Environmentally-friendly food processing

Related titles from Woodhead's food science, technology and nutrition list:

Novel food packaging techniques (ISBN 1 85573 675 6)
This comprehensive and authoritative collection summarises key recent developments in packaging. The book first discusses the range of active and intelligent techniques. It then summarises the major trends in modified atmosphere packaging. The final part of the book discusses general issues such as the regulatory context, packaging optimisation and consumer attitudes to novel packaging formats.

Food authenticity and traceability (ISBN 1 85573 526 1)
With recent problems such as genetically-modified ingredients in food, the need to trace and authenticate the content of food products has never been more urgent. The first part of this authoritative book reviews the range of established and new techniques for food authentication. Part 2 explores how these methods are applied to particular foods, whilst Part 3 reviews developments in traceability systems.

Rapid and on-line instrumentation for food quality assurance (ISBN 1 85573 674 8)
With its high volume of production, the food industry has an urgent need for instrumentation which gives rapid results and can be used on-line. This important collection reviews the wealth of recent research in the field. The first part of the book discusses product safety and the use of rapid techniques to identify chemical and microbial contaminants. Part 2 looks at techniques to analyse product quality.

Details of these books and a complete list of Woodhead's food science, technology and nutrition titles can be obtained by:

- visiting our web site at www.woodhead-publishing.com
- contacting Customer Services (email: sales@woodhead-publishing.com; fax: +44 (0) 1223 893694; tel.: +44 (0) 1223 891358 ext. 30; address: Woodhead Publishing Limited, Abington Hall, Abington, Cambridge CB1 6AH, England)

Selected food science and technology titles are also available in electronic form. Visit our web site (www.woodhead-publishing.com) to find out more.

If you would like to receive information on forthcoming titles in this area, please send your address details to: Francis Dodds (address, telephone and fax as above; e-mail: francisd@woodhead-publishing.com). Please confirm which subject areas you are interested in.

Environmentally-friendly food processing

Edited by
Berit Mattsson and Ulf Sonesson

CRC Press
Boca Raton Boston New York Washington, DC

WOODHEAD PUBLISHING LIMITED
Cambridge England

Published by Woodhead Publishing Limited
Abington Hall, Abington
Cambridge CB1 6AH
England
www.woodhead-publishing.com

Published in North America by CRC Press LLC
2000 Corporate Blvd, NW
Boca Raton FL 33431
USA

British Library Cataloguing in Publication Data
A catalogue record for this book is available from the British Library.

Library of Congress Cataloging-in-Publication Data
A catalog record for this book is available from the Library of Congress.

Woodhead Publishing Limited ISBN 1 85573 677 2 (book); 1 85573 717 5 (e-book)
CRC Press ISBN 0-8493-1764-9
CRC Press order number: WP1764

Cover design by The ColourStudio
Project managed by Macfarlane Production Services, Markyate, Hertfordshire
(e-mail: macfarl@aol.com)
Typeset by MHL Typesetting Limited, Coventry, Warwickshire
Printed by TJ International, Padstow, Cornwall, England

Contents

vi   Contents

x   Contents

Contributor contact details

Chapter 1

Dr B. Mattsson
The Swedish Institute for Food and
    Biotechnology (SIK)
PO Box 5401
SE-402 29 Göteborg
Sweden

Tel: +46 (0)31 3355600
Fax: +46 (0)31 833782
E-mail: bm@sik.se

Dr U. Sonesson
The Swedish Institute for Food and
    Biotechnology (SIK)
PO Box 5401
SE-402 29 Göteborg
Sweden

Tel: +46 (0)31 3355600
Fax: +46 (0)31 833782
E-mail: usn@sik.se

Chapter 2

Lic. Eng. J. Berlin
The Swedish Institute for Food and
    Biotechnology (SIK)
P.O Box 5401
SE-402 29 Göteborg
Sweden

Tel: +46 (0)31 3355600
Fax: +46 (0)31 833782
E-mail: jbe@sik.se

Chapter 3

Dr K. J. Kramer
Agricultural Economics Research
    Institute (LEI)
PO Box 29703
2502 LS The Hague
The Netherlands

Tel: +31 70 335 83 30
Fax: + 31 70 361 56 24
E-mail: klaasjan.kramer@wur.nl

Chapter 4

Dr L. Milà i Canals
Unitat de Química Física
Edifici Cn, Universitat Autònoma de
    Barcelona
E-08193-Bellaterra (Barcelona)
Spain

Tel: +34 93 581 21 64
Fax: + 34 93 581 29 20
E-mail: llorenc@klingon.uab.es

Dr G. Clemente Polo
Departament de Tecnologia
    d'Aliments
Universitat Politècnica de València
Camí de Vera s/n
46022 València
Spain

Chapter 5

Dr C. Cederberg
Swedish Dairy Association
SE-105 46 Stockholm
Sweden

E-mail:
    christel.cederberg@svenskmjolk.se

Chapter 6

F. Ziegler
The Swedish Institute for Food and
    Biotechnology (SIK)
PO Box 5401
SE-402 29 Göteborg
Sweden

Tel: +46 (0)31 3355654
Fax: +46 (0)31 833782
E-mail: fz@sik.se

Chapter 7

Dr W. Pelupessy
IVO
University of Tilburg
PO Box 90153
5000 LE Tilburg
The Netherlands

E-mail: pelupessey@uvt.nl

Chapter 8

Dr H. Dalsgaard and Dr A.W. Abbotts
COWI
Parallelvej 2
2800 Kongens Lyngby
Denmark

Tel : +45 70 10 1062
Fax: +45 70 10 10 63
Email : hda@cowi.dk, awa@cowi.dk

Chapter 9

Dr F. De Leo
University of Lecce
73100 Lecce
Italy

Tel: 00 39 0832 298755
E-mail: fedeleo@economia.unite.it

Chapter 10

Dr D. Dainelli
Cryovac Division
Sealed Air Corporation
Italy

Tel: +39 02 9332351
E-mail: Dario.dainelli@sealedair.com

Chapter 11

Dr V. Haugaard
The Royal Veterinary and
 Agricultural University
Department of Dairy and Food
 Science, Food Chemistry
Rolighedsvej 30
DK-1958 Frederiksberg C
Denmark

Tel: +45 3528 3349
Fax:+45 3528 3245
E-mail: vkh@kvl.dk

Dr G. Mortensen
Arla Foods Innovation
Roerdrumvej 2
DK-8220 Braband
Denmark
Tel: +45 8746 6771
Fax: +45 8746 6688
E-mail:
 grith.mortensen@arlafoods.com

Chapter 12

Dr M. Song and Dr S. Hwang
School of Environmental Science and
 Engineering
Pohang University of Science and
 Technology (POSTECH)
Kyungbuk, Pohang
South Korea

Tel: 054 279 2282
Fax: 054 279 8299
E-mail: shwang@postech.ac.kr

Chapter 13

Professor C. Hansen
UMC-8700
Utah State University
Logan
Utah 84322
USA
Tel: 435-797-2188
E-mail: chansen@cc.usu.edu

Professor S. Hwang
Environmental Bioprocess
 Engineering
B.E.S.T. Laboratory
School of Environmental Engineering
POSTECH
Kyungbuk, Pohang
South Korea

Tel: 054 279 2282
Fax: 054 279 8299
E-mail: shwang@postech.ac.kr

Chapter 14

Dr C. Simoneau and B. Raffael
European Commission Joint Research
 Centre
Institute for Health and Consumer
 Protection
Unit Physical and Chemical Exposure
TP 260
21020 Ispra
Italy

Tel: + 39 0332 785 889
Fax: + 39 0332 785 707
E-mail: catherine.simoneau@jrc.it

Dr R. Franz
Fraunhofer Institute for Process
 Engineering and Packaging (IVV)
Giggenhauser Strasse 35
85354 Freising
Germany

E-mail: Roland.Franz@ivv.fhg.de

Chapter 15

Dr B. Weidema
2-0 LCA Consultants
Borgergade 6, 1
DK-1300 Copenhagen
Denmark

E-mail: bow@lca.dk

Chapter 16

H. van Zeijts, G.J. van den Born and
    M.W. van Schijndel
Netherland Environmental
    Assessment Agency
National Institute for Public Health
    and the Environment (RIVM)

PO Box 1
3720 BA Bilthoven
The Netherlands

E-mail: henkvanzeijts@rivm.nl

Chapter 17

Professor B. Notarnicola
Department of Commodities Science
Faculty of Economics
University of Bari
Via C. Rosalba, 53
70124 Bari
Italy

E-mail: b.notarnicola@dgm.uniba.it

1

Introduction

B. Mattsson and U. Sonesson, The Swedish Institute for Food and Biotechnology (SIK)

The food industry is facing increasing pressure to improve environmental performance, both from consumers and regulators responding to consumer pressures. *Environmentally friendly food processing* has been designed to allow food manufacturers to understand better the effects their activities have on the environment and to take practical measures towards more sustainable production.

When discussing the environmental impact of food production it is important to use a holistic approach, a systems perspective. As an example, it is not efficient to reduce the emissions from a processing plant if it results in, for example, larger losses of raw material, which in turn increases the emissions from agriculture. As the food supply chain is complex, environmental impacts can occur in different places and different times for a single food product. Life cycle assessment (LCA), discussed in Part I, provides a way of addressing this problem. LCA is a means of assessing the environmental impact of a product over its entire life cycle, from raw materials to the point of consumption. LCA allows businesses to see their role as contributors to the overall environmental impact of the supply chain, and how they can work with suppliers to improve the environmental profile of products they manufacture. LCA also gives businesses the opportunity to anticipate environmental issues and integrate the environmental dimension into products and processes, rather than just manage the environmental impact of their existing operations. Part I includes chapters on the principles of LCA and its application to the production of vegetable, fruit, animal and seafood products as well as beverages such as coffee.

Important issues directly related to food processing are energy and waste management. Food production in general uses significant amounts of energy and produces relatively large amounts of waste, particularly packaging waste, both from secondary packaging and consumer packaging. The food sector, at least in industrialised countries, is the single largest user of consumer packaging. These issues are dealt with in Part II which concentrates on practical measures in improving environmental performance. It includes chapters on training and ways of improving energy efficiency, waste treatment and recycling. There is also a detailed comparison of integrated crop management and organic farming which allows businesses to make more informed decisions about how they source more sustainable raw materials. Given its importance, there is a detailed discussion of packaging, with chapters on ways of minimising packaging, methods of recycling, assessing the safety and quality of recycled packaging materials and the use of biobased packaging alternatives.

We wish to thank all the authors of this book for their time and effort to share their knowledge of different fields. We hope that the readers of this book will find it as interesting and inspiring as we do!

Part I

Assessing the environmental impact of food processing operations

2

Life cycle assessment (LCA): an introduction

J. Berlin, The Swedish Institute for Food and Biotechnology (SIK)

2.1 Introduction

Life cycle assessment (LCA) is a tool for evaluating the environmental impact associated with a product, process or activity during its life cycle. LCA is suitable for several purposes. LCA provides knowledge of a product and its related environmental impact. It also makes it possible to isolate which stages in the life cycle of a process or product make the most significant contribution to its environmental impact. Other reasons for undertaking an LCA study could be to assess improvements or alternatives, or to compare products, processes or services. Environmental communication such as environmental product declaration (EPD) can be based on LCA, and it can also be used as an instrument in environmentally adjusted product development.

Life cycle assessment is one of the tools included in the larger area of environmental systems analysis and is today one of the most commonly used tools within the subject. LCA has its roots back in the 1960s, when interest in energy requirement calculations started. During the oil shortage in the 1970s research was undertaken which included life cycle thinking for energy calculations and emissions released during energy production. However, after the oil crisis subsided, interest in LCA faded, but with the increased interest in the environment in the 1980s a revival of LCA occurred. Since 1990 LCA has expanded enormously, and the number of studies, publications, conferences and workshops is still growing (Lindahl *et al.*, 2001). Today LCA is recognised as an ISO standardised method (ISO 14040–14043, 1997–2000).

2.1.1   Environmental systems analysis

Environmental systems analysis is concerned with how to collect and assess information about a system's environmental impact. Several analytical tools have been developed. Findeisen and Quade (1997) describe the following common steps for performing an environmental systems analysis:

1. Formulating the problem
2. Identifying, designing, and screening possible alternative solutions
3. Forecasting future contexts or physical states
4. Building and using models to identify the range of potential outcomes of each solution from step 2
5. Comparing and ranking the alternative solutions.

Key components in such analyses include determining boundaries and constraints, data collection and analysis.

The following categorisation of types of environmental systems analysis can be made: flow models, monetary models, process models and risk assessment. These produce outcomes such as life cycle assessment, material flow accounting and substance flow accounting. LCA is further described below. Material flow accounting (MFA) describes all inflows and outflows and accumulation of a material, substance or element in a geographic area during a certain time period. Depending on the type of material studied, a further distinction of MFA is often applied. Bulk-material flow analysis studies flows of bulk materials, such as wood, iron or plastics, in a given region. Flows of substances such as nitrogen compounds and single elements such as cadmium or lead within a region are studied in substance flow accounting (Udo de Haes *et al.*, 1997). Cost–benefit analysis is used for assessing the total costs, including environmental costs, and benefits from a planned project. An example of a process analytical tool is 'Design for environment', which focuses on the environmental dimension of the design process. Risk assessment is a broad term and includes several different types of assessments. The focus can be on human health or environmental aspects. The risk can also vary from diffuse to specific and can be risk associated with natural operation or risk of accidents. The tools mentioned above are just a selection; for more information see Moberg (1999), Baumann and Cowell (1999) and Wrisberg *et al.* (2000).

2.1.2   Life cycle assessment

LCA is a method for assessing and evaluating the environmental performance of a product, process or service throughout its entire life cycle. The flow of material needed for the processing of the product or service is followed during the stages of the product's life cycle. At the same time input and output data such as emissions, waste, energy consumption and use of resources are collected for each unit process. This chapter provides an introduction to LCA. More information about LCA can be found by consulting ISO standard 14040–14043 (1997–2000) and Baumann and Tillman (2002).

2.2   The LCA process

The concept of life cycle assessment means that a product is followed and assessed from its 'cradle' all the way to the 'grave'. As shown in Fig. 2.1 the life cycle model starts with the acquisition of raw materials and energy that is needed for production of the studied object, the 'cradle'. The model follows the stages of processing, transportation, manufacturing, use and finally waste management which is considered as the 'grave'. The assessment is accomplished by identifying and quantitatively or qualitatively describing the studied object's requirements for energy and materials, and the emissions and waste released to the environment.

2.2.1   The LCA procedure

LCA is an ISO standardised tool (ISO 14040–14043, 1997–2000) and included in the standard is a working procedure, illustrated in Fig. 2.2 and described below. The LCA process consists of four phases (Fig. 2.2). In the first phase, *goal and scope definition*, the purpose of the study and its range are defined. In goal and scope definition important decisions are made concerning boundary setting and definition of the functional unit (i.e. the reference unit). During *inventory analysis*, information about the product system is gathered and relevant inputs and outputs are quantified. In *impact assessment*, the data and information from the inventory analysis stage are linked with specific environmental impacts so that the significance of these potential impacts can be evaluated. Finally, in the *interpretation* phase, the findings of the inventory analysis and the impact assessment are combined and interpreted to meet the

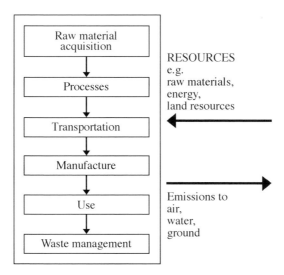

Fig. 2.1   The life cycle model (Baumann and Tillman, 2002). The arrows illustrate flow of energy and matter.

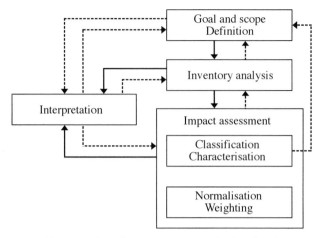

Fig. 2.2 Working procedure for an LCA. The unbroken line indicates the order of procedural steps and the dotted lines indicate iterations (ISO 14040, 1997).

previously defined goals of the study. Formalised, quantitative weighting methods are available for the aggregation of either inputs and outputs or environmental effects into one index.

An LCA starts with an explicit statement of the goal and scope of the study, the functional unit, the system boundaries, the assumptions and limitations and allocation methods used, and the impact categories chosen. The goal and scope includes a definition of the context of the study which explains to whom and how the results are to be communicated. The functional unit is quantitative and corresponds to a reference function to which all flows in the LCA are related, for example 1 kg of milk leaving the dairy. Allocation is the method used to partition the environmental load of a process when several products or functions share the same process.

In the inventory analysis a flow model of the technical system is constructed using data on inputs and outputs. The flow model is often illustrated with a flow chart including the activities that are going to be assessed and also gives a clear picture of the technical system boundary. The input and output data needed (resources, energy requirements, emissions to air and water, and waste generation for all activities within the system boundaries) are then collected. After that the environmental loads of the system are calculated and related to the functional unit, and the flow model is finished.

The inventory analysis is followed by impact assessment, in which the data are interpreted in terms of their environmental impact e.g. for acidification, eutrophication and global warming. In the classification stage, the inventory parameters are sorted and assigned to specific impact categories. The next step is characterisation, where inventory parameters are multiplied by equivalency factors for each impact category. Thereafter all parameters included in the impact category are added and the result of the impact category is obtained. For many LCAs, characterisation concludes this stage of LCA. Indeed, it is the last compulsory stage according to ISO 14042 (2000). However, some studies involve the further step of normalisation, in which the results of the impact categories from the study are compared with the total impact in the region. During weighting, the different environmental impacts are weighted against each other to get a single number for the total environmental impact.

The results from the inventory analysis and impact assessment are summarised during the interpretation phase. The outcome of the interpretation is conclusions and recommendations for action. According to ISO 14043 (2000) the interpretation should include:

- identification of significant environmental impact issues
- evaluation of the study for completeness, sensitivity and consistency
- conclusions and recommendations for action to reduce significant environmental impacts.

The working procedure of LCA is iterative as illustrated by the dotted lines in Fig. 2.2. The iteration means that information gathered in a later stage can affect an earlier stage. When this occurs the earlier stage and the following stages have to be reworked considering the new information. Therefore it is common for an LCA practitioner to work on several stages at the same time.

2.3   Key principles of LCA

There are some key principles besides the flow model and procedure of LCA that are important to know in understanding the concept of LCA. These are described in this section.

2.3.1   Functional unit

The functional unit is defined by the ISO standard as a quantified performance of a product system for use as a reference unit in a life cycle assessment study (ISO 14040, 1997). All data in the study are related to the functional unit, which means that all the inputs and outputs to the system are related to the unit. Therefore the unit must be defined and measurable. An example of a functional unit is 1 kg of potatoes leaving the farm or 1 litre of milk leaving the dairy. The chosen functional unit for a system is dependent on the goal and scope definition of the study (ISO 14040, 1997; Baumann and Tillman, 2002).

2.3.2   System boundary

The system under study is limited by a system boundary. All unit processes studied are within the system boundary. Tillman and Ekvall (1994) came to the conclusion that the boundaries need to be specified in several dimensions:

- Boundaries in relation to natural systems: the boundary between the technical system and the natural environment. The system's start and finish are specified in this dimension.
- Geographical boundaries: the area to which the system under study is limited.
- Time boundaries: the time perspective of the study, i.e. retrospective, present time or prospective.
- Boundaries within the technical system related to production capital, personnel, etc.: the activities that are needed in the life cycle of the studied object and are included in the study, together with a list of those excluded.
- Boundaries within the technical system in relation to other products' life cycles: when several products share the same processes the environmental load has to be shared between the products. This is further discussed below in Section 2.3.3.

The specification of the system boundaries takes first place in the phase of goal and scope definition. But the final boundary is decided when enough information has been collected during inventory analysis. If part of the life cycle is not investigated, this must be very clearly stated in the report. The technical system is preferably described by a flow chart of all unit processes included in the study.

2.3.3   Allocation

During the performance of LCA, allocation problems occur when the life cycles of different products are connected. Production of milk and cream as well as meat and milk are examples of connected life cycles. When such problems arise, ISO 14041 (1998) recommends expanding the system boundaries to include the co-products or to increase the level of detail in the life cycle. Increasing the level of detail involves detailed investigation of the process whereby the product under study is produced in order to identify the relevant data specific to the product.

If neither of the above approaches is applicable, an allocation method can be used to partition the environmental loads between the products or functions of the shared processes. The partitioning can be based on physical correlation, such that any quantitative changes in the produced products or their functions correlate with changes in the inflows and outflows of the system. Partitioning can also be based on economic allocation, that is, on the value of the produced products as reflected in their relative prices or their gross sales value. Baumann and Tillman (2002) give the example of a multi-output process producing gold, a valuable product, and zinc, a less valuable one. In such a situation it can be argued that economic allocation is preferable, since production of the valuable product is the reason for production in the first place.

2.3.4 Data quality and data collection

It is important to use data that suit the study's goal. Appropriate data quality increases the reliability of the results. A life cycle study is usually a summary of a large amount of data of varying quality, hence the transparency of data is crucial. It should be possible for the receiver of the study to trace the result back to the data used. The transparency is also important for the study's reliability. As an example, data can be collected from production companies directly or can be gathered from literature. The two ways of collecting data give different views of reality.

To minimise the variety of data quality, the data quality requirements should be set in the phase of goal and scope definition (Fig. 2.2) before the inventory starts. The following parameters concerning data quality requirements should be included according to the ISO standard (ISO 14041, 1998; Baumann and Tillmann, 2002):

- Time-related coverage: the age of data.
- Geographical coverage: the geographical area where the data is relevant.
- Technology coverage: the type of technology, e.g. best available, worst operating, weighted average of an actual process mix.
- Precision: the variance of the data values.
- Completeness: the percentage of the locations reporting primary data for each data category in a unit process.
- Representativeness: a qualitative assessment of the degree to which the data reflect the true value of the time-related coverage, geographical coverage and technology coverage.

When the study is completed the data used should be assessed with the same parameters to find out whether there are data that are crucial for the study that have to be improved. Von Bahr and Steen (2003) has suggested three criteria to measure data quality: relevance, reliability and accessibility.

2.4   LCA of food products

LCA studies have been used for many kinds of products. The first LCA studies of food products were performed at the beginning of the 1990s (Mattsson and Olsson, 2001). For every product under study there are questions that are unique for that particular area. The unique elements in LCAs of food products are described in this section.

2.4.1 Functional unit

All data are related to the functional unit of the study. As it is only possible to use one functional unit, it can be difficult to define a functional unit when the product under study fulfils more than one purpose. The mass of a specific product is commonly used for LCA studies of food products, e.g. 1 kg of cheese leaving the cheesemaking dairy, 1 kg of bread from a bakery, 1 kg of cod from the filleting industry, or 1 kg of apples from the greengrocer. But when selecting the functional unit the choice is not obvious. As described in both Baumann and Tillman (2002), Dutilh and Kramer (2000) and Andersson (1998), other functions that food products provide are, for example, nutritional value (nutrient or fibre content or calorific value), shelf-life, or sensory quality. An LCA can be related to only one functional unit, but the other functions are best described in qualitative terms in the interpretation stage of the LCA study.

2.4.2   System boundary
The boundary between the technical system and the natural environment is not clear when agriculture is considered, as the production takes place in the natural environment. Some examples of decisions that have to be made include whether or not the soil is to be included in the system. The time boundary is also not a clear choice. Should a crop rotation be included in the study? When animals are considered, a decision has to be made as to when the life cycle starts. As the choices are not obvious, it is important that system boundaries are clearly stated in the report.

2.4.3   Allocation
Allocation is recognised as a complex issue (ISO 14041, 1998). Several stages of a food product's life cycle may involve multifunctional processes, whether the agriculture phase, the phase of industrial production, retail distribution or consumption in the household. For instance, dairy cows produce both milk and meat, and a wheat crop gives both straw and wheat, which makes it difficult to divide the agricultural system into sub-systems. Many products are often produced at the same time, for example cheese, cream, milk powder and whey during dairy processing. The retailer sells an enormous number of products, including the particular product being studied. If the product is stored in the fridge or freezer in consumers' households, for example, it also shares the environmental impact of the freezer with other products. Different kinds of allocation methods can be used, but allocation criteria according to weight, volume or economic value are most commonly used in relation to food products.

2.4.4   Environmental impacts: land use and biodiversity
There is no general agreement as to how to handle the category of land use in LCA. Some guidance can be found in Udo de Haes *et al.* (2002). LCA is a method focusing on material flows. It is hard to connect it with the impact on biodiversity. Many food LCAs include just the area required for the agricultural production of the product under study with no connection to biodiversity. However, land use is a vital issue for LCA of foods, especially when agriculture is considered (Cederberg, 2002; Mattsson, 1999). A method for assessment of

agricultural land was tested by Mattsson *et al.* (2000). They concluded that the indicators included (soil erosion, soil, organic matter, soil structure, soil pH, phosphorus and potassium content of the soil and impact of biodiversity) gave a good picture of long-term soil fertility and biodiversity, but also that there is a need for a more simplified method.

2.5   Using LCA: some examples

Life cycle assessment can be used to answer questions that are interesting from an environmental point of view. For instance, it is possible to identify the sub-systems contributing most to the total environmental impact in a product system and it is possible to compare products or processes with the same function.

2.5.1   Organic versus conventional products

When comparing organic and conventional products, it should be remembered that production is on a different scale in the two methods. Som examples of case studies are given below. The organic and conventional production systems of two baby food products were investigated by Mattsson (1999). The major advantage of organic production was the ban on pesticides, while the major disadvantages were the lower crop yields and the difficulties of avoiding plant nutrient emissions from organic fertilisers. The pesticide use was the major drawback of the conventional cultivation systems, although the higher crop yields resulted in lower environmental impact per kilogram of product, even when the impact per hectare was the same as in organic production. However, a comparative study of organic and conventional farm milk production showed that the conventional system, with a high input of imported cattle feed, clearly has a larger environmental impact than an organic, more self-supporting production system (Cederberg and Mattsson, 2000).

2.5.2   Scale of production

Andersson (1998) has reported the results of a case study of bread. One of the objectives of the study was to compare the influence of scale of production. Home baking, a local bakery and two industrial bakeries of different sizes were studied. The home baking system showed a relatively high requirement for energy and water; otherwise, the differences between home baking, the local bakery and the small industrial bakery were negligible.

The scale of dairy milk production was studied by Høgaas Eide (2000). Three different Norwegian dairies were compared. The results showed that the environmental impact of the smallest dairy was significantly higher than for the other two dairies. The explanation was that the process equipment in the small dairy was cleaned more often, thus the energy use per kilogram of milk was higher.

It is often assumed that smaller enterprises cause less environmental impact than large companies. The two studies quoted above show that no such conclusion can be made. However, it is important to stress that in studies of existing companies not only the scale of production but also other subsystems differ among the subjects.

2.6   Future trends

The purpose of the study dictates the analytical tool to use. Sometimes there is no one tool that suits the purpose. In the circumstances Wrisberg *et al.* (2000) suggest that different tools might be combined to avoid this problem. Baumann and Cowell (1999) have suggested a framework for analysing tools. They observed that tools may be combined, for example through their consecutive use. The combination means that one tool acts as the input to the next tool. They also observed that some tools overlap each other. Successful case studies which combine tools include Sonesson and Berlin (2003) who combined material flow accounting, substance flow accounting and life cycle assessment in their study of future milk supply chains in Sweden. Berlin and Sonesson (2002) have also developed an environmental process model strongly influenced by LCA. A new trend in society when food is considered is the ethical and moral values. This will also probably influence LCA. Combining LCA and social values, such as the working environment and animal welfare, is still rare.

2.7   References

ANDERSSON K. (1998), *Life Cycle Assessment (LCA) of Food Products and Production Systems*, PhD thesis, School of Environmental Sciences, Department of Food Sciences, Chalmers University of Technology, Göteborg, Sweden.

BAUMANN, H. and COWELL, S. J. (1999), An evaluation framework for conceptual and analytical approaches used in environmental management. *Greener Management International, Journal of Corporate Environmental Strategy and Profile*, 109–122.

BAUMANN, H. and TILLMAN, A.-M. (2002). *The Hitchhiker's Guide to LCA,*, Chalmers University of Technology, Göteborg, Sweden.

BERLIN, J. and SONESSON, U. (2002) Design and construction of an environmental process management model for the dairy industry, submitted for publication in *Journal of Cleaner Production* (2002).

CEDERBERG, C. (2002). *Life Cycle Assessment (LCA) of Animal Production*, PhD thesis, Department of Applied Environmental Science, Göteborg University, Göteborg, Sweden.

CEDERBERG, C. and MATTSSON, B. (2000), Life cycle assessment of milk production: a comparison of conventional and organic farming. *Journal of Cleaner Production*, 8: 49–60.

DUTILH, C. E. and KRAMER, K. J. (2000), Energy consumption in the food chain. *Ambio*, 29 (2), March.

FINDEISEN, W. and QUADE, E. S. (1997), The methodology of systems analysis: an introduction and overview in *Handbook of systems analysis*, Vol. 1 edited by Miser, H. J. and Quade, E. S., John Wiley & Sons, Chichester UK.

HØGAAS EIDE, M. (2000), Life Cycle Assessment (LCA) of industrial milk production. *Int. J. LCA*, 3: 15–20.

ISO 14040 (1997). Environmental Management – Life Cycle Assessment – Principle and Framework (Stockholm, Sweden).

ISO 14041 (1998). Environmental Management – Life Cycle Assessment – Goal and Scope Definition and Inventory Analysis (Stockholm, Sweden).

ISO 14042 (2000). Environmental Management – Life Cycle Assessment – Life Cycle Impact Assessment (Stockholm, Sweden).

ISO 14043 (2000). Environmental Management – Life Cycle Assessment – Life Cycle Interpretation (Stockholm, Sweden).

LINDAHL, M., RYDH, C. J. and TINGSTRÖM, J. (2001). En Liten Lärobok om Livscykelanalys (A Minor Textbook about Life Cycle Assessment, in Swedish), Department of Technology, Kalmar University Kalmar, Sweden.

MATTSSON, B. (1999), *Environmental Life Cycle Assessment (LCA) of Agricultural Food Production*, PhD thesis, Swedish University of Agricultural Sciences, Department of Agricultural Engineering, Alnarp, Sweden.

MATTSSON, B. and OLSSON, P. (2001), Environmental audits and life cycle assessment, in *Auditing in the Food Industry*, edited by Dillon, M. and Griffith, C., Woodhead Publishing, Cambridge, UK.

MATTSSON, B., CEDERBERG, C. and BLIX, I. (2000), Agricultural land use in life cycle assessment (LCA): case studies of three vegetable oil crops. *Journal of Cleaner Production* 8, 283–292.

MOBERG, Å. (1999), Environmental Systems Analysis Tools – Differences and Similarities, Masters thesis, Stockholm University, Stockholm, Sweden.

SONESSON, U. and BERLIN, J. (2003), Environmental impact of future milk supply chains in Sweden – a scenario study, *Journal of Cleaner Production*, 11: 253–266.

TILLMAN A.-M. and EKVALL T. (1994), Choice of system boundaries in life cycle assessment, *Journal of Cleaner Production*, 2 (1): 21–29.

UDO DE HAES, E., VAN DER VOET, E. and KLEIJN, R. (1997), Substance flow analysis (SFA): an analytical tool for integrated chain management, in *From Paradigm to Practice of Sustainability*, paper presented at ConAccount workshop, 21–23 January 1997, Leiden, the Netherlands, pp. 32–42.

UDO DE HAES, H. A., FINNVEDEN, G., GOEDKOOP, M., HAUSCHILD, M., HERTWICH, E. G., HOFSTETTER, P., JOLLIET, O., KLÖPFFER, W., KREWITT, W., LINDEIJER, E., MULLER-WENK, R., OLSEN, S. I., PENNINGTON, D. W., POTTING, J. and STEEN, B. (2002), *Life-cycle Impact Assessment: Striving Towards Best Practice*. SETAL, Brussels, Belgium.

VON BAHR, B. and STEEN, B. (2003), Reducing epistemological uncertainty in life cycle inventory. *Journal of Cleaner Production*, in press.

WRISBERG, N., UDO DE HAES, H. A., TRIEBSWETTER, U., EDER, P. and CLIFT, R. (2000), Analytical tools for design and management in a system perspective, Centre of Environmental Science, Leiden University, Leiden, the Netherlands.

3

Life cycle assessment of vegetable products

K. J. Kramer, Agricultural Economics Research Institute, The Netherlands

3.1 Introduction

In Europe vegetable production is an important economic activity as measured by production and cultivated area. For the production of arable crops various amounts of inputs are necessary, such as labour, energy, fertilizers and pesticides. Due to consumer concerns for food safety, public concern about the use of pesticides is increasing. Human health plays a major role in food consumption, resulting in more consumption of fresh products such as vegetables. Besides paying attention to the use of pesticides, consumers are aware of the use of (synthetic) fertilizers for the production of foods. This trend can be illustrated by the growing interest in organic food products, where no chemical pesticides or synthetic fertilizers are permitted. This use of inputs in agriculture and horticulture could lead to negative impacts on the environment, for example through emissions of toxic compounds in the case of pesticide use, through emissions of nitrates and phosphates to soils and water, and through emissions of greenhouse gases resulting from energy use.

The methodology of environmental Life Cycle Assessment (LCA) can be used to determine the integrated environmental effects of crop production. This chapter highlights the use of LCA in protected as well as in open arable production. The results of the LCA are presented, as well as how LCA can be used to decrease the environmental impact of vegetable crop production. The following aspects are discussed:

- Methodologies to determine the environmental impact of pesticide use in LCAs.
- Case studies covering
  - Protected horticultural production

- Open arable production
- Comparison between organic and conventional agriculture.
- Future trends:
  - LCA and sustainability
  - Information systems.

3.2   Using LCAs: the case of pesticides

As mentioned in the introduction, the use of pesticides in the agricultural sector receives a lot of attention concerning their effects on human health and the environment. LCAs are aimed at an overview of the potential environmental aspects of products and consequently many LCAs cover a single moment of production. This means that there is no attention to the changing patterns of emissions over time or for site-specific issues. The inclusion of pesticide use in LCAs is therefore rather difficult and complex.

Some major studies have been undertaken to determine the effects of pesticides in LCAs and to develop toxicity equivalence factors. Heijungs *et al.* (1992) extrapolated toxicity data from the US Environmental Protection Agency (EPA) using data for maximum tolerable concentrations. Calculating characterization factors using this methodology results in factors that indicate only potential effects while ignoring the fate and nature of pesticides. After this first attempt to determine the toxic effect of pesticide use in LCA, several further studies were undertaken. Some of these are briefly described below.

3.2.1   Model-based initiatives

Raaphorst *et al.* (2001) tried to integrate the concept of the environmental yardstick in LCAs of horticultural products. The environmental yardstick is an instrument to determine the toxic effects of pesticide use on soils, surface water and groundwater. The environmental yardstick assigns environmental impact points for the risk to water and soil organisms of groundwater contamination for each pesticide (Leendertse *et al.*, 1997). These emissions depend on the vapour pressure of the pesticides and the method of application. The emitted pesticides are deposited on soil and surface water. The assumption is made that pesticides are exposed to decomposition in the atmosphere for 12 hours before they are finally deposited. So the amount of pesticides in the atmosphere depends on the period of exposure as well as the amount of pesticides applied. The calculated emissions were characterised with the aid of human toxicity and ecotoxicity data as described above. This method can be seen as a significant step to a more realistic interpretation of the toxic effect of pesticide use in agriculture. However, this method does not take into account exposure to pesticides by humans, animals, or plants.

Guinee *et al.* (2001) and Huijbrechts (1999) used models incorporating degradation and dispersal of pesticides to determine the human toxicological and ecological impacts of pesticides. This model is based on the Uniform System for the Evaluation of Substances 2.0 (USES). This USES model is able to calculate Prediction Environmental Concentrations (PEC) for air, water, agricultural and industrial soils resulting from a given constant emission of pesticides to air, water, agricultural and industrial soils. The USES-LCA model then calculates characterization factors from these releases for human toxicity, aquatic toxicity, sediment ecotoxicity and terrestrial ecotoxicity. The USES-LCA model has been used to determine ecotoxicity and human toxicity potentials for 181 pesticides.

3.3   LCA in horticultural production

LCA has been used in product design in the agricultural sector. In the Netherlands in the near future mandatory targets will be set for the use of energy, pesticides and fertilizers. In a research project carried out by the Research Station for Floriculture and Glasshouse Vegetables, an economic and environmental assessment was made of several greenhouse crops (Raaphorst *et al.*, 2001). The environmental impact of greenhouse production was determined with the help of an LCA. Figure 3.1 shows the environmental impact of 1 m^2 of tomato cultivation. The use of natural gas, necessary for heating the greenhouses and to provide the tomato plants with CO_2, is the most important element in the total environmental impact of tomatoes. More than 50% of the environmental impact is caused by this energy use. The use of electricity comprises about 15% of the total environmental impact. The use of pesticides and mineral fertilizers contributes less to this total environmental impact. The low impact of pesticides is partly due to the use of organic pesticides in greenhouse vegetable production.

The environmental impacts are aggregated using the Eco-indicator 1995 method, a distance-to-target approach. The underlying assumption is that there is a correlation between the seriousness of an effect and the distance between the current level and the target level. Since the 1995 method, the Eco-indicator 99

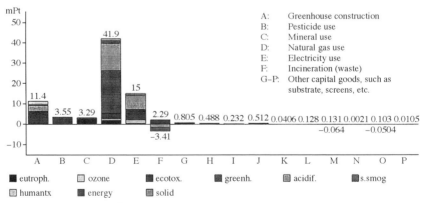

Fig. 3.1   LCA of 1 m^2 of tomato cultivation in 2000 (derived from Raaphorst *et al.*, 2001).

has been developed, a more damage-oriented impact assessment method. The Eco-indicator 99 scores are based on an impact assessment methodology that transforms the data of the inventory stage of an LCA into damage scores, which can be aggregated, depending on the needs and the choice of the user. To determine the environmental impact of pesticides, the method of the environmental yardstick for greenhouse horticulture is used, as described above.

Figure 3.1 shows that the impact of greenhouse construction is rather small at 15%. Through energy savings in production, the environmental impact of the construction materials will be relatively higher. Figure 3.2 shows the total projected environmental impact of tomatoes in 2010 compared to 2000. The production of 1 m^2 of tomatoes will be accompanied by a 20% lower environmental impact in 2010 compared to 2000 cultivations. Tomato cultivation in greenhouses in 2010 will differ from that in 2000. The greenhouses will be constructed from other materials, mainly to reduce the use of energy and heat losses. Another option to reduce the total use of energy is the application of fuel cell or other advanced techniques, in order to raise the efficiency of energy generation, possibly in combination with more efficient electricity delivery.

Pluimers has also carried out environmental research in the horticultural sector (Pluimers, 2001). The objective of that study was to identify technical options to reduce the environmental impact of greenhouse horticulture in the Netherlands and to evaluate their cost-effectiveness. The study focused on tomato cultivation and on the following environmental problems: global warming, acidification, eutrophication, dispersion of toxic biocides and the production of waste. The methodology of environmental system analysis was used to determine the effects of different technical options on environmental

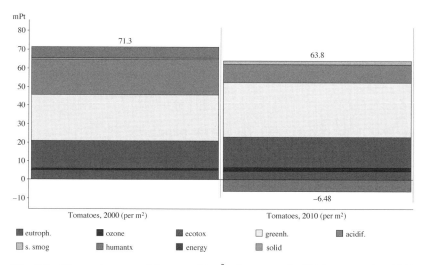

Fig. 3.2   The environmental impact of 1 m^2 of tomatoes in 2010 compared to 2002 (derived from Raaphorst *et al.*, 2001).

impact reduction in the Dutch horticultural sector. In a first step to define the system boundaries and to determine the components of the systems, a limited LCA for greenhouse tomato cultivations was carried out, with an emphasis on environmental aspects such as global warming, acidification and eutrophication. This limited LCA showed that a study of the environmental impact of Dutch tomato or horticultural cultivation needs to consider the CO_2 emissions as a result of the use of natural gas and the production of electricity, the emissions of nitrogen oxides (NO_x) (related to the use of natural gas and fertilizers and to the production of electricity and rockwool), and the losses of nitrogen and phosphorus from the use of fertilizers.

Besides these aspects, the emissions of pesticides and the production of waste were also included in the study. To determine the environmental impact of pesticides Pluimers describes two approaches. In the first approach, the methodology of the environmental yardstick for pesticide emissions from greenhouse cultivations was used. As described above, the emissions of pesticides depend on several chemical and physical properties of the pesticides. To determine the impact of these emissions, the deposited amount of pesticides is divided by the maximum allowable concentration (MAC) of pesticides for aquatic organisms, which are the most sensitive organisms to pesticides. The second approach is related to the use of pesticides. The amount of pesticides remaining after 12 hours is calculated with the aid of the same physical-chemical properties as with the environmental yardstick, followed by division by the MAC value. The result is the potential hazard of the environmental pollution, described as the amount of water polluted by more than the permissible concentration for the various pesticides.

Pluimers (2001) used the model to analyse cost-optimal strategies to meet national environmental targets for tomato cultivation in the Netherlands. With technological improvements at no extra cost the tomato growers can reach national environmental targets for CO_2-emissions and emissions of nitrates and phosphates. However, significant investment must be made to reach the environmental reduction target for energy efficiency (also concluded by Raaphorst *et al.*, 2001) and the emission reduction target of pesticides to air.

3.4   LCA for processed vegetable products

In the Netherlands, the Dutch Environmental Quality Label has been used for a steadily increasing number of agricultural products and foodstuffs since 1995, including potatoes, fruit and vegetables and ornamental plants. Products with this label meet the most stringent environmental criteria during their entire life cycle (Milieukeur, 2002). Environmental quality criteria are related to the use of resources and energy, pesticide and other emissions, waste, possibilities for recycling, and lifetime of products.

To define criteria for environmental quality labelling for processed vegetable products, LCAs were carried out to identify the relative environmental impact of the various parts of the life cycle. This analysis also showed the differences in environmental impact between vegetables with the Dutch Environmental Quality Label and conventional vegetables. The LCA was limited to four types of vegetables (carrots, French beans, spinach and red beets) as well as to four different types of vegetable processing (slicing, preserving in tins, preserving in jars and freezing). The following environmental impact categories were taken into account: the greenhouse effect, ozone depletion, carcinogenic emissions, human toxicity, eco-toxicity, acidification, minerals saturation, and use of mineral and fossil resources. The Dutch Environmental Quality Label is based on a complete life cycle analysis. However, only the relative contributions of the agricultural production stage and the vegetable processing stage were taken into account when environmental criteria were developed. It is not possible to define or control criteria for the consumption stage of these products, partly due to uncertainties as to how consumers prepare the vegetables and how, and for how long, these products are stored.

Figure 3.3 shows the normalized environmental profile of 1000 kg of tinned carrots (Roba/Gijsbers, 2000; Effting and Spriensma, 1999). The figure shows that pesticide use in the agricultural production stage is responsible for a relatively large share of the total environmental impact of the tinned carrots, followed by fossil fuel use during the production stage. Figure 3.4 compares the environmental impact of four different processed vegetable products. The total environmental impact of each product depends on the processing method and differs regarding product loss and the impact of packaging and energy use. The figure shows that the agricultural production stage is (mostly) dominant in the total environmental impact. The use of energy is also important for the storage of frozen vegetables, and has a higher contribution to the total environmental impact than pesticide use. These LCAs of processed vegetables resulted in some major findings: pesticide use in agriculture, the use of energy for

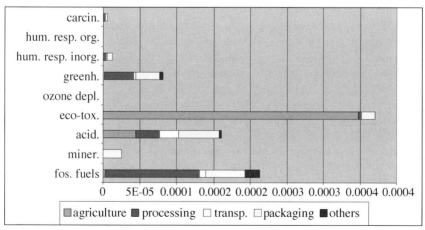

Fig. 3.3   The environmental impact of the various life cycle stages of 1000 kg of tinned preserved carrots (Effting and Spriensma, 1999).

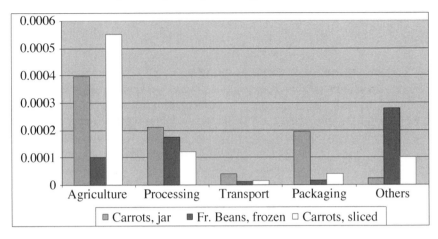

Fig. 3.4   The environmental impact of three types of processed vegetables (Effting and Spriensma, 1999).

vegetable processing as well as packaging are substantial in the total environmental impact of processed vegetables (Roba/Gijsber, 2000; Effting and Spriensma, 1999).

The Dutch Environmental Quality label scheme used these findings to define criteria for vegetable production and processing (Milieukeur, 2002). For the production of vegetables environmental criteria were developed for pesticide and fertilizer use, packaging and, in the case of greenhouse crop production, energy use. The most important environmental criteria for vegetable production are related to the use of pesticides. The Dutch Environmental Quality label distinguishes two types of criteria regarding pesticide use. The supply of pesticides is restricted to a maximum per hectare (kg of active compound per hectare). Furthermore, only a small selection of pesticides is allowed to be used in the production of a particular crop. These pesticides are judged according to their environmental impact, with the help of the environmental yardstick. The use of pesticides with high scores in these categories is discouraged by applying penalty points. To determine the effects of the main criteria on the environmental impact of the production of the crops, the results are monitored yearly. Figure 3.5 presents the results of production of carrots under the criteria of the Dutch Environmental Quality label scheme compared to the conventional growth of carrots. There are various indicators to determine the environmental impact of pesticide use: the amount of active ingredients used, emissions, and environmental impact points (EIP) for surface water, soil, and groundwater, calculated with help of the exposure risk indices. Figure 3.5 shows that the growth of carrots under the restriction of the Dutch Environmental Quality label result in a significantly lower environmental impact in the use of pesticides. Some indicators show a more than 80% reduction in environmental impact compared with the conventional growth of carrots (Milieukeur, 2002).

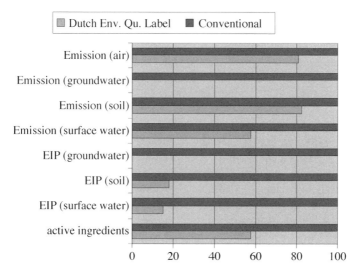

Fig. 3.5   The environmental impact in 2001 of pesticide use in conventional carrot production and carrot production under the criteria of the Dutch Environmental Quality label (Milieukeur, 2002).

Besides criteria to decrease the environmental impact in the primary production stage of the life cycles of arable crops, the Dutch Environmental Quality label scheme also defines criteria for other stages and other type of vegetable products. Figures 3.3 and 3.4 showed that the use of energy is an important aspect in the total environmental impact of processed, freshly washed and sliced or frozen vegetables. Consequently, the Dutch Environmental Quality label scheme has defined criteria to reduce the use of energy in the vegetable processing stage. Because it is difficult to allocate the use of energy to individual vegetable products (various vegetables are processed in the same plant), a major requirement is that the processing industry must have an energy-saving plan, in which they indicate how much energy is used and how the use of energy will be decreased in subsequent years. But the main environmental issue for the vegetable processing industry is related to the production of the crops. Figures 3.3 and 3.4 show that the use of pesticides is dominant in the total environmental impact of arable crops. Furthermore, Fig. 3.5 shows that this impact could be decreased significantly.

Besides a certification programme for washed and sliced vegetables, the Dutch Environmental Quality label scheme has also developed certification programmes for, among others, strawberries, apples and pears, and for some greenhouse vegetables. For the latter product category, an important criterion is the use of fossil fuel energy in greenhouse crop production, also shown by Fig. 3.1. The production of greenhouse vegetables requires restriction on the total use of energy to comply with the Environmental Quality label scheme (Milieukeur, 2002).

3.5   LCA for organic production

The previous section emphasized the impact of pesticide use on the total environmental impact of arable crops. Other than synthetic fertilizers, the use of pesticides is not allowed in the production of organic vegetable production. A few LCA studies have been undertaken on organic arable crop production.

3.5.1   Organic horticulture in the Netherlands

The former Research Station of Floriculture and Glasshouse Vegetables conducted studies on the economic and environmental impact of organic horticultural crop production compared to conventional crop production. Figure 3.6 shows the result of an LCA of 1 m^2 of tomato production, produced both conventionally and organically by seven producers (A–G) (Kramer et al., 2000). This figure shows some large differences in the total environmental impact of organic tomato production caused by the use of natural gas to heat the greenhouse. The organic tomato producing companies A–D use natural gas, while the other companies E–G do not use natural gas to heat the greenhouse. The total environmental impact of 1 m^2 of organic tomatoes grown without natural gas is 90% lower compared to conventional tomato production. It has to be emphasized that this type of tomato production is only possible during a small period of the year. The growth of organic tomatoes in heated greenhouses results in a 15–70% reduction in environmental impact per square metre compared to conventional tomato production in heated greenhouses. A lower environmental impact is caused by a lower energy use, resulting in part from the application of combined heat and power (CHP) generation. Figure 3.6 shows that large differences exist between the various organic tomato producing

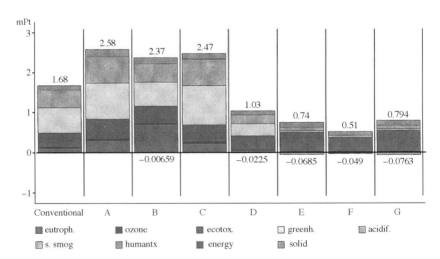

Fig. 3.6   LCA of conventional and organic tomato cultivation (1 m^2) (Kramer et al., 2000).

companies in energy use and consequently the total environmental impact. This is also the case in conventional tomato production. However, yields in organic production systems are lower than in conventional tomato production systems. When lower production rates are taken into account, another view of the total environmental impact of organic tomatoes is obtained. Tomatoes grown in unheated greenhouses have an environmental impact that is 40% lower per kilogram of tomatoes than the environmental impact of 1 kg of conventionally grown tomatoes. The total environmental impact of 1 kg of organic tomatoes and conventional tomatoes, when both are grown in heated greenhouses, is almost the same. However, organic horticultural greenhouse production is a relatively new activity in the Netherlands, and new developments are likely to result in higher yields and therefore should result in lower environmental impact per kilogram of organic tomatoes in the near future (Kramer *et al.*, 2000).

3.5.2   Organic farm systems

Geier and Köpke (1998) and Haas *et al.* (2001) used the methodology of LCA to analyse the differences in environmental impacts of various agricultural production systems. Geier and Köpke (1998) compared the environmental impact of a German region where conventional systems were converted into organic systems. Nine impact categories were selected for this comparison. Some impact categories were adapted in order to consider all important environmental impacts of agricultural systems, mainly wildlife conservation (biodiversity) and soil protection. The main results showed that, for seven impact categories, the LCA impact assessment showed the advantages of organic over conventional farming, with a 31–100% reduction in the environmental impact. Haas *et al.* (2001) adapted the LCA method to compare different farming (grassland) production systems. Like Geier and Köpke (1998), the LCA method was adapted to cover aspects such as impact on biodiversity, landscape image and animal welfare. The authors concluded that organic agriculture clearly showed ecological advantages compared with conventional farming, measured per unit of area.

3.5.3   Other resources

The influence of yield in the outcome of LCA is also shown by Nicoletti *et al.* (2001), who carried out an LCA of conventional and organic wine. The impact of organic wine production on the environment is lower. However, organic orchards produce much less grapes as compared with conventional orchards. Hence, after correction aspects for produced amounts quantities per hectare, the environmental benefits of organic wine disappear.

3.6   Future trends: LCA and sustainability

These case studies show that the methodology of LCA could be used in several ways. LCA can be used in the development of 'environmental' certification programmes. The methodology can also be used in product development or to compare various agricultural production systems, which could lead to measures to decrease the environmental impact of these systems. The illustrated LCA showed some interesting findings. In greenhouse crop production the use of energy is dominant in the total environmental impact of crops. Pesticide use makes only a minor contribution to this total environmental impact. However, in arable crop production the use of pesticides is dominant in the environmental load of the crops.

3.6.1   LCA, sustainability and chain information systems

LCA is a good methodology to determine the total environmental impact of production chain. Our society is increasingly demanding products and services produced in a sustainable way, whether defined in social, economic or environmental terms. To create a complete view of the level of sustainability is it important to take the entire life cycle of (agricultural) products into account. LCA can be seen as an instrument to analyse environmental sustainability. Such methodologies are not yet available for the economic and social aspects of sustainability. A lot of information or data is needed to execute an LCA. Much more information is required to determine the social and economic sustainability of agricultural chains. The Agricultural Economic Research Institute (LEI) currently conducts research into the possibilities of incorporating sustainability and chain information systems into LCA (Kramer *et al.*, 2002). A chain information system is defined as the way in which chain actors share information. Two different systems can be distinguished: a system in which information is transferred by the various actors in a chain; and a system which is placed outside the chain and acts like a certain type of database. In the latter system, all the actors in the chain have access to the system. Most current chain information systems are developed to trace and/or track products. Tracking provides information about the current position in a chain. Tracing provides information about where the products had been and under which circumstances. Food safety and legislation are important drivers in the growing concern for tracing and tracking within chain information systems. There are some indications that companies will use tracing and tracking in a more proactive way, for better control of production processes, for example. The LEI has investigated existing chain information systems in detail, what kind of information is gathered in these systems and whether there are some links with sustainability.

3.6.2   Case study

Groeinet is a Dutch company specializing in services such as crop registration, tracking and tracing, and benchmarking. Its registration programmes (GT.NET) allow for information on the use of pesticides, fertilizers and yield per crop. While the information system of Groeinet is not developed to provide indicators for sustainability, recorded information about pesticides and fertilizers could be used to determine environmental impacts. As an example, the production of these inputs is accompanied by the use of energy which can be very high (e.g. the fertilizer Calcium Ammonium Nitrate (CAN) is produced from natural gas in the Netherlands, and requires 0.7 m^3 natural gas for 1 kg). Databases with information about the energy contents of inputs are available to determine total energy use. The use of pesticides in arable crop production recorded by the Groeinet information system could also be used to determine the toxicological impact on water, soil and air. The information of Groeinet is useful in helping determine the environmental dimension of sustainability and in providing LCA studies with the necessary data for the determination of the environmental impact of arable crops.

The determination of aspects of social sustainability is more difficult. Chain information systems focus on products. While environmental aspects are also related to products, aspects of social and economical sustainability are more related to processes and people than to products. However, it is possible to relate social aspects to product, for example the number of hours of labour to produce a certain amount of product. Additional research is required to establish suitable benchmarks (Kramer *et al.*, 2002).

3.6.3   Concluding remarks

This chapter has shown various applications of LCA in arable and vegetable production systems. LCAs could be used to identify hotspots in production chains regarding environmental impact. Various options exist to reduce, with the help of the outcomes of LCA studies, this environmental impact. In addition, LCA could be a useful tool in setting targets for sustainability or social corporate performance.

3.7   Sources of further information and advice

For more information on LCA in agricultural production, see www.wageningen-ur.nl. For more information on economic and environmental research at LEI, see www.lei.wageningen-ur.nl.

3.8   References

EFFTING, S. and SPRIENSMA, R. (1999). *Verwerkte groenten. Screening LCA van verwerkte groenten afkomstig van de gangbare teelt of volgens Milieukeur*. Amersfoort, The

Netherlands.

GEIER, U. and KOPKER U. (1998). Comparison of conventional and organic farming by process life cycle assessment. A case study of agriculture in Hamburg. In: Ceuterick, D., *International Conference on Life Cycle Assessment in Agriculture, Agro-industry and Forestry.* Brussels, 3–4 December 1998.

GUINEE, J.B. (ed.), GORREE, M., HEIJUNG, R., HUPPES, G., KLEIJN, R., VAN OERS, L., WEGENER SLEESWIJK, A., SUH, S., UDO DE HAES, H.A., DE BRUIJN, J.A., VAN DUIN, R. and HEIJBREGTS, M.A.J. (2001). Life cycle assessment: an operational guide to the ISO standards. Centre of Environmental Science (CML), Leiden, The Netherlands

HAAS, G., WETTERICH, F. and KØPKER, U. (2001). Comparing intensive, extensified and organic grassland farming in Southern Germany by process life cycle assessment. *Agriculture, Ecosystems and Environment*, **83**, 43–53.

HEIJUNGS, R., GUINEE, J.B., HUPPES, G., LANKREIJER, H.A., UDO DE HEAS, H.A., WEGENER SLEESWIJK, A., ANSENS, A.M.M., EGGELS, A.M.M., VAN DUIN, R. and DE GOEDE, H.W. (1992). Environmental life cycle assessment of products. Guidelines and backgrounds. Centre of Environmental Science (CML), Leiden, The Netherlands.

HUIJBRECHTS, M.A.J. (1999). Priority assessment of toxic substances in the frame of LCA. Development and application of the multi-media fate, exposure and effect model USES-LCA. Interfaculty Department of Environmental Science, University of Amsterdam, Amsterdam, The Netherlands.

KRAMER, K.J., PLOEGER, C. and VAN WOERDEN, S.C. (2000). *Organic greenhouse vegetables production. Economic and environmental aspects 1998–1999.* Research Station for Floriculture and Glasshouse Vegetables, Naaldwijk, The Netherlands.

KRAMER, K.J., THORS, M. and WOLFERT, J. (2002). *Sustainability in chains.* Agricultural Economic Research Institute, The Hague, The Netherlands.

LEENDERTSE, P.C., REUS, J.A.W.A. and DE VREEDE, P.J.A. (1997). Meetlat voor middelgebruik in de glastuinbouw. Centre for Agriculture and Environment (CLM), Utrecht, The Netherlands.

MILIEUKEUR, (2002). The Dutch Environmental Quality Label. www.milieukeur.nl, The Netherlands.

NICOLETTI, G.M., NOTARNICOLA, B. and TASSIELLI G. (2001). Comparison of conventional and organic wine. In: *Proceedings of International Conference on LCA in Foods.* Göteborg, Sweden, 26–27 April 2001.

PLUIMERS, J. (2001). *An environmental systems analysis of greenhouse horticulture in the Netherlands the tomato case.* PhD thesis, Wageningen University, The Netherlands.

RAAPHORST, M.G.M., RUIJS, M.N.A., VAN WOERDEN, S.C., VAN PAASSEN, R.A.F., NIJS, E.M.F.M. and NIENHUIS, J.K. (2001). *Glastuinbouwbedrijfssystemen in 2010. Een studie naar toekomstige geïntegreerde bedrijfssystemen in de glastuinbouw in economisch en milieukundig perspectief.* Praktijkonderzoek Plant & Omgeving, sector glastuinbouw. Naaldwijk, The Netherlands.

ROBA/GIJSBERS MILIEUR B.V. (2000). *Rapport Verwerkte Groenten*, Deurne, The Netherlands.

4

Life cycle assessment of fruit production

**L. Milà i Canals, Universitat Autònoma de Barcelona, Spain and
G. Clemente Polo, Universitat Politècnica de València, Spain**

4.1 Introduction

This chapter builds upon other general references for LCA methodology applied to agriculture, mainly Audsley *et al.* (1997) and Weidema and Meeusen (2000). Chapter 3 gives a general description of application of LCA on food products. The objects of study in fruit production LCAs are fruit plantations, which are managed with the broad objective of producing fruit profitably for the farmer. This object must be defined properly for the LCA study to be consistent, and some specific comments need to be made in the case of fruit production. Different types of fruit may generally be distinguished, according to either the characteristics of the product or those of the plantation. The following groups of fruits are usually distinguished:

- Fresh fruit, including citrus, pip or seed fruit (apples, pears, etc.) and stone fruit (peaches, plums, etc.). Berries (soft fruit) are usually included within this category, and grapes may be included as well.
- Tropical fruit (bananas, mangoes, plantains, etc.).
- Grapes (and wine). They may be included with fresh fruit but, due to their economic importance in Europe, they are usually kept separate.
- Olives (and olive oil).
- Nuts.

Fresh fruits are the main fruit type from an economic point of view in Europe, representing 8% of agricultural production in the EU in 2000 (with apples accounting for half of total fruit tree production), ahead of 1.8% for olive oil and 5.5% for wines and must (Eurostat). In Europe, where over 12% of the world's fruit is obtained, production is concentrated in southern countries (Spain and

Italy produce two-thirds of European fresh fruit), and wine is produced mainly in France (with more than half of wine production in 2000) and Italy (26%). On a global scale, though, tropical fruit is very relevant as well, with bananas and plantains representing over 20% of total fruit production in 2002 (FAOSTAT). Oranges and apples are still very important globally, with 14% and 12% of total world fruit production in 2002. The main fruit producing regions in the world are spread around the warmer areas: China and India produce around 25% of the world's fruit, followed by other country groupings (e.g. the EU and South American countries with over 12% of world production each; FAOSTAT). Central America, the Middle East and south-east Asia are also relevant producing regions for some types of fruit. 'Emergent' suppliers such as China (on a global level) or Eastern European countries (for Europe) will affect production patterns within the course of this century.

From an LCA perspective, though, it is useful to classify fruit production according to the type of plantation, because that is what defines the operations performed in the field. These plantation types include production from fruit trees (permanent crops), from bushes, and from plants (e.g. berries). This chapter is focused on production of fruit from trees.

When defining the impact categories that are to be assessed within the LCA, it is very important to have in mind typical environmental problems that may arise with fruit. In temperate regions, fruit is usually produced in the sunnier areas, which often require irrigation in order to maximise and stabilise yields. This will usually lead to a concern on water scarcity and possibly to increased inputs and loss of nutrients and pesticides in the environment, which needs to be assessed within the LCA. On the other hand, most fruit types (chiefly fruit trees) are permanent crops that may be grown in conjunction with a vegetated and undisturbed soil. This type of crop represents an important potential for sequestering carbon into soil organic matter, and it may be relevant to address this issue in the impact assessment phase of the LCA. In tropical fruit, severe weather conditions may lead to locally relevant environmental problems, such as extreme erosion and yield drops due to intense rainfalls. Finally, fruit is commonly transported large distances before reaching the consumer, and this requires specialised storage and conservation treatments as well as a complex packaging. All these post-harvest treatments need to be thoroughly covered in the LCA.

4.2   Functional units and system boundaries

The functional unit is the reference unit to which the system's inputs and outputs are related. It is called 'functional' because it should be related to the function of the system in order to be able to make fair comparisons between different product systems performing the same function, or apparently similar products performing different functions. In most cases, the definition of a functional unit is a rather straightforward process, which takes a familiar unit such as 1 kg of product as a reference. Nevertheless, it must be noted that the definition of a functional unit has profound effects on the results of any LCA, and many aspects should be considered when analysing such results.

As agricultural systems are naturally multi-functional, the definition of a functional unit is not always a straightforward procedure. Cowell (1998) refers to this multi-functionality when discussing the definition of the functional unit. According to her, the function of agriculture can be related both to keeping the land to a definite shape and composition and to the production of products. Furthermore, agricultural products can be characterised by mass, energy, nutritional content, meal portions, etc., which renders the definition of a functional unit a complex and usually case-dependent process. Haas *et al.* (2000) provide an interesting example of the consequences of using different functional units on the LCA results.

In fruit production, the multi-functionality of agriculture is further complicated by the fact that different fruit qualities are usually obtained within the same orchard. Differences in quality characteristics (size, external and internal quality) should be included in the definition of a functional unit when this is considered to be relevant, and the allocation procedure to include these aspects should be consistent with the goal of the study. It is suggested by the authors that a mass-based functional unit (e.g. a tonne of fruit) is suitable for the agricultural stages of the life cycle. When the intended use of the fruit is included, more functional aspects may be brought into the functional unit, and then allocation between different fruit grades should consider relationships such as economic value, nutrient, sugar or juice content, etc.

4.2.1   Time-dependent system boundaries

Fruit trees are perennial cultures cultivated for several years. During the trees' whole lifetime, several stages may be differentiated that are relevant from an LCA perspective, which vary in duration depending on the crop:

1. Nursery (2–3 years): the trees are grown from seeds or seedlings.
2. Establishing stage (usually 1–3 years): trees are planted in the field, which has been previously prepared for good establishment of roots.
3. Stage of young trees and low yield (3–5 years): fruit is already harvested, and most operations are carried out, although in a somewhat less intensive way due to the smaller size of trees.
4. Stage of high yield (10–30 years): all usual operations.
5. Stage of old trees and low yield (0–5 years): all usual operations, but yield falls due to the trees' age;
6. Destruction stage (usually 1 year): the trees are removed and the field is prepared for future crops.

It is important that the time boundaries include all these stages, as they might have important contributions to the overall impacts. The reason is that many operations are performed during the different periods of the plantation that will affect the trees' whole lifetime. Specialised literature (see, e.g., Audsley *et al.*, 1997; Cowell, 1998) suggests that the inclusion of full crop rotations in LCA boundaries is a more correct approach, as they allow for the analysis of operations affecting different crops in the rotation. In a similar way, the whole lifetime of the trees should be included in order to assess the effects of such operations. This should generally be done in both descriptive and comparative studies; in the first case, the reason is obvious, as the main goal is to describe the environmental hotspots of fruit production, and in the latter case it is needed because the different stages might have different environmental consequences and durations in different production systems.

Only the high yield stage (stage 4) is usually analysed, because the common function required from a fruit production system is to produce fruit (measured in mass units). Nevertheless, it is not fair to overlook the fact that impacts per mass of fruit are generally higher in stages 1, 2, 5 and 6 and possibly also 3. Therefore, it is recommended that the environmental consequences caused in these stages be calculated and allocated amongst the product (fruit) that leaves the system throughout the years.

4.2.2   Physical system boundaries

The system's physical boundaries must be set in the whole orchard, including infrastructures directly used for fruit production, e.g. tree wind shelter, irrigation infrastructure, farm buildings, etc., as it is suggested in many studies that they can make a significant contribution to the LCA results. As an example, Audsley *et al.* (1997) suggest that machinery production plays a relevant role in agriculture, and report a share in total energy consumption of 13–37% in arable systems with different degrees of mechanisation. Milà i Canals *et al.* (2003) show that the contribution of machinery production to the overall impact of apple growing may be even higher in some specific impact categories, even though the share in the energy consumption indicator is somehow smaller: 7–15%.

On the vertical axis, soil is considered as part of the system as suggested in Audsley *et al.* (1997). Usually, the system's boundary is set at the depth of the water table, which in practice and for modelling purposes can be considered to be at 1 m (substances reaching this depth will probably reach the groundwater, and thus be a pollution threat). Substances crossing these boundaries will be considered as emissions to the environment. Also, it must be noted that soil leaves the system when it crosses the time boundaries (i.e., after the destruction stage 6). All differences in soil (presence or absence of substances, differences in soil structure, etc.) should be considered as impacts. This is particularly important for soil quality parameters (e.g. soil organic matter), and build-up of heavy metals. Agro-chemicals present in soil when this leaves the system should also be counted as emissions, but these will not generally be so important due to degradation.

4.3   Data collection: field operations

Fruit production is a very general term, which includes very different products. Seed or pip fruit trees (pear, apple, etc.) and stone fruit trees (peach, plum, etc.) will be considered in this chapter. Important differences in the production stage might appear in other fruit types (berries, grapes, etc.).

In the life cycle inventory (LCI) of fruit production, data must be gathered for the inputs and outputs crossing the system's boundaries. These inputs and outputs can broadly be classified into the following items:

- Field operations (directly affecting machinery use and energy consumption; see Sections 4.3.2–4.3.10)
- Nutrient balance (affected by the way some field operations are performed, and affecting the system's nutrient emissions; see Section 4.4)
- Pesticide field emissions (see Section 4.5)

These items are shown in Fig. 4.1 (adapted from Milà i Canals, 2003). This depicts the most typical field operations in fruit production, with their overall position in the crop calendar, and the main inventory aspects affected by each of them: nutrient balance, pesticide emissions, and energy consumption. Machinery use is of course affected by each one of these operations. The differences in soil quality (nutrient content, soil organic matter, toxic substances, etc.) when the soil leaves the system (after destruction stage or harvest) should be considered in the LCA study, and this is represented in Fig. 4.1 by SOIL' leaving the system.

This section deals mainly with the types of field operations typically found in fruit production on most sites. The main aim is to provide some sources of information for the typical inputs and outputs of a fruit plantation. These field operations will be translated into machinery use and energy consumption (further developed in Sections 4.3.2–4.3.10), and field emissions (both of nutrients, Section 4.4, and of pesticides, Section 4.5) which will also derive from the field operations. It is often necessary that detailed site-dependent data be

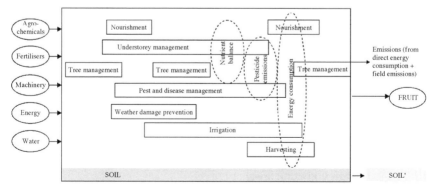

Fig. 4.1   Typical field operations and inventory items for fruit production (adapted from Milà i Canals, 2003).

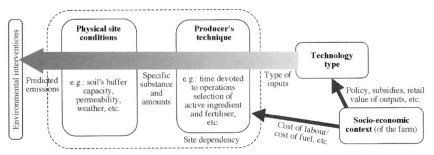

Fig. 4.2   Aspects influencing the LCI results in fruit production (Milà i Canals, 2003).

collected for the field operations, as farmers' practices may be the aspect most influencing the LCA results (Milà i Canals, 2003).

4.3.1   Site-dependency

Site-dependency is a key issue in agricultural systems, and particularly in fruit production: big differences exist between areas of production, due to climate and soil conditions and of course the different producers' practices. This has been discussed by Milà i Canals (2003), where the following items have been detected as the main contributors to the site-dependency in apple production (see Fig. 4.2).

Data for soil and climate conditions are relatively easy to obtain from competent bodies. The problem when developing an LCI in a specific area is to take into account the different producers' practices. In general, each farmer has his or her own production method. Thus, the best way to take quality data is to consult the farmer, though expert advice or information from a regulatory body can be valuable sources of information depending on the aim and scope of the study. Nevertheless, 'reference' practice may differ considerably from that of individual farmers, and this should be taken into account when deciding the source of data relevant for the goals of the study. In an integrated apple production study in New Zealand, Milà i Canals *et al.* (2003) find big differences (of up to one order of magnitude) in the contributions to environmental impacts between the reference values (suggested by expert advice and regulation body's records) and actual producers. These differences are caused by the selection and dosage of synthetic pesticides.

Indeed, data collection is facilitated when agricultural practices are controlled in order to optimise the use of resources, such as in integrated or organic fruit production schemes. The producers have to follow the guidelines of regulating bodies and they also have to inform them by handing in an annual report. Thus when developing an LCI in integrated or organic production, the data of producers' practices can be obtained from the regulating body (which can be considered as average practice in that area).

4.3.2   Understorey management

Fruit production systems generally have an understorey below the trees or bushes that actually produce the fruit. The understorey typically consists of herbage or bare soil, and its management is aimed at avoiding competition for water or nutrients with the trees, while preventing soil erosion and keeping a good structure for water infiltration. In plantations under integrated pest management (IPM) or organic farming the understorey has the additional and crucial function of providing a reservoir and shelter for pest predators. The main types of understorey management in fruit plantations are described below (Fernández, 1996):

- *Tillage:* This consists of the use of machinery (disc, plough or cultivator) between rows of trees in order to eliminate weeds (3–5 times per year over the entire surface). This practice is common in arid regions, where it is important to avoid competition for water, and there is little weed regrowth after spring discing because there is little rainfall.
- *No tillage with herbicide strip:* Herbicides are used to prevent weed growth. The period, the system and the dose of application depend on the herbicide and the climatic and soil conditions. Nevertheless, herbicides are generally applied two or three times during the year (Gil-Albert, 1995). Yagüe and Bolívar (2001) have listed the main herbicides used for fruit production. The production of such substances must be included in the LCI, as explained in Section 4.3.4. Toxicity impacts derived from herbicides must also be taken into account in the LCI, as referred to in Section 4.5.
- *Mulching:* A mulch (either organic like straw, rice shell or sawdust, or inorganic like plastic films or sand) is spread over the surface of the soil to control the weeds by choking (Gil-Albert, 1995). In this operation, machinery is needed, usually tractors equipped with a convenient device. Inorganic mulching is kept up for one year only, thus the impact of production, transportation and waste of inorganic mulching must be considered. Organic mulching has a contribution to the organic matter of the soil, which has to be considered in the nutrient balance (and in the carbon balance if this is included in the study).
- *Grassed understorey:* This is more common than mulching in fruit production. The soil is completely covered by artificially planted grass, which competes with the weeds by suffocating them. There exists an additional need for water in order to avoid the competition for it between the plant cover and the fruit trees. Machinery is needed for sowing the cover crop and for mowing it when it is convenient. The contribution of plant cover to the soil quality is important. When there is farm work, plant cover is mixed with soil (increasing nutrients and organic matter). If it is cut, there is an input into the soil as well, and the contribution to soil organic matter and nutrient can be significant (see Section 4.4). In other cases, the cover is perennial, such as in legume-based covers. In these cases, the fixation of atmospheric nitrogen has to be taken into account for the nutrient balance.
- *Mixed systems:* In these kinds of systems, two or more of those described above are combined. For example, one mixed system is the combination of the farm work during spring and summer and the maintenance of a plant

covering (natural or cultivated) during autumn and winter. It is also common to have a grassed understorey between the rows while keeping the tree lines free of vegetation either with a herbicide strip or using mulches together with the mechanical removal of weeds (e.g.. with a mower or weed-eater – a smaller machine, usually hand operated, used to reduce weeds' height).

4.3.3   Tree management

Tree management is the group of operations aimed at improving the system's productivity and efficiency by ensuring access to light and space for all fruits, prevention of pest and disease proliferation, and facilitating mechanical operations in highly mechanised systems (e.g. vine training). Pruning, thinning and tree training are the operations usually considered under tree management.

In fruit production, pruning is carried out with the aim of ensuring the provision of light and space for all the branches. In this way, all the fruit can develop satisfactorily, and thus the proportion of the crop that can reach the market (the 'pack-out') is higher. Pruning is usually carried out by hand using hand-operated pruners, loppers and saws and using manual ladders for elevation. Commonly used machine-operated equipment includes compressed air operated pruners and motorised hydraulic ladders (hydra-ladders) when trees are tall. Prunings are usually kept in the orchard by mulching them with a mulching mower, although some farmers remove them and burn them. While the mulching helps supply the nutrients and organic matter to the orchard soil, removing and burning prunings is usually faster, and therefore some producers prefer to do this. The problem with burning is atmospheric emissions from the combustion process. The fate of prunings should be considered in the nutrient balance of a fruit production LCA (see Section 4.3.10).

Fruit thinning is one of the key issues for most fruit growers. Its primary objective is to reduce the number of fruits per tree, increase fruit size and colour and hinder biennial bearing (i.e. alternate years of high and low fruiting). Furthermore, thinning results in fruit that is more evenly distributed within the tree canopy. This facilitates spray and light penetration, reduces potential insect feeding sites, and aids fruit drying after rain or irrigation, reducing fungal pathogen establishment. It is therefore a critical step in the production of high-quality fruit, and a big part of the 'art' of fruit growing that growers have to learn. In some fruits, there exist chemical fruit thinning agents that are used to facilitate this operation. When chemical thinning does not work well with a special fruit or variety (e.g. in peaches and nectarines), thinning is undertaken by hand, usually with the help of hydra-ladders. Chemical thinning agents are usually used only in conventional or integrated fruit production, and most of them are forbidden in organic plantations; this is one of the main reasons for the higher cost of producing fruit organically. When chemical thinning is performed, the fate of thinners should be included in the pesticide fate emissions, as they might be relevant from a toxicity point of view. In the case of manual thinning, this operation is usually the most labour-intensive one, and when mechanical aid is used (e.g. hydraulic ladders) it usually represents a highly significant part of the plantation's energy consumption.

Finally, some types of fruit trees may be trained in order to allow for mechanical harvesting (e.g. vines, new varieties of dwarf olive trees, orchard trees such as pear, apple, peach, etc.). This is mainly practised in highly mechanised fruit production systems, where high infrastructure investments are made with the aim of increasing productivity. When tree training is performed in such a way, all the infrastructure production (wires and poles to tie the trees to) should be included in the LCA together with the energy consumption for harvesting.

4.3.4   Pest and disease management

Pesticide production should be taken into account in an agricultural LCA, as it usually turns out to be a significant source of impacts: Milà i Canals *et al.* (2003) find that it represents up to 11–18% of total energy consumption in integrated apple production in New Zealand. To date, the most commonly referenced study for pesticide production impacts is Green (1987), although this is already quite old and most current active ingredients should actually be approached using Green's data. Audsley *et al.* (1997) give a useful procedure for approaching any active ingredient's production from the basic data in Green (1987). Nemecek and Heil (2001) provide a more recent source of data, even for some bio-pesticides. Data for sex-attractant pheromone production is available in Milà i Canals (2003).

There are different kinds of spray systems, from individual spray backpacks to systems attached to tractors. In addition, different techniques for pesticide spraying have different energy consumption requirements and have an effect on pesticide emissions. Examples of such differences are found when spraying at different concentrations, or at different times of the year (e.g., when trees are in full foliage or have no leaves at all). These effects should be taken into account in the LCI.

4.3.5   Irrigation

If there is not enough rainwater to supply the trees' needs, an irrigation system must be installed. There are different irrigation systems and the choice between them depends on many factors, described in Table 4.1 (Fernández, 1996):

- *Gravity irrigation (flooding and tracks)*: In these systems, the energy consumption is very low because water is distributed by gravity. The main problem is that the loss of water is high and this has to be taken into account in areas with water supply problems.
- *Spray irrigation (overhead sprinklers)*: In these systems the water is distributed in the form of rain. They are sometimes used in tropical fruit production and are not uncommon in places where frost fighting is needed.

Table 4.1   Factors affecting irrigation system choice (Fernández, 1996)

	Irrigation system			
	Flooding	Tracks	Spray	Drip
Slope limitations	<1%	<2%	No	No
Limitations of ground				
Infiltration (cm/h)	0.2–5	0.2–7.5	1.5–15.0	>0.05
Erosion danger	Moderate	Severe	Light	No
Salinity danger	Moderate	Severe	Light	Moderate
Water limitations				
Flux (l/s/ha)	1.6	1.2–1.6	1	<1
Climate				
Wind influence	No	No	Yes	No
Cost of the system				
Installation	Low	Low	High	High
Farm work	Moderate	High	Moderate to low	Low
Energy	Low	Low	High	Moderate
Irrigation efficiency (%)	40–80	40–70	70–90	80–90

Sprinklers, pipes (mainly plastic, PVC or polyethylene) and a pumping system are needed (Castañón, 2000). The energy consumption by the pumping system must be taken into account. The useful life of pipes is around 20 years and that of sprinklers and pumping systems around 10 years, so the production must be taken into account.

• *Drip irrigation*: These systems consist of the distribution of water by pipes to the point where it is needed under the tree canopy. They are fixed systems and there is no dependency on climate and soil conditions. A pumping system is also needed. The useful life of drips, pipes distribution in orchard and pumps is around 10 years, and that of pipes taking water to the orchard is around 20 years.

An efficient irrigation system must provide enough water to cover the farm's water necessities. Several references may be used to estimate these needs (Plaut and Meiri, 1994; Castañón, 2000; Martínez-Cortijo, 2001). Excessive supply of water has to be avoided in order to increase water use efficiency. This is very important in areas where water is scarce. The irrigation system will influence energy consumption. This influence will depend on the water source (surface water or groundwater) and on the irrigation type. In the case of groundwater, the extra energy consumption from the pump must be included. Low-density polyethylene (LDPE) pipes are used in drip and spray irrigation systems to distribute water. The amount of plastic used depends on the water needs, type of soil and plantation layout, and should be considered in the LCA inventory.

For instance, the typical plantation layout for citrus in the Valencian Community (Spain) is 5 × 4 m (i.e. one tree every 4 m in rows 5 m apart; this layout gives a density of 500 trees per hectare). In that case, a double line of

Table 4.2   Weight of LDPE pipes, depending on diameter and thickness (LDPE density = 0.941 kg/dm^3)

External diameter (mm)	Thickness (mm)	kg PE/m
12.0	1.0	$3.25*10^{-2}$
16.0	1.0	$4.43*10^{-2}$
16.0	1.3	$5.65*10^{-2}$
20.0	1.4	$7.70*10^{-2}$
20.0	2.0	$10.6*10^{-2}$
25.0	2.0	$13.6*10^{-2}$
32.0	2.0	$17.7*10^{-2}$
40.0	2.4	$26.7*10^{-2}$
50.0	3.0	$41.7*10^{-2}$
63.0	3.8	$66.5*10^{-2}$
75.0	4.5	$93.8*10^{-2}$
90.0	5.4	$135.0*10^{-2}$

distribution pipes, with one drip per metre is needed. For a 1 ha square orchard, this results in 4000 m of LDPE pipe per hectare for distribution in the field. Assuming 16 mm diameter pipe, the example would give 177.2 kg of polythene per hectare (see Table 4.2 for data on weight of LDPE pipes, depending on their diameter).

Bigger pipes are needed in order to transport the water to the field. In the example described above, 250 m/ha are needed (assuming 63 mm pipe, this is 166.25 kg polythene per ha). Finally, the amount of plastic calculated should be allocated throughout all the years it is functional in the field. The default value for the useful life of LDPE pipes can be considered to be 10 years for those within the orchard, and 20 years for those leading to it. Therefore, in the above example, 26 kg of LDPE per hectare per year would be used (177.2/10 + 166.25/20).

4.3.6   Nourishment

Soil cannot provide nourishment to plants indefinitely, and using fertilisers provides additional sources of nutrients. Fertilisers can be defined as organic or inorganic products containing at least one of the three main elements (nitrogen, phosphorus or potassium) (Domínguez, 1999). The production of fertilisers should be included in the LCI, and reference data for this production can be found in the literature (see, e.g., Audsley *et al.*, 1997; Davis and Haglund, 1999). Farmers monitor total amounts of fertilisers used, and it is not difficult to obtain these data from them or from agronomy books, where reference values are given.

Fertilisers can be applied to the soil in different forms: directly to the soil, mixed with irrigation water (fertirrigation) or directly to the tree by pulverisation. Depending on the form, the machinery used will be different and this will influence the LCI.

Normally, when an irrigation system (spray or drip) is installed the most usual way is to use fertirrigation for nourishment. This consists of administering the fertiliser dissolved in irrigation water. In this case, irrigation systems are being used to distribute nourishment (Domínguez, 1996; Moya, 1998), thus improving machinery efficiency. The only extra equipment needed is a tank where fertilisers are mixed with water. Sometimes, herbicides and fungicides are also applied with irrigation water.

Pulverisation to trees is used for specific treatments (amino acids, micro elements, etc.), which are needed for a special and precise requirement. The addition of nutrients to the system has a potential risk for causing emissions, to both water and air. These emissions (usually coming from the nutrient itself, e.g. NH_3, N_2O, NO_x, PO_4^{3-}, NO_3^-, but also CH_4 emissions from soil caused by fertilising) must be taken into account because they usually determine the LCA results. A good reference to address these emissions is Weidema and Meeusen (2000), and a proper tool to assess them is nutrient balance (see Section 4.4).

4.3.7 Weather damage prevention

Periods of extreme weather (frost, hail, excessive heat, etc.) may reduce the economic profitability of a plantation by destroying fruit or flowers, or otherwise damaging the trees. There are many examples of operations aimed at reducing weather effects. For example, frost fighting operations aim to avoid damage to the crop when temperatures fall below 0°C. Usually the idea behind these operations is precisely to prevent temperatures dropping below 0°C. The systems used for frost fighting are many; most of them are based on watering the trees, but other approaches such as the use of wind machines are important. Sunburn to the fruit can occur during the hottest days of summer. Activities to reduce sunburn damage are undertaken in order to prevent high temperatures causing effects similar to low temperatures.

The energy consumption allocated to these types of operations is usually high; Milà i Canals *et al.* (2003) describe an apple orchard where one-third of total diesel consumption was due to frost fighting. The reason is that extreme weather is treated as an emergency, and powerful machinery is involved in its treatment. Besides, the need for weather damage prevention may vary widely among plantations, as well as the means to perform this prevention. Usually, the more likely extreme weather events are, the more means will there be, and the richer is the socio-economical situation of the plantation. Consequently, differences in weather damage prevention may result in significantly different LCA results for different fruit production sites yielding the same product. Therefore, it is recommended that a careful description of the ways used to prevent weather damage prevention is included in the LCA, and that the effects of site- and location-dependency are considered in the analysis.

4.3.8   Harvest

Fruit harvesting is carried out by hand in many cases, but mechanical harvesting is possible for some particular (especially trained) crops. Also wide differences between particular locations are expected in this field operation, as they will reflect the effect of different productivities and tree management. Any use of machinery and energy consumption should be inventoried as in every agricultural LCA.

4.3.9   Post-harvest operations

Usually fresh fruits receive some post-harvest treatments before being commercialised. Fruits are ideally harvested when quality is at an optimum, and they may deteriorate after harvest because they are living systems. The main post-harvest operations are (Wills *et al.*, 1998):

- *Cold storage*: In this case, energy consumption of cold chambers will have an influence on LCI.
- *Storage atmosphere*: This is usually used in combination with low-temperature storage, thus energy consumption will also be considered in LCI.
- *Fungal treatments:* In order to control microorganism attacks, some chemicals are used as post-harvest fungicides, for instance benzimidazoles to control *Penicillium* in stone fruits or sulphur in order to control *Monilinia* or *Sclerotinia* in peaches.
- *Edible wax*: A layer of wax avoids loss of water, improves appearance and stimulates a controlled storage atmosphere.

The production, transport, method of application and emissions of these chemicals will have to be evaluated in an LCI of fruit production. Energy consumption during storage at a controlled temperature also has to be considered.

Transport of fruits is usually refrigerated. For instance, citrus transportation occurs in refrigerated trucks. The temperatures vary depending on variety and on the distance of transport. Before transportation, the fruits are pre-cooled, thus it is very important to design the packaging in order to optimise pre-cooling. Energy consumption for pre-cooling and for transportation must be included in the LCI. And if the transportation is carried out under refrigeration, energy consumption for maintaining low temperatures during transport will be considered. Also packaging tends to be complex in fruit transportation, usually because of the long distances involved and the fragility of the product. Many different materials may be used in the packaging, and this often makes it difficult to recycle them after use or in the event that a lot is spoiled.

4.3.10   Machinery production, energy consumption and emissions

Audsley *et al.* (1997) give a useful approach for calculating impacts caused by machinery production. In particular, data on the weight, composition (usually

Table 4.3 Characteristics of typical machinery used in fruit production (categories as defined in Audsley *et al.*, 1997)

Machine	Cat.[a]	Weight (kg)	Service life (years)	Total use in lifetime (hours)	Reference
Fertiliser Pendulum Spreader	C	125	10	800	Adapted from Hermansen (1998)
Air-blast sprayer (2000 l)	C	640	10	3000	Adapted from Hermansen (1998)
Herbicide sprayer	C	640	10	400	Adapted from Hermansen (1998)
Mower	C	450	8	400	Adapted from Hermansen (1998)
Weed-eater	C	50	8	300	Milà i Canals (2003)
Hydra-ladder, 12 hp	A1	700	10	10 000	Milà i Canals (2003)

[a] Category A1 is 'small tractors', and category C is 'other machines' (different from self-propelled machines and tillage machines) (Audsley *et al.*, 1997: 36).

simplified to steel and rubber), and total use in lifetime for each machine should be provided in order to complete the LCI. Values for total use in lifetime are usually very uncertain, and may heavily influence the results; therefore, special care should be taken with their collection. Table 4.3 provides information on typical machines found in fruit production.

Energy consumption is often difficult to estimate for field operations. Nevertheless, this generally represents one of the main sources of environmental impacts, and thus it should be estimated correctly. What is needed for the LCI is the energy consumption per hectare and year, but only the duration of operations can be commonly collected from farmers. This needs therefore a further transformation using fuel consumption per hour. If it is not possible to obtain data directly from the farmers, references from the literature should be used. Nevertheless, most arable crop operations can be found in the literature (see, e.g., Nielsen and Luoma, 2000; Vitlox and Michot, 2000; Cortijo, 2000; all of these are in Weidema and Meeusen, 2000), and fruit production operations have received less attention. Table 4.4 provides some examples of energy consumption for typical field operations in fruit (apple) production. It must be noted that fuel consumption per hour depends on many factors: the engine rate, field slope, driver's habits, etc.

An important aspect when an irrigation system is used is its energy consumption. The energy needed (E) for the irrigation of one hectare during a production campaign can be calculated by using the following equation:

$$E = (P * Q * t * \rho * g)/\eta$$

where P is the power needed for the irrigation system, Q is the volume of water provided by the pump, t is the length of time the system is working, ρ is the density of water, g is the acceleration due to gravity, and η is the output of the pump.

Table 4.4 Fuel consumption for different apple production operations using different tractors (Milà i Canals *et al.*, 2003)

Tractor rate and operation	Diesel consumption (l/h)	
	47 hp	60 hp
Tractor, mowing	4.0	5.0
Tractor, swing-arm mowing	6.0	9.0
Tractor, air-blast spraying	6.0	9.0
Tractor, spraying herbicides	4.0	5.0
Tractor, spreading fertilisers	4.0	5.0
Tractor, spreading mulching/compost	3.0	4.5
Tractor, collecting prunings	3.0	4.5
Tractor, mulching prunings	6.0	9.0
Tractor, harvesting	3.0	4.5

4.4   Data collection: nutrient balance

Nutrient balance is a useful tool for quantifying the flow of nutrients in agricultural production (Cederberg and Mattsson, 2000). It gives an indication of emissions derived from nutrient inputs. In addition to inputs from outside the system (fertilisers, both organic and inorganic), the nutrient balance must include the recycling of nutrients returning to the soil through tree leaves, prunings, fallen fruit, grass clippings, etc. Therefore, it is at least related to the operations of nourishment (for external sources of nutrients), understorey management (e.g. inputs from grass clippings), and tree management (prunings).

As an example, the results of a nutrient balance in the integrated production of oranges in the Valencian Community (Spain) are shown (Sanjuán et al., 2003). In this study, the soil's nutrient stock has not been considered. The average nitrate, potassium and magnesium contents in water used for irrigation have been obtained from public databases. Elements provided by rainwater have been taken from Carratalá et al. (1998) and an average annual rainfall in the zone of 500 mm has been considered. An average of 1.5% of organic matter in the soil was taken into account. It is assumed that organic matter content is stable. Organic matter mineralisation was computed using the method proposed by Henin and Dupuis (1945). The mineralisation coefficient has been taken from data from Tamés (1975) in Pomares (2001) for irrigation farming in eastern Spain. The C/N ratio in soil humus has been concluded to be 10, with 58% carbon content in soil (Brady and Weil, 1999; White, 1997; Rowell, 1994).

The nutrient extraction by the crop was obtained from Legaz and Primo Millo (1988). Nitrate leaching was calculated according to the results obtained by Ramos et al. (2002) for orange cultivation in the Valencian Community. To determine the N immobilised by microorganisms, the C/N ratio considered was 6 and its efficiency factor 1/3 (Brady and Weil, 1999; White, 1997).

The inputs of organic matter considered were sheep manure, prunes, and fruits and leaves that have fallen from the trees. It was supposed that the latter does not cause N immobilisation due to its high N content. Sheep manure does not cause N immobilisation according to the manure composition provided by suppliers. But it was considered that prunes cause N immobilisation. This was calculated according to data assessed by Moreno (2001). Nitrogen outputs by volatilisation and denitrification were calculated as proposed by Brentrup and Küsters (2000).

In Table 4.5 the results of the nutrient balance are shown. A, B, C and D are the four scenarios studied, depending on the origin of water, from a well or from a river, and on the fertilisation system, fertirrigation or solid fertilisation: A (well and solid fertilisation), B (well and fertirrigation), C (river and solid fertilisation) and D (river and fertirrigation).

In this balance, the reserves of the soil have not been included. This could explain the negative values in the N balance in Table 4.5 in an area such as Valencia where leaching is nowadays an environmental problem. The negative

Table 4.5   Results for nutrient balance (Sanjuán *et al.*, 2003)

INPUTS (kg/ha)		N	P	K
Mineralisation		31.32		
Fertirrigation (B and D)		291.59	28.32	111.87
Fertilisation (A and C)		291.00	39.33	74.70
Rain		2.52	0.00	0.00
Irrigation (groundwater)	A	52.68		9.69
	B	57.41		10.56
Irrigation (surface water)	C	11.12		16.72
	D	12.11		18.22
TOTAL A		**377.52**	**39.33**	**84.39**
TOTAL B		**382.84**	**28.32**	**122.42**
TOTAL C		**335.96**	**39.33**	**91.42**
TOTAL D		**337.55**	**28.32**	**130.09**

OUTPUTS (kg/ha)		N	P	K
Extraction		226.5	22.00	123.00
Lixiviation	A	124.58		
	B	126.34		
	C	110.87		
	D	111.39		
Immobilisation		44.63		
Volatilisation NH_3		4.97		
Denitrification N_2O		4.53		
Denitrification N_2		32.18		
TOTAL A		**437.39**	**22.00**	**123.00**
TOTAL B		**439.14**	**22.00**	**123.00**
TOTAL C		**423.67**	**22.00**	**123.00**
TOTAL D		**424.20**	**22.00**	**123.00**

BALANCE (kg/ha)	N	P	K
A	-59.87	17.33	-38.61
B	-56.30	6.32	-0.58
C	-87.71	17.33	-31.58
D	-86.65	6.32	7.09

values indicate a depletion of soil reserves, and the positive values indicate an accumulation.

When a grassed understorey is present, an important input of organic matter and nutrients (particularly N) may come from the herbage clippings, as suggested by Haynes and Goh (1980). They report a leaching loss of 33.1 kg $N\,ha^{-1}\,year^{-1}$ in an apple orchard in New Zealand receiving 71.6 kg $N\,ha^{-1}\,year^{-1}$ of fertilisers plus over 500 kg $N\,ha^{-1}\,year^{-1}$ of returns from the trees but mainly from the grass under the trees.

4.5   Data collection: pesticides

As noted in the introduction, LCI of fruit production systems should include site-dependent data. This is particularly important for the pest and disease management operations, as particular characteristics of the plantation may greatly affect the LCI results. A general model for the calculation of organic pesticide fractions reaching the environment is given in Hauschild (2000) and reviewed in Hauschild and Birkved (2002). Therefore, only some general comments are given below for the calculation of pesticide fractions reaching the environment:

- For the fraction drifting off the field with wind, Hauschild (2000) facilitates estimates from EPPO (1996), which are a worst-case estimate. Better data should be used if available, and considerations of the plantation limits such as the existence of windshields or any other type of fence should be included. Windshields usually act as barriers capturing part of the pesticide drifting off the field. Also the size and shape of the plantation should be taken into account, as only the pesticide that is being sprayed in close proximity to the field's edges is prone to wind drift (e.g., Praat *et al.*, 2000, suggest that only spraying of the first six tree rows has an effect on wind drift in an apple orchard).
- The amount of substance remaining in the field after subtracting the wind drift is divided into the fraction actually ending up on the plants (spray deposition) and the fraction on the field soil. The higher the foliage density at the moment of spraying, the bigger will be the amount of pesticide reaching its target on the plant. Thence, foliar density should be included in the analysis in order to detect effects on pesticide retention of both timing on pesticide application and tree distribution (note that the same spraying intensity will leave more pesticide on the ground if the trees have more space between them).
- Site-dependent models should be used to estimate pesticide leaching. The most relevant parameters affecting leaching are usually soil type (and particularly soil texture and organic matter content) and weather (amount of rainfall). Therefore, in order to detect potential risks in pesticide use, a model including these parameters should be used. Such models are increasingly available for different regions (e.g. the PESTLA model for Dutch conditions, PESTRISK in New Zealand, etc.).

Finally, pesticide emissions to air, surface- and groundwater and soil should be established from the modelling. Emissions to soil correspond to the fraction of pesticide present in soil after the plantation is abandoned or destroyed (when the system crosses the temporal boundaries); therefore consideration of the pesticide fraction degraded every year is needed. Usually, microbial degradation is the main process affecting pesticide residues in soil.

Some physical and chemical parameters of pesticides are needed in order to calculate these fractions: volatility, microbial degradation and photochemical degradation rates, partition between water and organic solvents (K_{OW}), etc. Most of these data may be found in a single reference: the *Pesticide Manual* of the British Crop Protection Council (e.g. Tomlin, 1995; newer editions are already available). Otherwise, the Material Safety Data Sheet from the pesticide producer usually gives detailed information as well. In the case of bio-pesticides, a good source of information is Copping (1998 and newer editions); nevertheless, as bio-pesticides' toxicity is usually negligible, modelling their fate is not of much concern.

The fractions of non-organic pesticides, such as copper, reaching the environment require a sound alternative approach. Audsley *et al.* (1997) give some general values for cereal crops, but these cannot be directly extrapolated to the case of fruit production. At least the consideration of fractions remaining in the soil (which might build up to highly relevant concentrations after the whole life cycle of the plantation), and those leaving the system to surface-water (runoff) and groundwater (leaching), should be included in the model. Also the fraction of heavy metals entering the plant and possibly ending up in fruit should be studied (the part reaching the fruit should be included as a potential risk for human health, while the rest either returns to the soil with litter or stays in the wood at the plantation destruction stage).

4.6   Assessing a LCA

As in any LCA, the impact assessment phase will determine how results can be interpreted, and whether these are relevant for the decision-maker. A fruit LCA should cover the 'usual' impact categories that are already developed and have a broad consensus within the LCA community. Besides, some specific categories are highly relevant specifically for fruit production (and agriculture in general). Firstly, human toxicity impacts caused by direct ingestion of pesticide residues in fruit will be a potential contribution to human toxicity. Secondly, the impacts caused in the working environment should be considered as relevant due to the high toxicity of some of the substances involved, and bad traditional practices. Other relevant environmental impacts associated with agriculture are those derived from land use and the use of water. In the case of land use, no consensus yet exists on the best available impact assessment methods, and there is not enough space in the present chapter for a proper dicussion of this issue. The reader is referred to Lindeijer *et al.* (2003) for a good review of the state of the art on land use impact assessment. Water consumption may lead to depletion in arid areas; in addition, some types of irrigation systems are actually designed to reduce water consumption, sometimes at the expense of higher energy consumption or infrastructure. Therefore, a sound evaluation of this impact should be included in any agricultural LCA, and thus also in the LCA of fruit production.

4.6.1   Human toxicity from pesticides

The three ways through which pesticides enter the human body are oral exposure by the mouth and swallowing, dermal exposure and absorption through the skin and eyes, and respiratory exposure (Bovey, 2001). A method that incorporates a fate-and-effect analysis of different substances is needed to assess their relative toxicity. Both physical–chemical and toxicological data on the pesticides will be needed to derive their characterisation factors, and these can be found in some references (e.g. Tomlin, 1995) or from the manufacturer (in the Material Safety Data Sheet). Some preliminary fate analysis for the calculation of the pesticide fractions reaching each environmental compartment has already been explained in Section 4.5.

4.6.2   Working environment

A special case for human toxicity in fruit production is the working environment. Consumers will be exposed to pesticides only by ingestion, but workers may be exposed by the three ways previously mentioned: ingestion, inhalation, and dermal contact. It is important for everyone working with agricultural chemicals to know the toxicity of the agro-chemical, both to himself and to other organisms (Bovey, 2001). Lundqvist (2000) suggests a procedure to include chemical risk of the working environment into LCA, after giving a review of the most important parameters for the working environment in a Swedish method. Some authors (Bovey, 2001; Zimdahl, 1999) give some recommendations for the safe use of herbicides. Most of these rules are common sense, but it is important to stress them in farmers' training and capacitating programmes.

4.7   Future trends

According to the aspects commented on in the present chapter and the previous findings in fruit LCA studies (e.g. Sanjuán *et al.*, 2003; Milà i Canals *et al.*, 2003), some research needs and future trends are suggested.

4.7.1   Producers' training

Farmers' practices have the capacity to change the environmental results of a fruit LCA analysis more than changes in production techniques for agro-chemicals or even farming machinery (Milà i Canals, 2003). Therefore, almost all of the improvement opportunities that may be suggested by LCA studies need the cooperation of farmers in order to succeed. This cooperation may be gained through training programmes.

New ways of communication should be developed to inform the farmers of the LCA results and the improvement opportunities. Indeed, it is difficult to communicate procedures that are usually against 'what has always been done', and still get the positive reaction that is needed if the environmental performance of agricultural production is to improve. In this respect, connecting the LCA results and the socio-economic needs of companies will be crucial for the farmers' acceptance. For instance, if they see LCA as a way to increase their competitiveness, to facilitate legislation compliance or the communication with the rest of the supply chain, there are more opportunities to get a positive reaction (Milà i Canals, 2003).

4.7.2   Consumers' education

The use of products by consumers is very important when developing an LCA of food products in general and of fruits in particular. Some of the fruits produced are not consumed because consumers are sometimes spendthrift (they throw away some of the product before consuming it). As a consequence, resources are consumed and emissions are produced without any benefit, and the LCA practitioner should consider this. Depending on the scope of the study, it might be better to consider a functional unit of 1 kg of *consumed* fruit rather than 1 kg of *produced* fruit. Apart from considerations of LCA scope, the authors suggest that the education of consumers in this way is very important in order to improve the environmental impacts of production systems.

4.7.3   Development of site-specific fate analysis models

One of the biggest difficulties when developing an LCI is to find models adapted to the local conditions of the study, in order to find suitable data or to evaluate the impact categories. Sometimes models or data from other conditions are used, which negatively affects the credibility of the LCI results and consequently on the LCA interpretation. There is a need to develop databases and models of impact assessment adapted to local conditions, taking into account weather and soil characteristics of each region. Data for soil and weather are locally available from public authorities.

4.7.4   Development of new impact categories for LCIA
Biodiversity and soil quality

Agriculture is one of the main sources of impacts on land, which affect both the intrinsic values of land (e.g. biodiversity) and the functional values (biotic production, nutrients and water cycling, etc.). These aspects have been poorly addressed by LCA applications until now, and should be further developed in order to increase the credibility of LCA studies. Lindeijer *et al.* (2003) offer an interesting review on this issue.

Water use

Water is basic for agricultural development. In some parts of the world water is a very scarce resource because of arid or semi-arid conditions. If groundwater is used to overcome lack of rainfall, the rate of renewal of aquifers should be considered, which generally raises potential depletion problems. Thus, the optimisation of water use is very important. But even in regions not having this problem (i.e. with wide availability of water), water consumption by agriculture is relevant because it generally reduces its quality, increasing salinity and the concentrations of toxic substances. Therefore, not only water quantity, but also quality aspects should be considered from an LCA perspective.

Toxicity of products and working environment
As discussed in Sections 4.6.1 and 4.6.2, the effect of chemical substances on life is very important. Humans, animals and plants are exposed to many toxic substances which have a negative effect on their health. The effect of these substances should be consistently studied, in order to include it properly in the LCA. Also research into their exposure paths and site-dependent models to include the fate of toxic substances within the LCA is needed. Application of such models and data in LCA might help in giving advice to achieve more environmentally sound alternatives of management, such as listings of the most problematic substances, application procedures to reduce risk of intoxication, etc.

4.8   References

AUDSLEY E (COORD); ALBER S, CLIFT R, COWELL S, CRETTAZ P, GAILLARD G, HAUSHEER J, JOLLIET O, KLEIJN R, MORTENSEN B, PEARCE D, ROGER E, TEULON H, WEIDEMA B and VAN ZEIJTS H (1997), *Harmonisation of Environmental Life Cycle Assessment for Agriculture,* Final Report. Concerted Action AIR3-CT94–2028. European Commission. DG VI Agriculture.

BOVEY R W (2001), *Woody Plants and Woody Plant Management. Ecology, Safety and Environmental Impact*, Marcel Dekker, New York.

BRADY N C and WEIL R R (1999), *The Nature and Properties of Soils,* Prentice-Hall International, Upper Saddle River, NJ, USA.

BRENTRUP F and KÜSTERS J (2000), 'Methods to estimate potential N emissions related to crop production', in Weidema B P and Meeusen M J G, *Agricultural Data for Life Cycle Assessments*, Agricultural Economics Research Institute (LEI), The Hague, Wageningen, The Netherlands vol 1, 104–113.

CARRATALÁ A, GÓMEZ A and BELLOT J (1998), Mapping Rain Composition in the east of Spain by applying kriging. *Water, Air and Soil Pollution*, 104: 9–27.

CASTAÑÓN G (2000), *Ingeniería del Riego. Utilización Racional del Agua, Paraninfo Thomson Learning*, Madrid, Spain.

CEDERBERG C and MATTSSON B (2000), 'Life cycle assessment of milk production – a comparison of conventional and organic farming', *J Clean Prod,* 8(1): 49–60.

COPPING L G (ed.) (1998), *The BioPesticide Manual* (first edition), British Crop Protection Council, Farnham, Surrey, UK.

CORTIJO P (2000), 'Conclusions: data on energy use and fuel emissions in agriculture', in Weidema B P and Meeusen M J G, *Agricultural Data for Life Cycle Assessments,,*

Agricultural Economics Research Institute (LEI), The Hague, Wageningen, The Netherlands, vol 1, 96–102.

COWELL S J (1998), *Environmental Life Cycle Assessment of Agricultural Systems: Integration into Decision-making*, Centre for Environmental Strategy, University of Surrey, Guildford, Surrey, UK.

DAVIS J and HAGLUND C (1999), *Life Cycle Inventory (LCI) of Fertiliser Production. Fertiliser Products Used in Sweden and Western Europe*, Report No 654, The Swedish Institute for Food and Biotechnology, SIK, Göteborg, Sweden.

DOMÍNGUEZ A (1996), *Fertirrigación*, Mundi-Prensa Madrid, Spain.

DOMÍNGUEZ A (1999), *Tratado de fertilización*, Mundi-Prensa Madrid, Spain.

EPPO (1996). Chapter 12: 'Air' in *Decision-making scheme for the environmental risk assessment of plant protection products* (draft version). European and Mediterranean Plant Protection Organisation, Council of Europe, 96/5543. Cited in Hauschild, 2000.

FERNÁNDEZ R (1996), *Planificación y Diseño de Plantaciones Frutales*, Mundi-Prensa, Bilbao, Spain.

GIL-ALBERT F (1995), *Tratado de arboricultura frutal. Vol IV Técnicas de mantenimiento del suelo en plantaciones frutales.* Ministry of Agriculture, Fishing and Food. Mundi-Prensa Madrid, Spain.

GREEN M B (1987), 'Energy in pesticide manufacture, production and use', in Helsel Z R (ed.), *Energy in Plant Nutrition and Pest Control*, Elsevier, Amsterdam.

HAAS G, WETTERICH F and GEIER U (2000), 'Life cycle assessment in agriculture on the farm level', *Int. J. LCA*, 5(6): 345–348.

HAUSCHILD M Z (2000), 'Estimating pesticide emissions for LCA of agricultural products', in Weidema B P and Meeusen M J G, *Agricultural Data for Life Cycle Assessments*, Agricultural Economics Research Institute (LEI), The Hague, Wageningen, The Netherlands, 64–79.

HAUSCHILD M Z and BIRKVED M (2002), *Estimation of pesticide emissions for LCA of agricultural products*, poster presentation, 12th Annual Meeting of SETAC Europe, Vienna, 6–10 May 2002.

HAYNES R J and GOH K M (1980), 'Distribution and budget of nutrients in a commercial apple orchard', *Plant and Soil*, 56: 445–457.

HENIN S and DUPUIS M (1945), 'Essai de bilan de la matière organique des sols', *Ann Agr,* 15(1): 161–172.

LEGAZ F and PRIMO MILLO E (1988), *Normas para la Fertilización de los Agrios.* Serie: 'Fullets Divulgació' nº 5–88, Direcció General de Innovació i Tecnologia Agrària, IVIA, Valencia, Spain.

LINDEIJER E, MÜLLER-WENK R, STEEN B (eds), BAITZ M, BROERS J, CEDERBERG C, FINNVEDEN G, TEN HOUTEN M, KÖLLNER T, MATTSSON B, MAY J, MILÀ I CANALS L, RENNER I and WEIDEMA B (contributors) (2003), 'Impact assessment of resources and land use', in Udo de Haes H A, Jolliet O, Finnveden G, Goedkoop M, Hauschild M, Hertwich E G, Hofstetter P, Klöpffer W, Krewitt W, Lindeijer E W, Müller-Wenk R, Olson S I, Pennington D W, Potting J and Steen B (eds), *Towards Best Practice in Life Cycle Impact Assessment. Report of the second SETAC-Europe working group on Life Cycle Impact Assessment*, Pensacola (USA), SETAC (forthcoming).

LUNDQVIST P (2000), 'Occupational health data in agriculture', in Weidema B P and Meeusen M J G *Agricultural data for Life Cycle Assessments*, Agricultural Economics Research Institute (LEI), The Hague, Wageningen, The Netherlands, 92–96.

MARTÍNEZ-CORTIJO F J (2001), *Introducción al riego,* Politechnic University of Valencia, Valencia, Spain.

MILÀ I CANALS L (2003), *Contributions to LCA Methodology for Agricultural Systems. Site-dependency and soil degradation impact assessment.* PhD thesis, Universitat Autònoma de Barcelona, Spain.

MILÀ I CANALS L, BURNIP G, COWELL S J and SUCKLING M (2003), 'Relevance of site-dependency in an apple production LCA', in preparation.

MORENO R (2001), *Caracterización analítica de residuos orgánicos para su posterior compostaje y aprovechamiento agrícola.* Final Project developed in the Valencian Institute of Agricultural Research (Instituto Valenciano de Investigaciones Agrarias, IVIA). Valencia, Spain.

MOYA J A (1998), *Riego Localizado y Fertirrigación,* Mundi-Prensa, Madrid, Spain.

NEMECEK T and HEIL A (2001), SALCA – Swiss Agricultural Life Cycle Assessment Database, Umweltinventare für die Landwirtschaft, Version 012. December 2001. FAL, Interner Bericht Teilprodukt Ökobilanzen, Produkt Öko-Controlling, 75 pp.

NIELSEN V and LUOMA T (2000), 'Energy consumption: overview of data foundation and extract of results', in Weidema B P and Meeusen M J G, *Agricultural Data for Life Cycle Assessments,* Agricultural Economics Research Institute (LEI), The Hague, Wageningen, The Netherlands, 48–63.

PLAUT Z and MEIRI A (1994), 'Crop irrigation', in Tanji K K and Yaron B (eds) *Management of Water Use in Agriculture,* Springer, New York, USA.

POMARES F (2001), *La Fertilización Orgánica. XIII Curso de Producción Integrada,* Federación de Cooperativas Agrarias Valenciana (FECOAV), Valencia, Spain.

PRAAT J P, MABER J and MANKTELOW D W L (2000), 'Managing spray drift on your orchard', *HortResearch Focus Orchard website* http://www.hortnet.co.nz/industry/ focus_orchard/managing_sd.htm 2000-07-26.

RAMOS C, AGUT A and LIDÓN A L (2002), 'Nitrate leaching in important crops of the Valencian Community Region (Spain)', *Env Poll,* 118: 215–223.

ROWELL D L (1994), *Soil science: Methods and applications,* Longman Scientific and Technical, London.

SANJUÁN N, CLEMENTE G and ÚBEDA L (2003), 'Environmental effect of fertilizers', in Dris, R (ed.) *Crop Management and Postharvest Quality Handling of Horticultural Crops,* Oxford & IBH Publishing Co, New Delhi, in press.

TAMÉS C (1975), 'Equilbrio del humus en los suelos cultivados. Fertilizantes Nitrogenados Nacionales', *Bol. Informativo,* 52(1): 1–48.

TOMLIN C (ed.) (1995), *The Pesticide Manual (tenth edition) Incorporating the Agrochemicals Handbook (third edition, of 1991),* Crop Protection Publications, Farnham, UK.

VITLOX O W C and MICHOT B (2000), 'Energy consumption in agricultural mechanisation', in Weidema B P and Meeusen M J G *Agricultural data for Life Cycle Assessments,* Agricultural Economics Research Institute (LEI), The Hague, Wageningen, The Netherlands, Vol. 1, 64–83.

WEIDEMA B P and MEEUSEN M J G (eds) (2000), *Agricultural Data for Life Cycle Assessments.* Agricultural Economics Research Institute (LEI), The Hague, Wageningen, The Netherlands.

WHITE R E (1997), *Introduction to the Principles and Practice of Soil Science,* Oxford, Blackwell.

WILLS R, MCGLASSON B, GRAHAM D and JOYCE D (1998), *Postharvest. An introduction to the Physiology and Handling of Fruit, Vegetables and Ornamentals,* CABI,

Wallingford, Oxon, UK.

YAGÜE J I and BOLÍVAR C (2001), *Guía Práctica de Herbicidas y Fitorreguladores*, BASF Española SA, Barcelona, Spain,.

ZIMDAHL R L (1999), 'Herbicides and the environment', in *Fundamentals of Weed Science*, Academic Press, San Diego, California, USA, 411–444.

5

Life cycle assessment of animal products

C. Cederberg, Göteborg University, Sweden

5.1   Introduction

Livestock production systems are generally less efficient than vegetable production systems. For example, Wirsenius (2000) shows that animal production systems accounted for roughly two-thirds of the total appropriation of phytomass, while their contribution to the human diet was about one-tenth. Van der Hoek (1998) calculated the global nitrogen (N) efficiency in animal production (defined as the ratio between N in animal output products and nitrogen input) to be slightly over 10%, resulting in 102 million tonnes N excretion yearly by all domestic animals. Livestock use 3.3 billion hectares of grazing land and one-quarter of the world's cropland that in total corresponds to 1.45 billion hectares (de Haan *et al.*, 1997). To feed the global livestock population, more than two-thirds of the world's agricultural land is needed for fodder production.

It is evident that it is important to assess animal products environmentally to gain knowledge on how to improve these production systems. Some common methodology problems in LCAs of animal products are discussed in Section 5.2. These methodology issues are, however, not unique for LCA but relevant to any environmental analysis. In Section 5.3, some results from LCA studies on consumer milk and pig meat are presented. Section 5.4 discusses how the results of LCAs of animal products can be used when improving the environmental performance of livestock production. Finally in Section 5.5, future consumption trends on milk and meat are presented.

5.2   LCA methodology and animal products

Livestock production systems are very complex biological systems with delicate interactions between animals, plants and soil. The producing domestic animal consumes fodder and emits biogenic emissions, especially ammonia and methane. It is often difficult to obtain good quality data on fodder consumption (i.e. resource use) and emissions since there can be large variation within an animal population and also in breeding and husbandry methods. Data collection and data quality is one problem to deal with when performing environmental analysis of animal products. Other methodology issues of importance in LCAs, of animal production are handling manure allocation, system expansion in milk LCAs and impact assessment of ammonia emissions and of land use, which are discussed in the following section.

5.2.1   Allocation of manure

Manure is produced in all animal production systems. It is of great interest to analyse the recycling and use of manure since the lion's share of the nutrients in the animal's feed intake ends up in the manure, for example 75–80% of the nitrogen (N) in a dairy cow's feed intake is found in the manure. The farmyard manure produced is often used in the crop rotation on the farm where fodder to the livestock is cultivated, and in this situation the manure stays inside the studied livestock system and is not an output product. However, it is also common that manure is exported from a livestock farm to an arable farm. The reason for this is usually that the livestock density is too high and that the surplus manure is not allowed to be used in the fields at the livestock farm. In this situation, the manure is an input in another system (open-loop recycling) and there is an allocation problem.

In the EU Concerted Action report on Harmonisation of Environmental LCA for Agriculture, there were several suggestions for this allocation situation but no consensus agreement was reached on how to solve it (Audsley *et al.*, 1997). In the case study in the EU report, farmyard manure was imported to fertilise an organic wheat crop. The final solution to the allocation issue was to include the burdens of nitrogen input via the manure input to the organic wheat system, by including the environmental burden arising from the equivalent nitrogen production using biological N fixation in a clover crop. In this procedure, inputs of phosphorus and potassium through the import manure were not considered.

Van Zeijts *et al.* (1999) suggest another way of dealing with the allocation problem in the open-loop recycling situation. Depending on the situation, if there is high livestock density and/or a market for organic fertilisers, animal manure can have an economic value at the livestock farm gate. It can be negative, which often is the case in Dutch agriculture, and livestock farmers pay to get rid of the manure. But it can also be positive, if there are arable farmers willing to pay for the manure. Somewhere in the process of storage and transport to the arable farms the negative economic value turns into a positive one. Van

Zeijts *et al.* suggest that if the manure is bought for arable farming the environmental impacts of storage and transport are allocated to the arable products.

In an LCA study comparing organic and integrated carrot purée production, the allocation method of the manure used in the organic system had a decisive impact on eutrophication and acidification (Mattsson, 1999). The manure used in the organic carrot cultivation was purchased from a livestock farm and, since the manure was regarded as a product with a value entering the carrot system, the ammonia emissions in the stables were allocated to the livestock while the storage and field emissions were allocated to the fertilised carrot crop. Since only synthetic fertilisers were used in the integrated carrot crop in this study, the ammonia emissions in the organic production were considerably higher, which resulted in a relatively bad outcome for the eutrophication and acidification categories of the organic system.

It is very important to observe that the choice of how to allocate the emissions from farmyard manure which are transferred from one production system to another will have influences on LCAs of the exporting livestock production system as well as the importing arable system. In a mixed farming livestock system in balance, the manure is used in the fodder crops only, i.e. within the system, thus all emissions belonging to the manure are allocated to the animal products of this system. In a mixed farming livestock system with high animal density, or in an industrial livestock system with a small land-base, it is common that the manure is exported to an arable farm. If some of the manure emissions are then allocated to the arable crops, and an LCA compares a balanced mixed-farming livestock system with an industrial livestock system, the latter system will most likely show better results in some of the impact categories, since the allocation procedure distributes some of the manure's emissions to an arable crop. Therefore, when comparing different livestock production systems, it is very important to assess how the environmental burden of the farmyard manure is allocated if manure is an output product (exported from the farm).

5.2.2   System expansion in milk LCAs

When analysing multifunctional production systems, it is important to determine to what extent the environmental burdens of the process should be allocated to the product investigated. Dairy production contains a number of multifunctional processes. In the production of concentrate feed for the dairy cows, co-products from vegetable oil, sugar and starch crops are important raw materials. In the dairy industry, depending on the fat content of the milk, cream is a co-product when consumer milk is manufactured. When producing cheese, whey is an important co-product, often used as feed in pig production. However, the most important co-product from milk production is beef and the choice of allocation between milk and meat has a big outcome for the result of an LCA. Milk and beef production are very closely interlinked and surplus calves and meat from culled dairy cows is an important base for beef production. In the EU, approximately 50% of beef products are derived from by-products of the dairy sector (Oomen *et al.*, 1998).

To be able to produce milk, a dairy cow must give birth to a calf. In Sweden, the average dairy cow produces 1.07 calves per year. The yearly replacement rate of the Swedish dairy cow population is 37%, and to replace these culled dairy cows, 0.37 heifer calf per dairy cow per year has to be raised as a replacement animal. This leaves a surplus of 0.7 calves per dairy cow per year to be used in beef production. What share of the environmental burden of a milk production system should the co-product meat from the culled cows and surplus calves carry?

Economic allocation is often used in LCAs. However, in the ISO standard for LCA (ISO, 1998), it is suggested that before allocation is performed reflecting some relationship (e.g. economic relationship) between the environmental burdens and the functions, allocation should be avoided if possible. System expansion is a method of avoiding allocation. When system expansion is used to avoid allocation of the environmental burden between milk and meat, the milk system is expanded also to include the alternative way of producing meat and calves. In the alternative beef-producing system, suckler cows are the mother animals. These cows produce one calf per year and their milk is not an output product but is used only to feed the calves. Since the two co-products (surplus calves and meat) from culled cows are always generated when milk is produced, the milk system is credited for the avoided burdens of these two products in the beef system when system expansion is performed. By this operation, system expansion gives a truer picture of the real relationship between the milk and beef producing systems than an allocation dividing the environmental burdens through some physical relation between milk and meat, for example through an economic relation.

Table 5.1 shows the share of the environmental burden that was attributed to milk when economic allocation and system expansion was used for dividing the environmental burden between milk and meat (Cederberg and Stadig, 2001). Due to the low income for meat and calves, as much as 92% of the income for the Swedish dairy farmer comes from milk. Thus when using economic allocation for dividing the environmental burden between milk and meat, 92% of the burden is assigned to milk with economic allocation. The study shows that

Table 5.1 Distribution of environmental burden to milk when comparing economic allocation and system expansion

	No allocation	Economic allocation	System expansion
Energy	100	92	87
Land use	100	92	66
Climate change	100	92	63
Acidification	100	92	60
Eutrophication	100	92	60

economic allocation between milk and meat favours meat. When meat production is a co-product of milk production this entails fewer animals and thus less manure handling and less fertilising of crops. The distribution of the important biogenic emissions from ruminants (methane and ammonia) and nitrogen losses due to land use and its fertilising (nitrate and nitrous oxide) shows the largest differences when system expansion is compared with traditional economic allocation.

When LCA is used in prospective studies, system expansion should be the prime option for handling allocation situations (Weidema, 2001). Milk and beef production is a good example of this. In the present EU agricultural policy, a milk quota system regulates total milk production, and with increasingly higher-yielding cows this will result in a lower dairy cow population. However, this will also lead to lower beef production as a co-product from milk production since fewer surplus calves are produced in the milk system. If beef consumption remains at a constant level, the pure beef production system will have to increase to compensate for a lower and more high-yielding dairy cow population. To foresee and make prognoses for the overall environmental effect of future milk production, changes in beef production systems must also be considered. Modelling those two interacting systems with system expansion is a good method for understanding how the overall environmental impact will alter if future milk production systems change.

5.2.3   Impact assessment of ammonia emissions

Emissions of ammonia (NH_3) represent a considerable loss of nitrogen from agriculture and are largely responsible for a poor N-balance in animal production. The deposition of NH_3 and NH_4^+ can cause acidification and eutrophication, to an extent that depends on the sensitivity of the area where the emitted NH_3 is deposited. The dry deposition velocity of NH_3 is high, which leads to relatively high dry deposition of NH_3 close to sources (<1 km) (Asman, 2001). In comparison with NO_x and SO_2, NH_3 is transported shorter distances before deposition.

Intense livestock production located on a small area can generate high local ammonia emissions. Table 5.2 shows ammonia emissions per tonne of milk and per hectare of arable land from a case study assessing conventional and organic milk in Sweden (Cederberg and Mattsson, 2000). The total NH_3 emissions were 2.5 times greater per hectare on the conventional farm in comparison with the organic farm. The NH_3 emission per tonne of milk (functional unit) was, however, only 15% higher on the conventional farm, so by regarding the NH_3 discharges only in relation to the product unit, the difference between the two systems was small. The main cause for the discrepancy was the significantly higher milk production per hectare of arable land on the conventional farm, which produced 7.4 tonnes of milk per ha per year compared with 3.3 tonnes of milk per ha on the organic farm. When assessing NH_3 emissions from animal production, discharges per unit of

Table 5.2   Ammonia emissions on conventional and organic dairy farms

	Conventional dairy farm	Organic dairy farm
NH$_3$-N emissions, kg N/ha	61	24
NH$_3$-N emissions, kg/tonne milk	7	6.1

product (functional unit) do not give sufficient information, because the environmental damages on a local scale can be high since a large portion of the emitted ammonia can be dry-deposited very close to the source. The risk of negatively affecting nitrogen-sensitive flora is thus greater close to area-intensive livestock farms.

Livestock production is responsible for the greater part of anthropogenic ammonia emitted into ecosystems. Ammonia can cause effects on a local as well as a regional basis. One way to describe the regional effects of ammonia emissions was presented by Potting *et al.* (1998) who used the RAINS model to relate the site of the emission to the impact on its deposition area. With calculations based on the RAINS model, Potting *et al.* have set land-based acidification factors for SO$_2$, NO$_x$ and NH$_3$. These factors relate the regions of emissions to the impact on their deposition area. However, this model is not sufficient for assessing the whole potential impact of ammonia since the ammonia is emitted very close to the source. In an LCA of animal products, ammonia losses should be accounted for per product unit (functional unit) as well as per hectare to give information on the spatial concentration of the emissions and thereby the risk of local damage.

5.2.4   Impact assessment of land use

Land use has so far seldom been incorporated in the impact assessment of LCA. A major reason is that although a number of methods have been suggested, it has been difficult to make this impact category operational. Also, the problem of collecting data for assessment of land use impacts is an obstacle (Cowell and Lindeijer, 2000). The great need for land use impact assessment is obvious since land is the basic resource for all agriculture. In LCAs of animal products, information on land use assessment is very important. Depending on the type of fodder (i.e. grass, cereals, soymeal), the influence on land use quality can vary significantly.

A method for impact assessment of land use was tested in a case study of cultivated vegetable oil crops (Mattsson *et al.*, 2000). Three sub-categories were chosen: (1) soil fertility, (2) biodiversity, and (3) landscape values. Indicators were selected to characterise these sub-categories. Land use was investigated thoroughly for rapeseed cultivation in Sweden, soybean cultivation in Brazil and oil palm cultivation in Malaysia. This study revealed that for some of the indicators suggested it was not possible to obtain adequate data concerning the crops' production impact. The information on land use impact assessment in this study was a mix of quantitative and qualitative data, which makes it extremely difficult to aggregate the data in an acceptable way. Accordingly, the assessment of the three vegetable oil crops was more descriptive and a step closer to Environmental Impact Assessment than are traditional LCA results. The results of such an assessment can still be useful to decision-makers. As an example of the type of results yielded by this method, a comparison between the protein feedstuffs rapeseed meal and soymeal was undertaken. Although the results were not aggregated, they showed that the rapeseed meal was clearly a better choice from the perspective of land use (Mattsson *et al.*, 2000).

In an LCA of Swedish consumer milk with 1.5% fat content, the yearly land use was approximately 1.4 m^2 to produce 1 litre of medium-fat consumer milk: see Fig. 5.1 (Svensk Mjölk, 2001). Grassland and grain were cultivated at the dairy farm site or close to it. Crops for production of protein feed, mainly rapeseed and soybeans, were to a large extent imported to the Swedish feed industry and processed as concentrate feed. A total land use assessment should include indicators such as soil erosion, soil structure, soil organic matter and biodiversity. The assessment showed that soil erosion and loss of biodiversity could be a problem in some of the imported crops, especially soybean cultivation.

The potential environmental impacts related to land use are very important in agricultural LCAs and especially in LCAs of animal products since the different fodder crops have varied impacts on the quality of land use. At the very minimum, the sub-categories soil fertility and biodiversity should be included. LCAs of animal products should at least contain a descriptive survey of land use so that decision-makers pay attention to the effects caused by land use for fodder production.

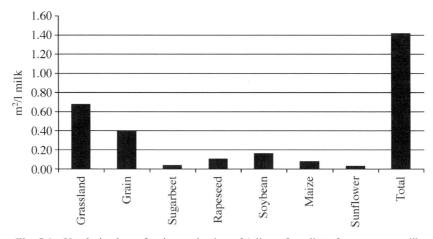

Fig. 5.1   Yearly land use for the production of 1 litre of medium-fat consumer milk.

5.3   LCA in practice: the cases of milk and pig meat

The environmental impacts of seven food items were analysed in Sweden using LCA methodology according to ISO 14040–14043 (LRF, 2002). The purpose of the study was to identify which parameters are most important for environmental impact (hotspots) in the food chain, from the production of raw material to the final consumption in the consumer's home. The knowledge gained was to be used in the food sector in environmental work and communication with consumers and customers.

Detailed and specific data on farm production, transport, industry processes and packaging were collected from farms and industries, including the packaging industry. The retail and consumer segment was not studied in detail, but was instead based on certain assumptions. It was assumed that 42% of consumers use cars for grocery shopping trips and that 58% cycle, walk or take public transport to the store. Consequently, only 42% of grocery shopping trips have an environmental impact in this study. It was also assumed that, on average, 15 kg of goods were bought per shopping trip, which covered a distance of 5 km (round-trip) to and from the retail outlet.

5.3.1   LCA of medium-fat milk

The production system of 1 litre of consumer milk was divided into six sub-systems:

1.   Production at the farm
2.   Transport to the dairy
3.   Production at the dairy
4.   Packaging (including waste treatment)
5.   Distribution to the market
6.   Market and consumer.

The functional unit was *1 litre of chilled and packaged medium-fat milk*, ready to consume in the customer's home. As described earlier, the milk chain contains many allocation situations. The division of environmental impact between main products and by-products was based mainly on economic allocation. When the milk study was conducted there were not enough data about alternative production of the co-products to use system expansion in order to avoid allocation. Also, the goal and scope of the study was not a prospective one, but rather to identify the hotspots of today's milk chain. System expansion was used when assessing the impact of the waste management, resulting in heat recovery from milk packages in incineration processes being credited to the milk system.

Generally, the 'Farm production' subsystem was responsible for the largest impact in the milk's life cycle. However, energy use was the exception where approximately 50% of the energy was used at the farm site, including the production of concentrate feed and fertilisers (Fig. 5.2). About 25% of the

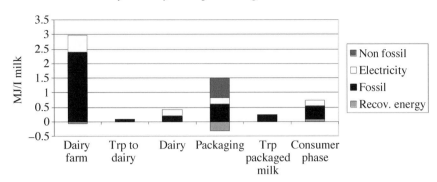

Fig. 5.2   Energy use in the life cycle of medium-fat consumer milk.

energy cost was due to packaging. Owing to heat recycling of the package in incineration processes, some energy was recovered in this system (shown as a negative in Fig. 5.2). The most important use of energy in the consumer phase was the car transport used in the shopping trips. This transport used more energy than the dairy's two transports of milk taken together.

Important environmental impacts caused by animal production and consumption are climate change, eutrophication and acidification. More than 90% of the greenhouse gas emissions are discharged in the 'Farm production' sub-system (Fig. 5.3). Emissions of greenhouse gases from biological systems play a larger part than CO_2 emissions from the use of fossil fuel. Methane (CH_4) from the ruminant's enteric fermentation is the most important greenhouse gas emission. Emissions of nitrous oxide (N_2O) are connected to the nitrogen cycle on the farm (losses from soils) and N_2O emissions from synthetic fertiliser production.

The most important emission of acidifying substance is ammonia (NH_3) from the cattle's manure, corresponding to more than 90% of the acidifying emissions. Ammonia emissions are also responsible for the eutrophication potential of milk together with nitrate (NO_3^-) which leaches from the fields where the cow's fodder is grown. Principally all of the nutrifying emissions in the milk's life cycle occur in the 'Farm production' sub-system.

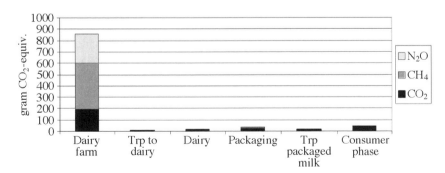

Fig. 5.3   Emissions of greenhouse gases in the life cycle of 1 litre of medium-fat consumer milk.

5.3.2 Pig meat
The production system for pig meat was divided into six sub-systems:

1.   Production at the farm
2.   Transport to the slaughterhouse
3.   Slaughterhouse
4.   Packaging (including waste treatment)
5.   Distribution to the market
6.   Market and consumer.

The functional unit was *1 kg of bone-free pig meat*, transported to the customer's home. Similarly to milk, the 'Farm production' sub-system was responsible for the largest environmental impact in the life cycle of pig meat. Roughly 70% of the total energy was used in the primary production, including the production of imported feed and fertilisers. Sub-systems also important for the total energy use were manufacturing of package material and processes at the slaughterhouse. Transports were only a minor fraction of the total energy use (Fig. 5.4). When cooking the meat in the consumer's home, an additional electrical energy use of approximately 3.5 Mj is used in the life cycle.

In the life cycle of 1 kg of bone-free pig meat, 4.8 kg of CO_2 equivalents were emitted (Fig. 5.5). Almost 90% of the greenhouse gases were emitted in the 'Farm production' sub-system. Nitrous oxides from soil processes when manure and fertilisers are transformed in the arable land are the most dominating greenhouse gas emission. The most important methane emission is due to manure management (slurry).

Similarly to milk and other dairy products, emissions of nitrate from the arable land where the fodder crops are grown, and of ammonia emissions from the farmyard manure, are responsible for almost all the discharges of nutrifying substances in the life cycle of pig meat. Ammonia from manure is also the dominating acidifying substance that is emitted in the life cycle.

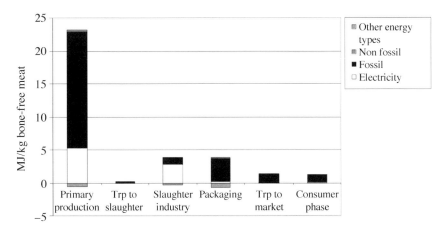

Fig. 5.4   Energy use in the life cycle of 1 kg of bone-free pig meat.

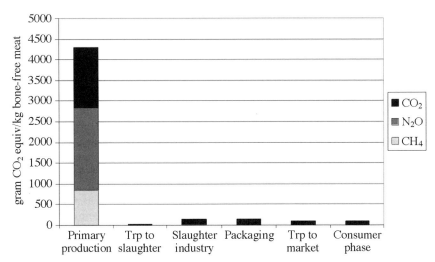

Fig. 5.5   Emissions of greenhouse gases in the life cycle of 1 kg bone-free pig meat.

5.4   Using LCA to improve production

The most important purpose of LCAs is to gain better knowledge on where in the life cycle it is most efficient to focus the environmental improvement measures. LCA studies of milk and meat show that the primary production at the farm is the most crucial stage in the route of the animal products and that the handling of the nitrogen cycle on the farm should be the focus of attention. The choice of fodder and the methods of fodder production can have a considerable influence on the outcome of several impact categories. Intensive livestock production can be very resource-efficient per unit of product, but because the livestock is concentrated in a small area, there can be large local environmental problems caused by such area-intensive production.

5.4.1   The farm's N-flow in focus

When the environmental impact of animal products is to be reduced, improvement measures are most beneficial at the farm level. Losses of reactive nitrogen compounds (NH_3, NO_3^-, N_2O) are important not only for eutrophication, acidification and climate change but also for the use of energy, since production of synthetic N fertilisers is responsible for a considerable share of the total energy use in primary production. Measures that lead to higher N efficiency in agricultural production should have high priority. For example, in Sweden there is now an on-going extension project aiming to increase farmers' knowledge about nutrient cycling, losses and use. The project is partly financed by taxes on synthetic fertilisers. Farmers are helped with analyses of the nutrients flows and losses on their farms and the analyses are the basis for improvement measures. Excessive use of nitrogen is often a result of lack of knowledge of the N content in manure and protein content in farm-produced fodder. Nitrate leaching can be reduced by postponed cultivation in the autumn, use of catch crops and modified handling of manure. More efficient use of the manure nitrogen means that less synthetic fertilisers are needed, which reduces the emissions of N_2O in the production of commercial fertilisers. It is very important to work with a whole systematic approach at the farm level when solving the problem of losses of harmful nitrogen compounds.

5.4.2   Importance of fodder production

A large share of the economic cost in animal production is due to fodder production. This is parallel to an 'environmental cost' caused by fodder production and this becomes obvious when environmental analyses of animal products are carried out. The origin of an animal product can vary significantly, especially for beef and milk. On the one hand, beef and milk can be produced in systems where grazing is the method of feeding the cattle, which implies low input of energy, fertilisers and pesticides. On the other hand, beef and milk can be produced in intensive systems where the feeding is stable-based, large amounts of grains and other concentrate feed are used, and roughage fodder (e.g. grass/maize silage) is harvested by machine. In the developed world, pig and poultry are mostly raised with cereals and different kinds of protein sources, in which soymeal is the most important. It is evident that the organisation of the animals' feeding system and the type of fodder are very important for the environmental performance of the final products.

The source of protein in the feed ration seems to be an important factor to observe when analysing the environmental consequences of fodder production. Fish meal is a protein feed with high energy cost and its production exploits the fish stocks in the seas of northern Europe. Soybean cultivation in the tropics can lead to soil erosion and losses of biodiversity. Cederberg and Mattsson (2000) showed that relatively large amounts of toxic pesticides – not allowed or with restricted use in Sweden – were used in the cultivation abroad of imported raw material for concentrate feed production in the Swedish feed industry. Increased use of domestically produced grass/maize silage and rapeseed meal in Swedish milk production, instead of importing a lot of concentrate feed from the tropics, is a way of reducing the total impact of fodder production. An experience from LCA studies is that the environmental aspect of protein feed production should be more carefully considered than is the case today.

5.4.3   Intensification vs extensification

Organic milk production is a promising way to attain more environmentally friendly dairy products (Cederberg and Mattsson, 2000). The main positive effect resulting from organic farming principles is the reduction of pesticides, which is substantial. For other environmental impacts that are connected to the use and emissions of nutrients, the positive effects from organic milk production are very much related to extensification: since inputs of fertilisers and feed are limited in organic farming, this requires a larger farm-based area for growing fodder crops and thereby lower milk production per hectare. This production system can be a good option for countries where the resource land is relatively ample and cheap. There are other ways besides extensification that can achieve milk production with lower environmental impact. Experience from the 'De Marke' experimental farm in the Netherlands, obtained in the period 1992–98, indicates that an average intensive commercial Dutch farm can halve inputs of fertilisers and feed without the need to reduce milk production per hectare or to export slurry (Aarts, 2000). The reduced N surplus from 'De Marke' has led to a significant decrease of nitrate concentration in the upper groundwater. A British study of different dairy farm systems shows the same result, that nutrient surplus and thereby potential losses can be substantially decreased in intensive farming systems. In the British study, phosphorus (P) surplus in a dairy farming system was lowered considerably by reducing fertiliser and/or feed input without limiting either milk or crop production (Withers *et al.,* 1999). Oomen *et al.* (1998) recommend a long-term solution for all EU agriculture in which a reintegration of crops and animals, on the farm level as well as on the regional level, would lead to a substantially lower N surplus and more efficient resource use.

5.5   Future trends

5.5.1   Production trends

A trend in global livestock production today is that industrial livestock systems grow much faster than more traditional mixed farming systems. Seré and Steinfeld (1996) define industrial livestock production systems as those in which less than 10% of the feed is produced within the farm unit. Industrial livestock production according to this definition represents as much as 43% of global meat production and two-thirds of global egg supply. Landless livestock systems are increasingly common for pigs and poultry but occur only rarely for ruminants. Industrial livestock systems are very dependent on an outside supply of feed and energy. A large amount of farmyard manure is produced on a concentrated area, and if this manure is not recycled to larger areas of arable land, the local effects can be substantial.

In mixed farming systems, crops and livestock are integrated on the same farm. This system dominates European livestock production and especially milk production. In this farming system, the manure is used in a crop rotation that normally produces fodder for the animals. There is, however, a broad spectrum as to just what extent crop and livestock production is in balance in mixed farming systems. In the developed world, mixed farming systems are very often connected with nutrient surplus, which is a result of an excess of manure due to high livestock density.

Table 5.3 Projected consumption trends of milk and meat to 2020

	Total 10^6 tonnes 1997	Total 10^6 tonnes 2020	Per capita kg, 1997	Per capita kg, 2020
Developed world				
Meat	98	114	75	84
Milk	251	276	194	203
Developing world				
Meat	111	213	25	35
Milk	194	372	43	61

Notes: Consumption is direct use as food, uncooked weight and bones. Meat includes beef, pork, mutton, goat and poultry. Milk is milk and milk products in liquid milk equivalents.
Source: Delgado *et al.* (2001).

5.5.2 Consumption trends

The International Food Policy Research Institute (IFPRI) forecasts food consumption trends. From 1997 to 2020, IFPRI projects the aggregate consumption growth rate of both milk and meat to be 2.9% per year in developing countries compared to 0.7% and 0.6%, respectively, in developed countries (Table 5.3).

Livestock production requires a large share of global agricultural land for producing fodder. In comparison with vegetable production, livestock production has a lower nutrient efficiency and is today responsible for a large share of the losses of reactive nitrogen from the agricultural sector. The future extent of milk and meat consumption as well as how future livestock production will be organised will have a major impact on the whole agricultural sector's environmental performance.

5.6 Sources of further information and advice

Food consumption patterns have a decisive impact on disruptions of the global N cycle. Azzaroli Blekken and Bakken (1997) have calculated and analysed the N flow in the whole food chain in a developed society (Norway). This study shows very clearly that the 'nitrogen cost', defined as the N input in primary production in relation to the N output in the food, varies a lot between different food items. Production of animal protein creates reactive nitrogen along the food chain to a much larger extent than vegetable. This is a good pedagogical paper in helping to understand the flow of N from primary production to final consumption.

The international organisation IFPRI publishes research and reports on future food consumption and has especially focused upon future livestock production and meat consumption in developing countries: see their website www.ifpri.org. Forecast environmental impact caused by agricultural expansion during the next

50 years is analysed by Tilman *et al.* (2001). This analysis forecasts that the pressure on resources such as land and increasing N emissions will be heavy due to the growing demand for food by a wealthier and 50% larger global population. For example, 10^9 hectares of natural ecosystems would be converted to agriculture by 2050.

The question of how to divide resource consumption and emissions between the output products from multifunctional processes is an important one in LCA. Weidema (2001) gives a number of good examples on how different allocation situations can be handled by system expansion. As discussed in Section 5.2, the food chain includes several allocation situations, a fact that is particularly important for animal products. In prospective environmental studies of food production and consumption, it is important to understand how different production processes are integrated in each other. By using the technique of system expansion, the true environmental effects of the use of co-products are better modelled and understood.

5.7   References

AARTS, H. F. M. 2000. *Resource management in a 'De Marke' dairy farming system.* De Marke Report No. 26, Research Institute for Animal Husbandry, Lelystad, The Netherlands.

ASMAN, W. A. H. 2001. Modelling the atmospheric transport and deposition ammonia and ammonium: an overview with special reference to Denmark. *Atmospheric Environment*, 35: 1969–1983.

AUDSLEY, E., ALBER, S., CLIFT, R., COWELL, S., CRETTAZ, P., GAILLARD, G., HAUSHEER, J., JOLLIET, O., KLEIJN, R., MORTENSEN, B., PEARCE, D., ROGER, E., TEULEON, H., WEIDEMA, B. and VAN ZEJTS, H. 1997. *Harmonisation of Environmental Life Cycle Assessment for Agriculture.* Final Report of Concerted Action AIR3–CT94–2028, Silsoe Research Institute, Bedford, UK.

AZZAROLI BLEKKEN, M. and BAKKEN, L. R. 1997. The nitrogen cost of food production: Norwegian society. *Ambio*, 26 (3): 134–142.

CEDERBERG, C. and MATTSSON, B. 2000. Life cycle assessment of milk production a comparison of conventional and organic farming. *Journal of Cleaner Production* 8: 49–60.

CEDERBERG, C. and STADIG, M. 2001. System expansion and allocation in Life Cycle Assessment of milk and beef production. In: *Proceedings from International Conference on LCA in Foods*, pp. 22–27. SIK-Document 143. Swedish Institute for Food and Biotechnology, Göteborg, Sweden.

COWELL, S. J. and LINDEIJER, E. 2000. Impacts on ecosystem due to land use: *biodiversity, life support and soil quality*. In: *Agricultural Data for Life Cycle Assessment* (ed. Weidema, B. P. and Meeusen, M. J. G.), Report 2.00.01, vol 2, pp 80–91, Agriculture Economics Research Institute (LEI), The Hague, The Netherlands.

DE HAAN, C., STEINFELD H. and BLACKBURN H. 1997. *Livestock and the Environment: Finding a Balance.* Report of a study coordinated by the Food and Agriculture Organisation of the United Nations, the United States Agency for International Development and the World Bank. Brussels: European Commission Directorate-

General for Development.

DELGADO, C., ROSEGRANT, M. and MEIJER, S. 2001. *Livestock to 2020: The Revolution Continues*. Paper presented at the annual meeting of the International Agricultural Trade Research Consortium (IATRC), Auckland, New Zealand, January 18–19, 2001.

ISO 1998. Environmental management – Life cycle assessment – Goal and scope definition and inventory analysis. ISO 14041:1998 (E). International Organisation for Standardisation, Geneva, Switzerland.

LRF 2002. *Mat och Miljön. Livscykelanalys av sju livsmedel (Food and the Environment. LCA of seven food items)*. Lantbrukarnas Riksförbund, 105 33 Stockholm (in Swedish).

MATTSSON, B. 1999. *Environmental Life Cycle Assessment of Agricultural Food Production*. Doctoral Thesis, Agraria 187, The Swedish University of Agricultural Sciences, Alnarp, Sweden.

MATTSSON B, CEDERBERG C. and BLIX L. 2000. Agricultural land use in life cycle assessment (LCA): case studies of three vegetable crops. *Journal of Cleaner Production* 8: 283–292.

OOMENS, G. J. M., LANTINGA, E. A., GOEWIE, E. A. and VAN DER HOEK, K. W. 1998. Mixed farming systems as a way towards a more efficient use of nitrogen in European Union agriculture. *Environmental Pollution*, 102(S1): 697–704.

POTTING, J., SCHOEPP, W., BLOK, K. and HAUSCHILD, M. 1998. Site-dependent life-cycle assessment of acidification. *Industrial Ecology*, 2 (2): 63–87.

SERÉ, C. and STEINFELD, H. 1996. *World livestock production systems. Current status, issues and trends*. FAO Animal Production and Health Paper 127, FAO, Rome, Italy.

SVENSK MJÖLK. 2001. *Milk and the Environment*. Swedish Dairy Assocation, 105 46 Stockholm.

TILMAN, D., FARGIONE, J., WOLFF, B., D'ANTONIO, C., DOBSON, A., HOWARTH, B., SCHINDLER, D., SCHLESINGER, W. H., SIMBERLOFF, D. and SWACKHAMER, D. 2001. Forecasting agriculturally driven global environmental change. *Science* 292: 281–284.

VAN DER HOEK, K. 1998. Nitrogen efficiency in global animal production. *Environmental Pollution* 102(S1): 127–132.

VAN ZEIJTS, H., LENEMAN, H. and WEGENER SLEESWIJK, A. 1999. Fitting fertilisation in LCA: allocation to crops in a cropping plan. *Journal of Cleaner Production*. 7: 69–74.

WEIDEMA B 2001. Avoiding co-product allocation in life-cycle assessment. *Journal of Industrial Ecology* 4(3): 11–33.

WIRSENIUS, S. 2000. *Human Use of Land and Organic Materials*. Doctoral thesis. Department of Physical Resource Theory, Chalmers University of Technology, Göteborg, Sweden.

WITHERS, P. J. A., PEEL, S., MANSBRIDGE, R. M., CHALMERS, A. C. and LANE, S. J. 1999. Transfers of phosphorus within three dairy farming systems receiving varying inputs in feeds and fertilisers. *Nutrient Cycling in Agroecosystems*, 55: 63–75.

6

Environmental impact assessment of seafood products

F. Ziegler, The Swedish Institute for Food and Biotechnology (SIK)

6.1 Introduction: the need for a sustainable fishing industry

The chapter starts by briefly describing the development and current situation of world fisheries. The environmental aspects connected to the two main seafood production systems, capture fisheries and aquaculture are then introduced. A section introducing environmental assessment of seafood products follows and the last part of the chapter deals with the future and how the two seafood production systems can be improved for increased sustainability. Finally, some sources for more information in this field are given. Technical terms from biological and fisheries sciences are marked in italics and explained in the Appendix the first time they are used.

Around a century ago, it was considered as more or less impossible that fisheries affected fish *stocks* and marine ecosystems to any significance (see references in Botsford *et al.*, 1997 and Auster and Langton, 1999). However, world fisheries have undergone rapid development in fishing capacity and gear, propulsion and navigation technology, especially since the end of World War II. Trawling became a common fishing method together with the use of diesel engines as the most common vessel propulsion technology in the 1950s. The introduction of new fishing methods as well as a general growth of fishing fleets and increasing vessel sizes led to dramatically increased fishing capacity in the decades after World War II. Figure 6.1 shows the development of total catches in capture fisheries and production of aquaculture. During the 1990s, reported landings have stabilised around 90–100 million tons caught per year, while aquaculture has continued to increase dramatically. In 2001 total production of capture fisheries was 91 million tons with the main fishing nations China, Peru, Japan, United States and Chile landing the greatest amounts (Anon., 2002b). However, these figures have been

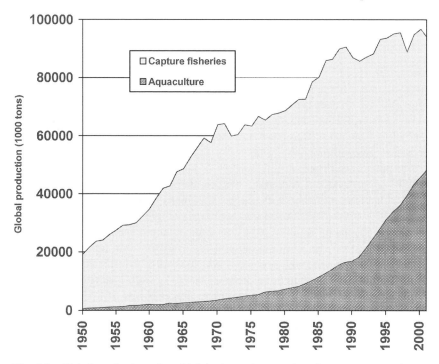

Fig. 6.1   Global production of world fisheries and aquaculture between 1950 and 2001 (from FAO statistics).

questioned. Some countries have consistently over-reported their catches (Watson and Pauly, 2001) and discards and illegal, unreported catches are not accounted for in the official statistics. It is therefore more probable that total landings have been decreasing for a number of years (Pauly *et al.*, 2002). A change that has taken place in the fisheries of many countries, is a switch from mainly landing fish for human consumption (species which are found higher in the *food webs*) to an increasing proportion of total landings being constituted by small *pelagic* fish (lower in the *food webs*) used for fish meal production which is used to produce feed for agriculture and aquaculture.

Almost half of the exploited fish stocks in the world are considered to be fully exploited and another 18% over-exploited (Botsford *et al.*, 1997; Anon., 2002b). Even though there are only moderately exploited or underutilised fish resources, it is hardly possible to increase the total landed volume of capture fisheries. Catches of many of the over-exploited commercial species have to be decreased considerably in order to reach sustainable stock levels, and this is the main reason why aquaculture of commercially important species is sometimes seen as the solution to the problem of supplying enough raw material to the ever-growing seafood market.

The human population has also increased during this period and it will continue to increase in the future. One estimate is an increase from 6 to 9 billion between 2000 and 2050 (Wurts, 2000). A large proportion of the human population (65%) live in coastal regions and as much as 20% of the total human protein consumption may originate in seafood (Botsford *et al.*, 1997). The proportion of humans living in coastal regions of the world is expected to increase further in the future: from 65 to 75% in the next 30 years (Anon., 1997), with growing demand for seafood products and increased pressure on marine ecosystems as likely results. In developing countries, people living in the coastal zone often depend on fishing and aquaculture for their own protein supply or for trading to generate important income.

For public health reasons, authorities in many countries promote the health benefits of seafood and encourage the consumption of more marine products. Such campaigns, together with the recent image problems of the meat and poultry industry in health, animal welfare and environmental issues, have probably contributed to the increasing demand for seafood products. In the developed part of the world, a general tendency towards more healthy and 'natural' food like seafood is evident and today the main limiting factor for seafood-producing companies is the availability of high-quality raw material. Even though our knowledge about marine resources is limited, we know enough about the complexity of marine ecosystems to realise that productive capacity will not keep up with an ever-growing demand (Pauly *et al.*, 2002). Therefore, it is important to develop effective methods to assess the environmental impact of seafood production.

6.2   The role of aquaculture

The growing demand for seafood, coupled with the limited possibility of increasing production of capture fisheries, have created high expectations of the development of aquaculture, i.e. man-controlled farming of commercially attractive fish, crayfish, molluscs and algae, in order to bridge the gap between supply and demand for seafood products. Aquaculture means that aquatic plants or animals are collected at some stage in their life-cycle and cultivated at a specially designated site until they have reached the right size for harvest. Later the entire life-cycle of the species can be managed in culture, e.g. for salmon. Mobile animals, such as fish, need to be caged; while sessile organisms, such as mussels and algae, are grown on special constructions in designated areas. After a period of growth, the species is harvested and the products go into the processing and marketing chain. Aquaculture is more similar to agriculture than fishing in the sense that it is a controlled production in a limited system, sometimes with input of nutrients and other substances. An advantage of aquaculture over agriculture is that production and harvesting can be done continuously and is normally not restricted to specific seasons. As evident in Fig. 6.1, world aquaculture has undergone spectacular development during the last two decades. In 2001 total aquaculture production was 38 million tons with China, India, Japan, the Philippines and Indonesia as the main producing nations

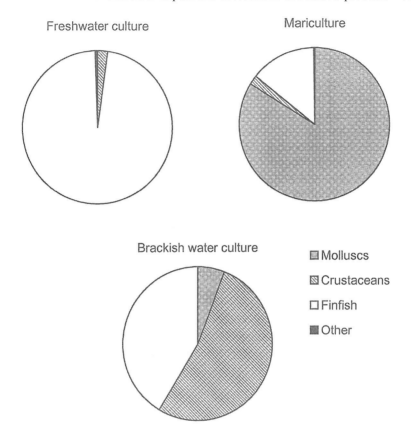

Fig. 6.2   World aquaculture production: proportions of species groups by environment in 2001 (from FAO statistics).

(Anon., 2002b). Aquaculture can take place both in freshwater, brackish and marine water and the typical types of organisms cultivated are shown in Fig. 6.2.

Limitations to the development of aquaculture are in part biological. Many commercially important species cannot be farmed because there are no methods for hatching and raising them in the laboratory. The early stages of larval development are very sensitive and the crucial combination of factors regarding feed and environmental requirements remains unknown for many species. Moreover, some species cannot be kept at densities high enough to make aquaculture profitable. Considerable research effort is necessary before it is possible to farm a new species commercially. In Norway eel (*Anguilla anguilla*) and cod (*Gadus morhua*) are species for which farming methodology is being developed and the future possibilities for farming look promising. In Chile, farming of several evertebrate species and red algae is growing. Introduction of a new farmed species on the market always has to be preceded by careful marketing in order to make often conservative consumers appreciate the new product, especially when foreign species are concerned.

In developing countries, the main organisms in aquaculture are *herbivorous* fish and shrimps of the genus *Penaeus sp.* Selective breeding and genetic engineering of shrimps is sometimes considered as a means to overcome biological limitations and increase production to supply export markets with more products (Khan and Bhise, 2000). Yet another type of biological limitation is the prevalence of diseases and parasites which spread easily among the dense populations of organisms in aquaculture units, especially in intensive ones (D'Souza and Colvalkar, 2001).

Another important constraint to aquaculture development is the availability of suitable land or sea area. Marine-based systems, such as mussel or fish farms, are localised in protected water with sufficient water exchange and at certain water depths. Such areas have other potential uses, such as for tourism, coastal fisheries or shipping, which are often in conflict with aquaculture. Land use conflicts in the coastal zone are frequent, especially close to urban areas where prices of building sites can make it difficult to establish a profitable aquaculture business. In tropical shrimp farming, mangrove areas are used for pond construction resulting in the deterioration of those and surrounding areas. The ponds in intensive farming can only be used for a couple of years and are then abandoned.

Mussel farming is limited by the temporary occurrence of toxicity in mussels due to blooms of harmful algae. Since *plankton* is the food of the mussels, they accumulate toxins when filtering the water during a bloom of toxic algae, and even though this does not seem to disturb the mussels themselves, they cannot be marketed during periods of high toxin levels, which can last for weeks or months. During such blooms, sea-based fish farms may also be severely affected (Karunasagar and Karunasagar, 1998; Gjosaeter *et al.*, 2000). Some species of microalgae are not toxic, but are covered with spines that can clog fish gills, which can also kill them (Hallegraeff *et al.*, 1995).

Stricter environmental legislation regarding nutrient emissions to aquatic environments can in the future also be expected to be a limiting factor for aquaculture development (Wurts, 2000; D'Souza and Colvalkar, 2001). The dependence of aquaculture on fish meal-based feeds to farm *carnivorous* species can also be seen as a limitation, since the resource used (small, pelagic fish) is a limited natural resource, like all wild fish stocks.

Section 6.4 briefly introduces aquaculture and its environmental impact.

6.3   The environmental impact of fishing

6.3.1   Effect on target, by-catch and discard species

The most evident and direct environmental effect of fishing is the removal of biomass of the target species as well as of by-catch species. Fishing can have one or several target species, which are the main purpose of the fishing activity. By-catch is the part of the catch, apart from the target species, that is also landed and sold, but is not the main driving force of the fishermen to go out fishing

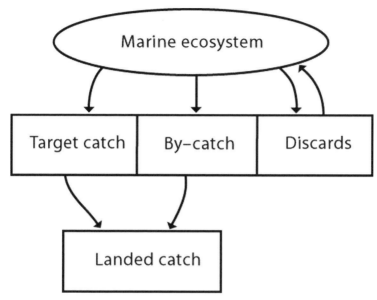

Fig. 6.3   The catch consists of target catch, by-catch and discards.

(Fig. 6.3). Sometimes, the by-catch in one type of fishing can be the target species of another type. *Demersal* fisheries along the Swedish west coast e.g. target both cod (*Gadus morhua*) and Norway lobster (*Nephrops norvegicus*) and land by-catch species such as plaice (*Pleuronectes platessa*), saithe (*Pollachius virens*) and brown crab (*Cancer pagurus*). In the eel trap fishery in the same area, however, eel (*Anguilla anguilla*) is the target species and cod is the main by-catch species, of which a part is landed.

The discard, in contrast, is not landed but thrown overboard. Discards can consist both of undersized specimens of commercial species, of fish and evertebrate species with a current low economic value, of damaged fish and sometimes even of fully marketable commercial fish. The latter occurs when there are limiting quotas and fishermen can 'save' their quotas for a more valuable catch – a phenomenon called upgrading or high-grading. A large proportion of the discards do not survive the treatment of being fished, taken on-board, handled and then thrown back into the sea (Alverson *et al.*, 1994) and therefore discarding must be seen as a waste of limited resources. Discarding undersized specimens of commercial species is not only a waste of biological resources, but also of economical resources, since a part of the discards, if left in the sea, would have grown to commercial size. Discarding is generally not reported, which increases the sources of uncertainty in preparing management plans. The true *fishing mortality* is simply unknown.

The large scale removal of target and by-catch species by fishing the worlds' oceans has caused changes in the populations of these species. Very generalised, intensive use of a fish species leads to markedly lower mean size both in the catches and in the natural fish stocks. The size at a certain age and genetic diversity in the population also decreases. How fast such changes occur, under what fishing pressure, and the degree of reversibility depends on the sensitivity of the species. Species with a high reproductive age, low growth rate and low *fecundity* (for example sharks and rays) are more at risk of being over-exploited than small, fast-growing species with a fast reproduction cycle, such as herring (*Clupea harengus*). For management purposes, the *spawning* biomass in relation to the *fishing mortality* and *recruitment* rates is evaluated for a stock in order to determine whether it is within so called safe biological limits or not. This means that the *spawning* biomass needs to be over a certain threshold level and the *fishing mortality* under a defined level. If not, lower *fishing mortality* recommended, resulting in lower quotas. For many by-catch species, there are no quotas or other regulations. They are especially vulnerable in times when increasing fishing pressure is directed towards them, due to the decrease in catches of the main target species, since we know very little about these stocks both with and without fishing pressure.

6.3.2   Ecosystem effects

Marine species do not live in isolation, but are tightly connected to each other in marine *food webs*. Therefore, a large-scale decrease of one species can indirectly impact on other species which are either *prey*, *predators* or competitors (Ramsay *et al.*, 1998; Anon., 2000; Pauly *et al.*, 2002). There are theories that fishing down one species, such as cod in the Baltic Sea, has led to an increase in the main prey species which are herring and sprat (*Sprattus sprattus*) (Anon., 2000). Indirect effects can occur also further away in the *food web*. In the above mentioned example with cod and herring/sprat one could for example expect *zooplankton* abundance to decrease since it is the main food item for herring and sprat and *phytoplankton* to increase, since zooplankton are the most important phytoplankton grazers (Fig. 6.4). However, such indirect effects are most often hard to follow and connect to a single human activity as there are many other simultaneous ways humans interfere with the marine environment, e.g by discharging waste containing nutrients into the oceans, causing local or regional eutrophication which can also affect phytoplankton abundance and growth. Additional factors can make the picture even more complex, such as the fact that sprat feeds on cod eggs (Fig. 6.4), the survival of which is a bottleneck for the successful reproduction of cod in the Baltic. Marine ecosystems are generally so complex, with inter-connected food chains, natural variation in the abundance of species and variation in environmental conditions, that clear cascading effects are seldom observed (Pauly *et al.*, 2002) and therefore it is in most cases difficult if not impossible to establish cause- effect-chains. What is clear is that fishery can affect entire *food webs* and that fishing down stocks and simplifying *food webs* leads to increasing vulnerability to natural factors such as environmental variation. Exploited stocks, hence, are more susceptible to environmental changes (Pauly *et al.*, 2002). Severe over-fishing can cause complete ecosystem shifts. When for example cod, the former main *predator* in

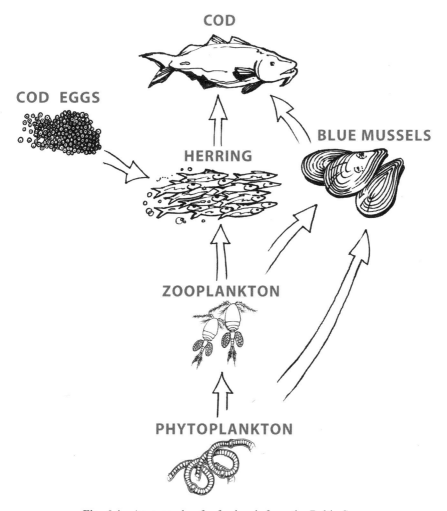

Fig. 6.4   An example of a food web from the Baltic Sea.

New England ground fisheries, was fished down to a minimum level and reproduction failed during a period (probably due to natural variation), other species such as the lower-value Arctic cod (*Boreogadus saida*) and some ray and shark species, took its place. After more than 10 years of stopping cod fishing completely, cod has not re-established its position as the main *predator* (Myers *et al.*, 1997).

6.3.3   Seafloor effects of towed gear

When fishing for *demersal* species, seafloor impact is inevitable. Towed gear like dredges and trawls, which are actively pulled along the seafloor in order to obtain the catch, affect a greater area than do passive gear like gillnets, long-lines and traps, which are left in the sea for a period (Jennings and Kaiser, 1998). A typical otter trawl used for cod trawling in Sweden normally consists of a nylon net, 50–100 m wide and several hundred metres long, connected with iron chains to two heavy iron otter boards (around 450 kg each) which make sure the trawl is dragged along or just above the seafloor (depending on the targeted species) and that the net is kept open (Fig. 6.5). Chains and ropes connect the otter boards to the fishing vessel. The trawl is pulled with a speed of 2–3 knots for a couple of hours and then hauled, emptied and set again.

The passage of fishing gear can cause physical, biological and chemical changes on and in the seafloor. Physical structures of biological origin such as corals, sponges or reef-building organisms are very sensitive to this type of disturbance. They can be crushed by the gear itself or, since they filter-feed on *planktonic* organisms, be damaged by the increased resuspension of sediment after the passage of a trawl. Too much sediment can clog their filtering organs and kill them. Patchy environments tend to become more uniform when such biogenic structures disappear and this reduced habitat complexity leads to reduced overall species diversity in fished areas (Thrush et al.,1998; Kaiser *et al.*, 2000).

Benthic organisms can be directly killed or injured by the passage of the gear. The 'digging' of gear to catch crayfish living in the seafloor causes turbation which exposes *benthic* organisms to *predators* as well as directly killed and injured organisms. This sudden availability of food favours *scavenger* species such as certain species of molluscs, crayfish and starfish, which have been reported to increase in abundance (Lindeboom and de Groot, 1998; Ramsay *et al.*, 1998; Demestre *et al.*, 2000), while long-lived, sessile and fragile marine

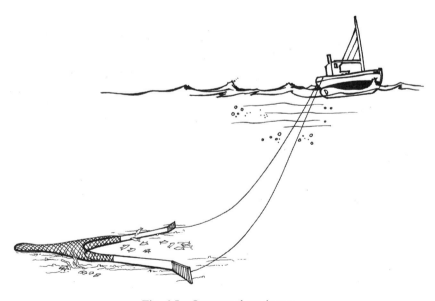

Fig. 6.5   Otter trawl net in use.

species often decrease in abundance at trawled sites (Olsson and Nellbring, 1996; Hall, 1999; Anon., 2000; Bergman and van Santbrink, 2000).

The biological impact of *demersal* trawling depends on the frequency of the activity and the habitat type impacted. The geographical distribution of fishing *effort* in different habitat types and thereby the intensity in fishery-related seafloor impact is rarely known (Auster and Langton, 1999; Jennings *et al.*, 2000). For regions where seafloor maps exist and fishing effort is reported with high geographical resolution, analysing these data in a Geographical Information Systems (GIS) could be valuable to quantify the overall fishing intensity in an area or in a specific habitat type (Nilsson and Ziegler, 2002).

6.3.4   Fuel consumption and emissions

In the industrialised part of the world, fishing vessels are nowadays almost exclusively run by marine diesel engines. Fuel consumption depends on the size and the hydrodynamic properties of the vessel hull, but also on the speed and engine age as well as the fishing method used. Generally spoken, passive gear like gillnets, seines, traps and long-lines require less fuel per unit catch landed than do active gears like trawls and dredges (where the gear is actively pulled) (Meltzer and Bjørkum, 1991; Bak, 1994; Lillsunde, 2001; Ziegler and Hansson, 2002), but there are also exceptions to this general rule (Tyedmers, 2002). Pelagic trawling generally requires less fuel than *demersal* trawling per kg of catch. This is due to the greater force required to pull the trawl along the seafloor than when it is just dragged through the water. Pelagic species often form schools and catches can therefore be enormous, which additionally reduces the fuel consumption per landed catch compared to *demersal* fisheries in which more effort has to be invested in finding the fish. Compared to other food types, capture fisheries is often an energy-intensive way to obtain food (Pimentel and Pimentel, 1979; Cederberg, 2002), especially *demersal* fisheries for popular seafood species like cod, haddock, plaice, shrimps and crayfish. However, seafood originates in an environment unaltered for food production and many of the other environmental issues connected to other food production systems are not applicable to fisheries.

Emission levels of carbon dioxide and sulphur oxides are directly related to the fuel consumption and sulphur content of the fuel respectively. Emissions of nitrogen oxides, hydrocarbons and carbon monoxide depend mainly on the engine age and load (Ziegler and Hansson, 2002). The engine load depends on the fishing method used and during a fishing trip, the engine is normally used under a range of different loads, e.g. during steaming, setting and hauling of the gear, during trawling and during manoeuvring in the port, resulting in highly different emission levels. The typical load profile during gillnet fishing has been shown to lead to higher emissions than during trawl fishery due to the, on average, low engine load with less efficient fuel combustion as the result (Ziegler and Hansson, 2002). However, the higher fuel consumption in trawl fishery normally overshadows this difference.

6.3.5   Anti-fouling

It was mentioned in the previous section that the hydrodynamic properties of the hull influence fuel consumption. The hull surface is kept free from growth of marine organisms by applying anti-fouling paints. Such paints contain toxic agents which prevent organisms like barnacles, algae and mussels from settling and growing on the hull. They are applied once or twice a year and the active substance slowly leaks out into the surrounding water, efficiently preventing fouling. Toxicity of these substances is broad and many marine organisms that are not targeted are affected. The earlier most common substance, tributyltin (TBT), has been phased out globally from 2003 due to its documented negative impact on the marine environment. In many countries it has already been replaced by e.g. mixtures of copper and the herbicide Irgarol. However, in order to avoid these broadly toxic substances in the future, research is ongoing trying to find 'smarter', i.e. more specific ways to target the species which cause the fouling problems, or nontoxic ways of keeping the hull free from marine organisms. Making the hull surface unattractive for marine larvae to settle on is one possible way. Mechanical cleaning and docking in freshwater every now and then are traditional ways of getting rid of marine organisms on ship hulls.

6.4   The environmental impact of aquaculture

6.4.1   Intensive or extensive aquaculture

There are numerous methods for aquatic farming. Shrimp farming can e.g. be done in basins on land or in coastal constructed ponds in former mangrove swamps where shrimps are hatched and grown until they reach commercial size. All feed is added, excretion products removed and medicines used when needed. Such systems are called intensive farming systems (D'Souza and Colvalkar, 2001). A contrast is provided by extensive-traditional farming systems, where wild shrimps are fished or trapped during high tide and then cultivated in natural ponds in mangrove areas where they are left to grow and reproduce without, or with very little, addition of feed. Extensively farmed shrimps feed on planktonic organisms just like wild shrimps do and they are not treated for diseases or to increase growth. Organism densities are much lower in extensive systems and therefore, the need for disease control is lower than in intensive farming systems. There are also many stages in between intensive and extensive farming. More intensive methods normally mean higher yields and more input of energy and chemicals and resulting emissions.

Mussel farming is often done in extensive systems by putting a clean surface such as a rope, pole or raft in the water at the time when mussel larvae abundance is high, when the larvae will settle on any free surface. The mussels are then left growing, feeding on the planktonic organisms passing with water currents for a year or two, until they are harvested. Another way of mussel farming, common in Denmark and the Netherlands, is farming on the seabed. Small, wild mussels are fished and then placed in suitable areas in high densities where they can be easily recollected with dredges after a growth period. Harvesting of mussels can often be done continuously and is only restricted by blooms of harmful algae which periodically can make mussels toxic. Farming of finfish is normally done in intensive farming systems based either on land, in seas or lakes, which is described in the following section.

6.4.2   Land-based or marine farming/aquaculture

Aquaculture can be undertaken in artificial ponds or basins on land. Freshwater or seawater (depending on the species to be farmed and on the life-cycle phase of the species) is pumped to the facility. Farming can also be carried out in the sea where fish or crayfish need to be caged in some way. In the case of land-based units, it is easier to keep track of discharge of residual water and its content of nutrients and other emissions, since there is usually a single wastewater tube. It is also easier to control the spread of infections and escapes. Sea-based systems require less energy for pumping water and building of ponds or basins, but are more difficult to control regarding nutrient emissions, disease spreading and escapes since the farming is done in the seawater, with only a net or cage separating the farmed fish from the wild stocks (when farming domestic species).

Shrimp farming is done in ponds, either natural ones or constructed ponds in former mangrove areas. In the case of intensive farming, where mangroves are cut down and artificial ponds constructed, land use will be an important environmental factor as the ponds only can be used for a couple of years, after which they are abandoned and new sites have to be occupied. Land use per yield is greater in extensive systems. On the other hand, the land can be used for a longer time and is not left as devastated as former intensive shrimp ponds. An additional problem is the salinisation of surrounding soils by sea water pumped through aquaculture ponds and leaking out, especially as aquaculture is practised further and further away from the coast e.g. in Thailand (Mungkung, 2003).

6.4.3   Species cultured

The species to be cultivated is also an environmental issue, since feed demands and sensitivity to disease and environmental settings differ between species. One environmental risk of aquaculture of domestic species are escapees who can survive and reproduce their genes in the wild stocks. The introduction of hybrids can change ecological links in marine *food webs* such as competition and *predation*. Diseases and parasites naturally spread more easily in fish farms due to the high densities of individuals in the cages and fish farms can therefore spread diseases and parasites to wild stocks. Examples are escapees from salmon (*Salmo salar*) or rainbow trout (*Salmo trutta*) farms who can spread lice and diseases to the wild salmon and trout stocks. When non-domestic species are farmed, there is a risk that new species, both the farmed one and species associated to it, are introduced to a new environment where it does not naturally occur. They may well be able to survive and colonise it with ecological changes as a result. The intentional or unintentional introduction of so called alien invasive species is considered one of the major threats to the marine environment today.

6.4.4   Feed and eutrophication

Most fish species considered for aquaculture in northern, temperate oceans are *carnivorous*, whereas some fish species farmed in tropical countries are *herbi-* or *omnivorous*. When farming *carnivorous* fish species, the feed is normally based on fish meal, which is based on capture fisheries of small pelagic fish species. Around 1–2 kg of dry feed are needed to produce one kg of farmed fish. To produce fishmeal for the feed of one kg of farmed salmon 3.2–6.6 kg of fish resource is used (Tyedmers, 2000), depending on the type and quality of the raw material. Vessels targeting small pelagic species are often large-scale, landing enormous catches using fishing gear with a very small mesh size. One problem in these fisheries is by-catch of undersized specimens of more high-value species in the small meshes. Many people argue that it is a waste of marine resources to use high-value fish protein to produce feed for agri- and aquaculture. On the other hand, the species in question have a low value on the market for human consumption and some argue that producing feed is the best way to use this resource. Nevertheless, these so called industrial fisheries have a bad environmental image among consumers. Over-fishing is also threatening pelagic stocks. As a result, the aquaculture industry would like to decrease its' dependency on fisheries for feed production. Alternatives are e.g. soy-based protein leading to other types of environmental impact.

Optimising feed dosage and feeding technique is crucial in order to minimise loss of feed which ends up as nutrient emissions from the fish farm. Excretion products from the animals cultured do escape and, in order to avoid local eutrophication problems around the production facility, wastewater treatment as well as proper location of the marine farm (in areas with sufficient natural water exchange) is very important. Local eutrophication can lead to very low oxygen levels and formation of hydrogen sulphide in the bottom-near water layers, altering the *benthic* community beneath and surrounding the aquaculture facility. Extensive farming, when no feed is added and the species feeds on planktonic organisms, can contribute to decreased eutrophication by removing biomass from the water column. Mussel farming has even been suggested as a method to improve water quality in a eutrophied area (Haamer, 1996) and as a wastewater treatment method.

6.4.5   Drugs and chemicals

The use of antibiotics to combat bacterial diseases in fish and shrimp farming is very common, but the amounts used have been reduced in many countries. Antibiotics are not always broken down easily in nature and are potentially bioaccumulating. The spreading of anti-bacterial substances in nature also creates a risk of development of resistant bacterial strains and of spreading of resistance to other bacteria than those originally targeted. Vaccines are today used on a routine basis to combat common salmon parasites, such as salmon lice. Various chemicals are also used either as dipping treatments or as feed additives. The environmental effects of these chemicals are not fully known (Anon., 2002a). To prevent settling and growth of marine organisms on surfaces. equipment is often treated with copper solutions.

6.5   Environmental assessment of seafood products, sustainable fishing and aquaculture

Environmental assessment of seafood products is primarily done for three target groups:

1. Companies working at some stage in the seafood production chain who want to learn more about the environmental characteristics of their own products or products they purchase in order to improve them or deal with consumer concerns.
2. Authorities responsible for developing the fishery sector towards increased sustainability.
3. Environmentally conscious consumers who want to make informed choices when purchasing seafood by buying eco-labelled seafood products.

Life Cycle Assessment (LCA) is a method for assessing the environmental impact of products or services from a life-cycle perspective, i.e. from raw material production to processing, packaging and consumption. When applying LCA to seafood products originating in capture fisheries, a number of fishery-specific types of environmental impact have to be considered as discussed in Section 6.2. Resources like land use and extraction of biological resources are relatively new in LCA methodology and development is still ongoing. Both of these issues are relevant when studying the environmental impact of seafood production. A system difference when studying fishery compared to other production systems is that the production is wild, i.e. not controlled by man like e.g. in agri- and aquaculture, which implies that the sustainability in the use of the limited resource base (the fish stock) should be evaluated when studying the environmental impact of such a system.

Characterisation methods are under development for ecotoxicological effects. Some researchers work on developing regional impact assessment indexes, which would be very useful for e.g. eutrophication and toxicity in the marine environment. Current methods are based on data from freshwater ecosystems, which often differ considerably from marine ecosystems in response to nutrient emissions, depending on the nutrient status of the recipient. There are currently no characterisation methods for the extraction of biological resources (landed and discarded catches), seafloor use and impact on target, by-catch and discard species. Due to the lack of data and methodology, these impact categories can at present only be assessed qualitatively in seafood LCAs.

Two life-cycle studies of seafood products indicate that fishery is the environmentally most important step of the life-cycle both regarding energy consumption and resulting green-house, acidifying and eutrophying gaseous emissions (Ritter, 1997; Ziegler et al., 2003). For both products studied; Danish pickled herring and Swedish frozen cod fillets, fishing was by far the most energy-consuming step in the life-cycle and there was a clear difference between fishing methods, with seining and gillnetting requiring less energy per unit of fish landed than trawling.

Ecological Footprint Analysis (EFA) and Energy Analysis are other methods for environmental assessment. Ecological footprint analysis quantifies the area of primary production required to sustain the production of a product. Energy analysis means quantifying input of different types of industrial energy into a production system with defined end points (in fact it can be considered being a partial LCA). Salmon farming and fishing in British Columbia, Canada, have recently been analysed with these methods. It was concluded that farming salmon required considerably more resources than fishing (Tyedmers, 2000), a conclusion also drawn by Pimentel and co-workers (Pimentel et al., 1996). A difference between fishing methods was also demonstrated with purse seining of salmon more resource-efficient than gillnetting and trolling. The results from these studies are likely to be case-dependent and therefore too general conclusions should not be made from the few case studies available in this field.

6.5.1   Sustainable fishing

Management

Fisheries management regulates amounts of catches, by-catch and discards in different fisheries for some species, based on biological advice and political considerations. It also regulates which gear types are allowed and the periods when fishing is allowed in certain areas. With the alarming state of many commercial fish stocks today, interest in avoiding catching undersized specimens by using more selective gear types is increasing. The establishment of management plans for important stocks as well as multispecies stock assessment with an ecosystem perspective are steps towards increased understanding of the dynamics of marine *food webs* and improved management schemes (Botsford et al., 1997). Fisheries management can also regulate the amount of fishing effort in an area and the way fishermen report effort. As mentioned earlier, high-resolution effort data would be valuable both in order to estimate the catch per unit of effort (*CPUE*) and to make reliable estimations of stock size, and also to study the seafloor effects of fishing. A problem facing fisheries managers is the so called burden-of-proof. Due to the large uncertainties in predicting environmental effects of human actions, politicians prefer not to argue for restrictions which have social and financial consequences. A more general application of the precautionary principle in the future would turn this problem around. The users of marine resources would have to prove that their activities did not affect the long-term sustainability of marine ecosystems.

Environmental effort

The smaller the stocks, the more effort has to be used for fishing them. This connection has been discussed in two studies of the environmental impact of Swedish cod fisheries (Ziegler and Hansson, 2002; Ziegler *et al.*, 2003). The effort can be seen as an investment with the prospect of getting a certain amount of catch paying for the investment. Decreasing stocks often lead to decreasing catch per unit of effort (*CPUE*) and thereby increased environmental impact per kg of catch landed (and increasing costs). If more resources and energy are used to obtain a certain amount of catch, we can say the 'environmental effort' increases. Optimally, fisheries managers in the future will consider the environmental effort of a fishery and try to minimise it, just as fishermen themselves are becoming more aware of these relationships and of their importance. From an economic point of view, the optimum level of harvest is even lower than the biological limits and therefore economically optimal fishing would also be less resource-demanding.

Marine Protected Areas

Fishing has so far only been a sustainable activity when there was an inability either technologically or geographically to fish in the entire spatial range of a species (Botsford *et al.*, 1997; Pauly *et al.*, 2002). With more and more advanced gear, vessel and navigation technology, these areas have been reduced to almost zero, increasing the risk of over-fishing stocks. It has become even more important to create marine protected areas with both no-take-zones and zones with limited fishing activity in order to maintain or re-establish both habitat integrity and viable stocks. In areas where conflicts between fishing using different gear (often fixed gear and mobile gears) exist, these conflicts can be solved by applying zoned management allowing different gears in different zones of the fishing grounds. Zoned management, besides resolving the initial conflict, protects an area from fishing disturbance by e.g. towed gears (Kaiser *et al.*, 2000). Improved mapping of the seafloor disturbance of fishing should be a high future research priority so that predictive habitat management becomes possible. With better knowledge about the distribution of marine habitats and fishing effort as well as effects of certain levels of fishing pressure in different habitats, management would be much more precise. The precautionary principle would have to be used less (Auster, 2001), reducing costly restrictions on the fishing sector.

Product waste in the supply chain

In environmental assessments of food products, the early stages of the life-cycle have been shown to be the most important regarding environmental impact (Mattsson, 1999). This has been shown to be true also for seafood products, a block of frozen cod fillets (Ziegler *et al.*, 2003) and a jar of pickled herring (Ritter, 1997). The most important environmental issue after landing was to maximise quality and minimise product losses. For losses occurring after landing, more fish have to be fished at high environmental costs. Therefore, improving product logistics, packaging and continuous cooling to minimise losses is important also from an environmental point of view.

Engine and fuel technology
Considerable improvements in fuel consumption and emission levels were identified in a study of Swedish cod fisheries (Ziegler and Hansson, 2002). By modernising engines, optimising vessel speed for low fuel consumption and increasing the use of passive fishing gear slightly, fuel consumption and emissions could be decreased considerably. Likewise, SO_x emissions would be reduced to 2% if diesel of the best environmental quality was used. Even though these calculations were made for the Swedish fishing fleet, the order of magnitude might apply also to the fleets of other countries. In other fisheries; e.g. the British Columbian salmon fishery, decreasing gillnetting and trolling and increasing the use of traps, weirs and fishwheels would lead to lower energy consumption per landed ton of catch (Tyedmers, 2000).

Consumers demand and eco-labelling
Consumers are already aware of environmental issues and worried about the impact fishing and aquaculture have on the environment (Anon., 1998a; Anon., 1998b). A demand for more sustainably produced seafood products, with adequate eco-labelling, will contribute to the future development of the sector towards increased sustainability by creating market incentives for more sustainable seafood production (Botsford *et al.*, 1997; Pauly *et al.*, 2002). There is an array of qualitative labelling schemes today, e.g. the MSC and the Seafood Watch of the Monterey Bay Aquarium (www.msc.org; www.mbayac.com). In the future such schemes could be combined with more quantitative types of environmental assessments like LCAs or EFAs to provide a scientific baseline for eco-labelling criteria. It is also desirable that aspects other than environmental ones are included in labelling schemes, e.g. working conditions and socio-economic factors.

Environmental training
Some of the improvement options that were concluded from environmental assessments of seafood products are not possible without the active cooperation of fishermen, e.g in optimising product quality from the moment of catching the fish and in optimising vessel speed to reduce fuel consumption. Such changes have to be built on growing consciousness of environmental issues only achievable by environmental training.

6.5.2   Sustainable aquaculture
The problem of intensive aquaculture is the failure to recognise the linkage to the resource base supporting the activity. Considerable effort has been done to increase sustainability in salmon farming in Norway by decreasing the use of antibiotics, optimising feed dosage and feeding techniques (Joshi, 2001). In

Finland, the possibilities of decreasing the nutrient load caused by aquaculture by improving feeding techniques and using locally produced fish feed have been discussed (Ruohonen and Mäkinen, 1991). Improving sustainability in shrimp farming includes the maintenances of buffer strips of mangrove around pond areas and the minimisation of discarding of fish and shrimp fry when collecting the wild brood stock (today around 50 fry of other species are discarded for every tiger prawn fry collected (Shiva and Karir, 1996)). Residual sediment deposits can be converted to manure after proper treatment and salinisation of nearby agricultural fields can be prevented by the construction of drainage canals around the ponds (D'Souza and Colvalkar, 2001)

   However, more basic changes are probably needed before aquaculture of fish and shrimps can be called sustainable. Less intensive systems are less resource-demanding and hence more sustainable. The most important input to intensive aquaculture systems, both in terms of material flow and energy consumption is the feed and therefore the composition of the feed is a key to decrease overall energy consumption and environmental impact of such systems. Livestock- and fishery-based ingredients represent high-energy ingredients even if they only constitute a small percentage of the feed. Often, waste products from other types of production (e.g. fish waste from a filletting factory) is used and in such cases an allocation should be done between the main product and the by-products. The major part of resource use and environmental impact needs to be allocated to the main product (e.g. the fish fillet) and a much smaller part to the by-product (i.e. fish waste used for fish meal production). Minimising feed use, especially overfeeding, and changing the composition of the feed to more agricultural products or waste products from agricultural production are important measures to decrease the resource consumption of intensive aquaculture. To farm species which do not need feed input, but feed on planktonic organisms in the water (i.e. filter feeders like mussels and *zooplanktivorous* fish) is a way to avoid the problem. A key to increased sustainability in fish farming is also to farm several species, using different food sources, together. This approach is called polyculture, integrated farming system or mixed cropping system and it means that algae, mussels, zooplankton or other filter feeders are farmed together with fish, removing suspended solids, dissolved organic matter and nutrients from the water before it reaches the surrounding environment. At the same time, other seafood products or raw materials for feed production are produced. In order to minimise energy conversion losses from one trophic level to another, cultivating lower-trophic-level organisms would be helpful. Such organisms are e.g. algae, filter feeders like mussels and *zooplanktivorous* or *herbivorous* fish.

6.6   Conclusions and future trends

Currently, most seafood is produced at high energy and environmental costs compared to other types of food. Many of our wild resources in the sea are harvested at an unsustainable rate in the long term. With the increasing human population and its growing demand for sustainably produced seafood in mind, it is of utmost importance to change our seafood production systems towards increased sustainability. For fishing such changes comprise improved fisheries management with e.g. gear and effort regulation and the establishment of management plans for important stocks. Management plans and quotas for more species are necessary. The establishment of marine protected areas as refuges for exploited fish stocks is another critical issue to ensure both habitat integrity and viable stocks in the long run. Improved gear, engine and fuel technology can contribute to lower resource use and increased sustainability in fisheries. The most important environmental issues after landing are to maintain high quality and minimise product loss in order to minimise the fishing effort needed to provide consumers with seafood products.

The latter is true also for seafood products originating in aquaculture. Continuous improvement of disease prevention, e.g. by keeping the animals in lower densities, is important as well as changes in the composition in the feed. Reducing the share being represented by fishery- and livestock-based ingredients would reduce the overall environmental impact of aquaculture. The planning of mangrove buffer strips and drainage channels around shrimp ponds would contribute to increased sustainability. The deliberate or unintentional introduction of alien invasive species should be minimised. Risk assessment of the introduction of new genetic material to wild fish stocks is desirable, at least in eco-labelled production. Other improvement options for aquaculture include the development of 'smart', integrated production systems copying nature by farming several species simultaneously which can benefit from each other. Farming in less intensive systems and farming species from lower trophic levels than what is common today (e.g. filter feeders) are some of the most important improvement options in order to make aquaculture activities more sustainable.

6.7   Sources of further information and advice

The Food and Agricultural Organisation, FAO (of the United Nations): www.fao.org
The International Council for the Exploration of the Sea, ICES: www.ices.dk
The Marine Stewardship Council (MSC label), MSC: www.msc.org
Monterey Bay Aquarium (eco-labelling of seafood): www.mbayaq.org

6.8   Acknowledgements

I am grateful to Mikkel Thrane, Per Nilsson and Johan Modin who helped me improve previous versions of this chapter. My little son Maurits has helped by sleeping while I was doing the last changes to this chapter.

6.9   Appendix

Technical terms from biological and fisheries sciences used:

Benthic	Marine organisms living on or in the seafloor
Carnivorous	Feeding on animals
CPUE	Catch per unit of effort (measured as tons caught per hour trawled or per day at sea e.g.)
Demersal	Organisms living on or close to the seafloor
Effort	Technical 'strength' of a fishery e.g. measured as number of hours spent fishing, mandays at sea or engine power of a fishing vessel or fleet
Fecundity	Capacity to reproduce (number of eggs surviving to reproductive age)
Fishing mortality	Percentage of a fish stock being killed by fisheries each year
Food web	Interconnected food chains of prey and predator species in an ecosystem; starting with primary producers and ending with top predators
Herbivorous	Feeding on plants
Omnivorous	Feeding on both plants and animals
Pelagic	Marine organisms living in the free water mass or at the surface
Phytoplankton	Freefloating, photosynthetisising organisms, often microscopic
Plankton/-ic	Freefloating, often microscopic, aquatic organisms
Predator/Predation	Any organism that catches and kills other organisms for food
Prey	Any animal being caught and killed for food by other animals
Recruitment	Number of juveniles in a stock surviving to a certain age
Scavenger	Feeding on dead animals or organic refuse
Spawn	To deposit eggs
Stock	Group of individuals/Population being geographically and/or genetically isolated from others of the same species
Zooplanktivorous	Feeding on zooplankton
Zooplankton	Freefloating small animals or larval stages of marine organisms, often microscopic

6.10   References

ALVERSON D L, FREEBERG M H, MURAWSKI S A, POPE J G (1994), *A global assessment of fisheries bycatch and discards*, FAO, Rome, 233pp.

ANON. (1997), *Hav och kust. Miljö och naturresurser i u-ländernas havs- och kustområden (Ocean and coast. Environment and natural resources in coastal areas of developing countries)*, Stockholm, Swedish International Development Cooperation Agency (Sida), 32pp.

ANON. (1998a), *The Nordic Fisheries in the new consumer era – Challenges ahead for the Nordic Fisheries sector*, Nordic Council of Ministers, TemaNord, 1998: 526, 100pp.

ANON. (1998b), *Forbrugerholdninger til baeredygtigt fiskeri og oekologisk fisk (Consumer attitudes towards sustainable fishery and ecological fish)*, Nordic Council of Ministers, DIVS report 1998:810, 32pp.

ANON. (2000), *Report of the working group on ecosystem effects of fishing activities*. Copenhagen, ICES Advisory Committee on the Marine Environment (ACME), 93pp.

ANON. (2002a), *Environmental Report 1999–2002 Food fish and brood stock*, Fjord Seafood Midt Norge AS.

ANON. (2002b), *The state of the world fisheries and aquaculture (SOFIA) – 2002*, FAO, Rome, ISBN 92-5-104842-8.

AUSTER P J (2001), 'Defining thresholds for precautionary habitat management actions in a fisheries context', *North American Journal of Fisheries Management,* 21, 1–9.

AUSTER P J, LANGTON RW (1999), 'The effects of fishing on fish habitat', *American Fisheries Society Symposium,* 22, 150–187.

BAK F (1994), *Brancheenergieanalyse og standardlösningar for fiskeriet (Sector energy analysis and standard solutions for fishery)*, DTI Energi Motorteknik, Aarhus, Denmark.

BERGMAN M J N, VAN SANTBRINK J W (2000), 'Fishing mortality of populations of megafauna in sandy sediments', in Kaiser M J, de Groot S J, *Effect of fishing on non-target species and habitats*, Blackwell Science, 49–68.

BOTSFORD L W, CASTILLA J C, PETERSON C H (1997), 'The management of fisheries and marine ecosystems', *Science,* 277, 509–515.

CEDERBERG C (2002), *Life Cycle Assessment (LCA) of Animal Production*. Ph.D. thesis, Dep. of Applied Environmental Science, Göteborg University.

DEMESTRE M, SÁNCHEZ. P, KAISER M J (2000), 'The behavioural response of benthic scavengers to otter-trawling disturbance in the Mediterranean', in Kaiser M J, de Groot S J, *Effect of fishing on non-target species and habitats*, Blackwell Science, 121–129.

D'SOUZA J, COLVALKAR N (2001), 'Shrimp farming and its impact on the environment', *Journal of environment and pollution,* 8(1), 19–34.

FAO Databases downloadable at: www.fao.org/fi/statist/FISOFT/FISHPLUS.asp

GJOSAETER J, LEKVE K, STENSETH N C, LEINAAS H P, CHRISTIE H, DAHL E, DANIELSSEN D S, EDVARDSEN B, OLSGARD F, OUG E, PAASCHE E (2000), 'A long-term perspective on the Chrysochromulina bloom on the Norwegian Skagerrak coast 1988: a catastrophe or an innocent incident?' *Marine Ecology Progress Series,* 207, 201–218.

HAAMER J (1996), 'Improving water quality in a eutrophied fjord system with mussel farming' *Ambio* 25(5), 356–360.

HALL S J (1999), *The effects of fishing on marine ecosystems and communities*, Blackwell Science.

HALLEGRAEFF G M, ANDERSON D M, CEMBELLA A D (1995), *Manual on harmful marine microalgae*, UNESCO, IOC Manuals and Guides No.33, Paris, 551pp.

JENNINGS S, KAISER M J (1998), 'The effects of fishing on marine ecosystems', *Advances in Marine Biology*, 34, 201–352.

JENNINGS S, WARR K J, GREENSTREET S P R, COTTER A J R (2000), 'Spatial and temporal patterns in North Sea fishing effort', in Kaiser M J, de Groot S J, *Effects of fishing on non-target species and habitats*, Blackwell Science.

JOSHI R (2001), *Ocean harvest. Norwegian fishing, aquaculture & seafood*, Horn Forlag.

KAISER M J, SPENCE F E, HART P J B (2000), 'Fishing-gear restrictions and conservation of benthic habitat complexity', *Conservation Biology*, 14(5), 1512–1525.

KARUNASAGAR I, KARUNASAGAR I (1998), 'Effects of eutrophication and harmful algal blooms on coastal aquaculture and fisheries in India', in Natarajan P, Dhevendaran K, Aravindan C M, Kumari R, *Advances in aquatic biology and fisheries*, Trivandrum, India, Prof. N Balakrishnan Nair Felicitation Committee, 163–173.

KHAN T A B, BHISE M P (2000), 'Genetic improvement of farmed shrimp for sustainable aquaculture', *INFOFISH International*, 2000(1), 41–44.

LILLSUNDE I (2001), *Havs- och kustfiskets miljöeffekter i Skärgårdshavet och Bottenhavet (The environmental impact of coastal and open sea- fisheries in the Archipelago Sea and the Bothnian Sea)*, Åbo, Jord- och skogsbruksministeriet, Fisk- och viltavdelningen, 33pp.

LINDEBOOM H J, DE GROOT, S J (1998), *IMPACT II The effects of different types of fisheries on the North Sea and Irish Sea benthic ecosystems*, Netherlands Institute for Sea Research (NIOZ), 404pp.

MATTSSON B (1999), *Environmental Life Cycle Assessment (LCA) of agricultural food production*, Ph.D. thesis, Swedish University of Agriculture (SLU), Alnarp, 55pp.

MELTZER F, BJØRKUM I (1991), *Kartleggning av avgasutslipp fra fiskeflåten (Emissions from the fishing fleet)*, Trondheim, MARINTEK/SINTEF, 37pp.

MUNGKUNG R (2003), Personal communication about aquaculture practices in Thailand (e-mail:r.mungkung@surrey.ac.uk)

MYERS R A, HUTCHINGS, J A, BARROWMAN N J (1997), 'Why do fish stocks collapse? The example of cod in Atlantic Canada', *Ecological Applications*, 7, 91–106.

NILSSON P, ZIEGLER F (2002), unpublished data on seafloor effects in different habitats of the Kattegat (e-mail:fz@sik.se).

OLSSON I, NELLBRING S (1996), *Fiske och vattenbruk- ekologiska effekter (Fishery and aquaculture-Ecological effects)*, Naturvårdsverket report 4247, 181pp.

PAULY D, CHRISTENSEN V, GUÉNETTE S, PITCHER T J, SUMAILA U R, WALTERS C J, WATSON R, WELLER D (2002), 'Towards sustainability in world fisheries', *Nature*, 418: 689–695.

PIMENTEL D, PIMENTEL M (1979), *Food, Energy and Society*, London, Edward Arnold.

PIMENTEL D, SHANKS R E, RYLANDER J C (1996), 'Bioethics of fish production: Energy and the environment', *Journal of agricultural and environmental ethics*, 9(2), 144–164.

RAMSAY K, KAISER M J, HUGHES R N (1998), 'Responses of benthic scavengers to fishing disturbance by towed gears in different habitats', *Journal of Experimental Marine Biology and Ecology*, 224, 73–89.

RITTER E (1997), *Livscyklusvurdering for marineret sild i glas (Life Cycle Screening of marinated herring in glass jars)*, Hirtshals, Denmark, DTI Miljö and DIFTA.

RUOHONEN K, MäKINEN T (1991), 'Potential ways to diminish the environmental impact of mariculture on the Baltic Sea', *Finnish Fisheries Research*, 12, 91–100.

SHIVA V, KARIR G (1996), *Towards a sustainable aquaculture* – Chenmmeenkettu, New Dehli, Research Foundation for Science, Technology & Natural Resource Policy, 126pp.

THRUSH S F, HEWITT J E, CUMMINGS V J, DAYTON P K, CRYER M, TURNER S J, FUNNELL G A, BUDD R G, MILBURN C J, WILKINSON M R (1998), 'Disturbance of the marine benthic habitat by commercial fishing: Impacts at the scale of the fishery', *Ecological Applications*, 8(3), 866–879.

TYEDMERS P (2000), *Salmon and sustainability: The biophysical cost of producing salmon through the commercial salmon fishery and the intensive salmon culture industry*, Ph.D. thesis, University of British Columbia.

TYEDMERS P (2002), *Energy consumed by North Atlantic fisheries*, manuscript, University of British Columbia.

WATSON R, PAULY D (2001), 'Systematic distortions in world fisheries catch trends', *Nature,* 424, 534–536.

WURTS W A (2000), 'Sustainable aquaculture in the twenty-first century', *Reviews in Fisheries Science,* 8(2), 141–150.

ZIEGLER F, HANSSON P-A (2002), 'Emissions from fuel combustion in Swedish cod fishery', *Journal of Cleaner Production*, 11(2003), 303–314.

ZIEGLER F, NILSSON P, MATTSSON B, WALTHER Y (2003), 'Life Cycle Assessment of frozen cod fillets including fishery-specific environmental impacts', *International Journal of LCA*, 8(1), 39–47.

Part II

Good practice

7

Environmental issues in the production of beverages: global coffee chain

W. Pelupessy, Tilburg University, The Netherlands

7.1 Introduction

Coffee is not the most environmentally damaging tropical crop. The cultivation of food crops such as maize and beans, as well as many vegetables and fruits, imposes a much higher environmental burden. However, it is necessary to consider the environmental impacts of coffee because of its widespread global presence in environmentally sensitive areas, the scale of production and the inevitably long-distance transport of most of the intermediate produce. There is also the economic pressure to apply potentially contaminating fertilisers and pesticides to increase returns and control diseases such as leaf-rust in coffee-growing countries (Varangis *et al.*, 2003: 56). The growing awareness of sustainability among final consumers in the main importing countries makes the treatment of the problem both an environmental and an economic necessity. At present, sustainable coffee sales worldwide represent less than 1% of the market (Giovannucci, 2001: 25). To treat the related environmental issues adequately, it will be essential to understand the two most outstanding economic problems, which are the tendencies to global overproduction and the limited share of the returns from coffee production for coffee growers in developing countries.

Coffee is an important income generator for more than 125 million people in 52 tropical developing countries.[1] About 25 million people, mostly small-holders, cultivate this perennial crop on 11.8 million hectares of arable land, to produce 6.6 million tonnes of consumable coffee annually. These are the huge

---

[1] When not mentioned otherwise, data from the website of the International Coffee Organization (ICO) are used.

numbers behind our daily cup of coffee. About a quarter of this quantity is consumed in the countries of origin, while three-quarters are traded globally, to be imported in the 25 most developed countries and other destinations. After oil, coffee is the second largest internationally traded commodity.

ʹCoffee growing forms part of specific global commodity chains which involve value-added networks of producers, traders and service providers (Pelupessy, 1999; Fitter and Kaplinsky, 2001; Ponte, 2002). In these chains the agricultural raw material of coffee berries is hand picked, then transformed into exportable green coffee. After being imported in the developed countries the green coffee is roasted, ground and packed, and then purchased by consumers. The networks cross borders between developing and developed countries, include at least two main transformation processes and a series of sequential coordinated or imperfect markets, and are affected by interventions of governments and private associations. This all makes the global coffee chain a so-called long chain with a complicated and internationally fragmented pattern of creation and distribution of income and environmental impacts. The present enduring international crisis of coffee prices, low farmers' income and overproduction is very much a consequence of this extended chain structure.

The purpose of this chapter is to discuss the creation and distribution of economic value and environmental effects in the global coffee chain and to examine the possibilities to enhance its sustainability. The analysis will be developed in five sections. After this introduction we will discuss in Section 7.2 the importance of coffee for developing economies, taking into account the structure and dynamics of the global chains. The market failures that lead to overproduction and the free fall of prices will be analysed in Section 7.3. Section 7.4 will provide a detailed treatment of the environmental effects in the different chain segments. The international distribution and the measures taken to reduce these effects are presented in Section 7.5. The alternatives to improve the environmental impact of coffee activities, considering future consumption trends, will be examined in Section 7.6.

7.2   Development issues

The two commercially most exploited coffee species are *arabica*, which originates from the Ethiopian Massif, and *canephora*, more commonly known as *robusta*. The first grows in uplands at over 600 m altitude with a temperate climate, but at 1000 m and higher gives better quality. Robusta thrives in hot humid lowlands, bears more fruits, is resistant to epidemics like leaf-rust and has a caffeine content of 2.0–2.5% or more. Arabica has a content of 1–1.5% and ripens more quickly than the other species. The taste of robusta is more acid and bitter and this coffee is considered of a lower quality.

Natural advantages could explain the initial geographical spread of the two species, but the differences are no longer so clear cut. The major arabica country, Brazil, is increasing its production and exports of the other species, while Vietnam, with a predominantly robusta culture, is introducing arabica. Anyhow, a whole series of different varieties or subspecies have been developed by the traditional producer countries of each species. Brazil developed Caturra and Mundo Novo from arabica's Tipica and Bourbon; Colombia and Costa Rica had their Caturra, Catuai and Catimorra later; there is Blue Mountain in Jamaica, and many others. From canephora the variety Kouilon was developed in the Ivory Coast and tradable Robusta has been selected in Indonesia and Congo (de Graaff, 1986: 21–22).

A further commercial classification into Naturals and Milds has taken place in accordance with the nature of the first processing techniques. The main traded coffee varieties, with market prices going from lower to higher values, are grouped in Robustas (35% by volume of world exports, 1997–2002), Brazilian Naturals (24%), Other Milds (27%) and Colombian Milds (14%). It is interesting to note that in most cases the research, development and distribution of seeds and seedlings of coffee varieties and sub-varieties has been controlled by national public and private institutions of the producer countries. This is quite different from the cases of other crops, where development and provision of seeds are mostly controlled by multinational companies. The five main producing and exporting countries in 1997–2002 (Table 7.1) were Brazil (Brazilian Naturals), Vietnam (Robusta), Colombia (Colombian Milds), Indonesia (Robusta/Arabica) and Mexico (Other Milds). Globally just five countries (fewer than 10% of those worldwide) harvested more than 60% of the crop.

The next five countries in Table 7.1 produced together hardly 18%. The export volumes and incomes went more or less in the same order. However, the share of the crop in foreign exchange is on average higher for the second group, and there are even smaller producers like Uganda and Burundi, which depend mostly on coffee for their exports with 60% and 70% respectively. Exports for almost 80% are the first processed or green coffee. The rest are the industrialised decaffeinated, roasted and soluble coffees and coffee extracts. Almost a quarter of all exports are re-exported by non-producer countries, among which Germany, the US, Belgium, Singapore and Italy are the most important. Germany is quantitatively the number four coffee exporter of the world with a considerable share of the more refined products (ICO, 2002).

For most producer countries coffee is more than just a foreign exchange earner. A considerable part of the arable land and rural labour is employed in this activity. In Central America coffee occupies 53% of the permanent crop land and 28% of the rural labour force (CEPAL, 2002).

As observed in Table 7.1, productivity varied greatly between countries. The highest levels were achieved by the big robusta country Vietnam. The main arabica producers showed lower yields but still considerably above the world average. Ethiopia, India and Guatemala showed the higher yields in this group. In most countries, cultivation is done by smallholders and is an important monetary source for subsistence or the acquisition of basic services. The 1997–2001 average export prices received by each country follow more or less the

Table 7.1   Global coffee production, 1997–2001/2 (averages)

Grower countries	Production (%)	Productivity (kg/ha)	Share in national exports[a]	Export price (US cents/lb)
Brazil	31.0	752	3.8%	87
Vietnam	9.4	2142	5.2%[b]	41
Colombia	10.0	756	11.0%[c]	116
Indonesia	6.5	482	0.8%	62
Mexico	4.5	473	0.4%	111
India	4.5	900	0.7%	80
Guatemala	4.0	900	15.1%	93
Ivory Coast	3.7	175	6.7%	54
Ethiopia	3.0	909	28.9%	117
Honduras	2.4	684	14.2%[c]	100
Rest	21.0	540		83
Total %	100	594		84

Total (1000MT)6575

[a] Goods and services.
[b] 1997.
[c] 1997–2000.
Sources: Estimates based on CEPAL (2002: cuadro 2); ICO and World Bank databases.

order of the quotations of the commercial coffee varieties. An exception is Ethiopia, which received with Brazilian Naturals the highest price, which may be an indication of high quality. There are considerable differences within the same variety. For instance, the lowest price is realised by Vietnam's Robusta, while Indonesia received over 50% more for the same variety.

The effects of price declines may be differentiated depending on the production structure (ICO 2002: 2). In the cases of well-developed technology and low production costs as in Brazil, the adverse effects could be observed in reduced farmer spending and decreasing employment rates. Where coffee is the cash crop in subsistence farming as in African and some Asian countries, the basic livelihood expenditures (medicines, education, etc.) have been affected. Where farmers depend almost entirely on coffee for their income there were reduced food purchases, increased indebtedness and abandonment of land. This resulted in famine in some coffee areas of Nicaragua and Guatemala at the start of the twenty-first century. For all these reasons, declining coffee prices will often cause supply increases instead of reductions by developing countries. Finally, it should be mentioned that unlike an export crop such as bananas, coffee has also generated processing, service and input provisions linkages and historically has supported the advent of capitalist classes and other social institutions (Williams, 1994: 197–233).

7.3   Market trends and their environmental and social impacts

There is no such thing as a coffee market where consumers meet growers. In global chains they are operating at the extreme ends and direct interchange of market signals between these parties does not exist. In between are value-adding actors and a sequence of interrelated markets of a mostly imperfect nature, as we will see later.

Table 7.2 gives an overview of the dynamics of the most important consumer markets. A dozen countries account for more than 70% of world coffee consumption. Only the number one producer, Brazil, appears as the second biggest consumer country. All other significant consumer markets are located in developed countries. Demand elasticities are low, between -0.2 and -0.4, except for Japan and the UK which are mainly tea-drinking countries. This means that in all other cases price fluctuations will hardly affect consumption of coffee, and income will do so in only a limited way. This may be a consequence of the habit-forming nature of the product (Olekalns and Bardsley, 1996). Based on the consumption per capita, we have indicated the degree of saturation of these markets and the trends at the end of the 1990s. The markets are saturated and show declining trends in Germany, the Netherlands and Sweden. Together they comprise 14% of global consumption, with a per capita average of four

Table 7.2   Main consumer markets for coffee

Country	World consumption (%) 1996–2001	Consumption per capita (kg) 2000–2002	Trend in 1990s
Germany	10.6	6.8	decreasing
France	5.9	5.4	stable
Italy	5.5	5.4	increasing
United Kingdom	2.6	2.3	decreasing
Spain	3.4	4.4	stable
Netherlands	2.2	6.7	decreasing
Sweden	1.4	8.3	decreasing
Total EU	37.5	5.4	stable
Poland	2.1	2.4	increasing
Russia	1.4	0.7	increasing
USA	20.4	4.0	stable
Brazil	13.9	4.9	increasing
Japan	7.0	3.2	increasing
Rest	23.6		
Total[b]	100	4.6	stable

[a] Saturated: >7 kg/cap., Mature: 5–7 kg/cap., Emerging: 2–5 kg/cap., Underdeveloped: <2 kg/cap.
[b] Data from Brazil included.
Sources: Database ICO and USDA.

cups or more a day. The mature markets of 5–7 kg per head are France, Italy and Brazil, where the first two are stagnating. The consumption per person in the most important markets is generally not increasing. Emerging markets of 2–5 kg are growing, while the underdeveloped markets such as Russia (0.7 kg) and Vietnam (0.3 kg) have a lot of growth potential. However, there may be ample scope for substitution among the varieties and their blends. In the high-income markets there is a trend to consume more differentiated and higher quality coffees (Lodder, 1998; Fitter and Kaplinsky, 2001). Gourmet coffees have already captured 17% in volume and 40% in value of the US market (Giovannucci, 2001: 7). While global demand for coffee is growing at 1.5% annually, the rates for gourmet and organic coffees are estimated at respectively 8% and 20% (ACPC, 2002).

The coffee consumption markets are supplier markets with high concentrations of roasters and retailers and sometimes vertical integration between them. Four large multinational companies provide more than half of all coffee in the 25 biggest consumer countries (excluding Brazil): Jacobs/Kraft General Foods, Nestlé, Procter & Gamble, and Sara Lee/DE. The first three control 73% of the US market, while in the EU the four largest roasters have half of the market. The principal coffee markets are dominated by small numbers of firms with differentiated products, which is typical of monopolistic competition where non-price instruments such as advertising, bonus systems, etc., rather than prices are used to compete. These companies may be considered as the leading and coordinating forces of the coffee chain (Pelupessy, 2001b: 22–24). They are protected by escalating tariffs on more elaborated coffees and non-tariff barriers. Taxes on coffee consumption give additional income to governments of importing countries.

One of the instruments for competition is the use of a pyramid of blends to adjust to consumers' preferences, where each variety has its own place (Pelupessy, 2001a: 80). Colombian and high-quality Other Milds (e.g. from Mexico and Guatemala) serve as tastemakers of a small, highly priced segment. A larger part is complementary which consists of Other Milds and Unwashed Arabica (Brazil), while the remainder is the filler of Unwashed Arabica and Robusta (Uganda, Vietnam). The precise composition of a blend is a secret and a barrier to entry. Substitution with cheaper varieties is done as long as the taste of the blend will not change or at least the consumer will not notice it (Sellen and Goddard, 1997: 134). Blending makes the roaster independent from the specific origins. A new technological development of the big roasters is the steaming of Robusta to reduce bitterness, which increases the possibility of replacing more expensive varieties. As a consequence there are great variations in the composition of coffee imports of all consumer countries.

The next link to the growers is through the international coffee markets. On the spot markets two-thirds of the demand comes from the two biggest economic blocs, the EU and US, while the supply is equally concentrated in the main producer countries. However, the real trade is executed between a very concentrated number of private agents from consumer countries and numerous providers from exporting countries. Only seven trading companies handle more than 50% of the global imports, while the top ten reaches more than 60% (Fitter and Kaplinksy, 2001). Among these are the big five: Rothfos, E.D. & F. Man, Volcafe, Cargill and Aron. They are clearly operating buyers' markets. But spot prices are highly influenced by the futures markets in New York (Arabicas) and London (Robusta). These are financial markets used for risk reduction and speculation, often by parties not related to the coffee business. The fact that the volumes of futures transactions for Arabicas and Robusta are respectively nine and five times their production figures is an indication of the impact on the spot price and its volatility. The influence of producer countries on these prices is limited, and even the intervention of associations of producer and consumer countries has not given the expected results (Pelupessy, 2001b).

The next sequence of markets is located within the growers' countries and consists basically of the producers' markets for coffee berries and those for processed or green coffee. In the past these markets were heavily influenced by government regulations and policies (Akiyama, 2001: 83–120). The objectives were to make coffee export profitable, to collect taxes and to guarantee minimum incomes to the grower. Today these markets are to a high degree liberalised because of inefficiencies of the old system and pressure from the international finance institutions. The market for coffee berries is a buyers' market of an oligopsonistic nature, because the very numerous smallholders are providers to a small number of middlemen and coffee millers. In the second half of the 1990s market shares of the top ten coffee exporting firms varied from 50 to 90% in the East African countries. Most of the growers manage small to very small properties, varying from on average 0.5 hectare in Ethiopia to 1.6 hectare in Central America (CEPAL, 2002: 21). Their very restricted economic capacity is also evident from the low supply elasticities in producer countries. In most countries these are lower than 0.3 for the short run. In the long run elasticities may be increasing, depending on the availability of arable land or forests.

The markets for green coffee in producer countries are often bilateral oligopolies where millers trade with exporters. Sometimes these two may be vertically integrated, which increases their market power towards the growers. In most cases there exists the possibility of new entrants in these green coffee markets, especially by foreign capital agents. This gives the market the nature of an unstable oligopoly (Pelupessy, 2001b: 18–19). From this analysis it may be clear that the possibility is great for a global mismatch between the supply of coffee berries by smallholders and the demand for roasted and ground coffee by Western consumers. Especially when the great variations in quality are considered, consequences for prices may be serious.

At the beginning of 2003 the average international price of all coffee categories was only about half of the 1997 benchmark after a five-year period of almost continuous decline. The average export prices were in nominal terms 30–60% lower than those at the beginning of the 1980s. According to ICO the composite export price of around 50 US cents per pound of green coffee is the lowest in real terms for 100 years. As a result of the operation of markets in the chain, an oversupply has developed on the global export market. Coffee production in the last five years has been rising at 3.6% annually, while total demand grew by only 1.5%. The production expansion in Vietnam and Brazil, the continuous productivity rise in Vietnam, and the overvalue of the dollar, are among the underlying factors for the continuous supply increase (Gilbert and Zant, 2001: 1). However, in many countries coffee plots have been abandoned, not maintained properly or not harvested. Obviously this could mean that the crisis will be solved in a 'natural' way by underavailability of quality coffee in the future, but at huge social costs.

The balance may also be restored by diversion of significant quantities of low-quality coffee from the market during a number of years. Each 60 000 tons removed may increase the market price by 2 US cents per pound of green coffee (Gilbert and Zant, 2001). The diversion means that the production of about one million smallholders will be affected globally in the first year of the measure and a half a million yearly during the following four years. Through successive global Coffee Quality Improvement Programmes the ICO has put into effect minimum grading standards and maximum moisture content for coffee exports (ICO, 21 August 2002: 3). Sub-standard coffee should be eliminated from the market and supply reduced. It seems that coffee-growing countries and farmers will be paying the bill for global supply control.

There has been a strong negative distributional impact of the price decrease, which has hit developing countries disproportionately. Where in the early 1990s producing countries received on average 40% of the retail values in importing industrialised countries, this share has fallen to almost 8% today. A comparison of prices paid to coffee growers and retail prices reveals enormous differences

Table 7.3   Minimum and maximum monthly price averages for main grower and consumer countries (US cents/lb)[a]

Country		1997	1998	1999	2000	2001
Colombia Milds						
Colombia	Grower	94–172	88–125	76–97	68–87	53–66
Sweden	Retail	350–506	381–489	321–394	272–332	238–287
Other Milds						
Guatemala	Grower	78–121	74–147	71–90	49–89	31–60
USA	Retail	330–467	345–403	334–350	321–368	291–322
Brazilian Naturals						
Brazil	Grower	66–102	62–105	49–67	27–61	13–34
Germany	Retail	438–511	479–528	385–501	315–383	303–339
Robusta						
Vietnam	Grower	37–70	57–70	40–67	16–36	12–18
Italy	Retail	526–570	536–589	484–568	412–484	413–454

[a] Prices by green coffee equivalents.
Source: ICO database.

which tend to increase with falling world market prices (Table 7.3). Farm prices follow the known variety differentials, which seems not to be the case with final consumer prices.[2] The minimum–maximum differences seem to be relatively less for retail prices in consumer countries compared to those paid to growers. The latter follow the international spot market price fall, while this decline is only partially passed on to consumers. The gap between consumer prices and green coffee prices has been increasing (Morisset, 1998; Bettendorf and Verboven, 2000; Feuerstein, 2002). Market imperfections between different chain segments, taxation and high sunk costs in the consumer countries could be implicated. Both the general imbalance and the asymmetric international distribution effects may have had consequences for the environmental impacts of the coffee chain. Already before the crisis of 1998 there had been an unequal distribution of environmental effects of the production, trade and long-distance transport processes, which made it more difficult to reduce the negative effects significantly.

7.4   The environmental impact of the coffee supply chain

The commodity chain approach has the potential to consider both value-creating activities and externalities in an integrated way. In many discussions about the environmental effects of international trade, a product-oriented assessment is applied, where only the final use impact of traded goods is considered (see *Special Issues on Trade and the Environment of Ecological Economics*, 1994, no. 9). In an interesting study by Nestel (1995: 165–178) of Mexican coffee activities, the socio-economic and ecological systems are still considered largely separately and connected only by the farmers' activities. Nevertheless, all value-generating activities in an input–output system could have environmental effects because of the use of non-renewable resources or the emission of polluting substances. Therefore it will be necessary to map out the main environmental impacts in each of the segments of the coffee chain.

This process-oriented approach opens the way to combine the life cycle assessment and global commodity chain methods, in order to apply an integral tool for sustainable management of specific coffee chains. However, this will need other, more detailed efforts in quantification and measurement of the economic and environmental effects, which is beyond the scope of this chapter.

We will give a first indication of those activities within the chain where the reduction of negative environmental effects would be needed and how this could be realised. However, measures to decrease or eliminate certain environmental effects may disproportionately reduce the competitiveness of specific links, affecting the position of actors within the chain. Because of the imperfect markets and the asymmetric distribution of generated values,

---

[2] However, one should remember that we are not consuming coffee from one origin because of the blends.

explained in Sections 7.2 and 7.3, there are unequal capacities to internalise the costs of these externalities. Environmental policies and other interventions in the chain may affect both income distribution and choice of technologies, which may in turn influence the treatment of the environmental effects of production and trade.

7.4.1   Coffee growing

As a perennial crop the cultivation stage needs basically maintenance and harvesting activities after the first investment in land preparation and planting with seedlings or seeds. The farm produces the coffee berries or parchment coffee when it integrates the first drying process. It takes three to five years to the first profitable harvest and up to 30 years before renewal of the plantation will be necessary. The cultivation practices are tillage and weeding, mulching, nurturing, pruning and disease control. The main environmental inputs are soil, water and non-renewable energy to produce the agrochemicals, while the outputs are erosion and residues to soil and water. There are also positive effects of coffee plantations in terms of biodiversity protection, carbon capture and oxygen generation. It is estimated that coffee plants give an equivalent of about 35% of the positive effects of carbon dioxide capture of the same area of woodlands (Anacafé, 2001: 249–255). However, the real environmental impact depends on the different agricultural technologies implemented which are related to the coffee variety, the inputs used and the presence of shade. A wide range of options may be available which can be grouped into four basic production systems (see Table 7.4).

Table 7.4   Alternative coffee growing technologies

Input–output	Traditional	Shaded	Sun-grown	Organic
Plant density (plants/ha)	1000–3000	2500–4000	> 5000	1000–3000
Coffee variety	Tipica, Bourbon	Tipica, Bourbon	Catuai, Catimora, Caturra	Tipica, Bourbon
Labour[a]	36 days/ha	72 days/ha	118 days/ha	Intensive
Agrochemicals	Little	Increasing	High	None
Shade	60–90% area	Declining (up to 50% area)	None	Intensive > 50%
Area	< 14 ha	14–35 ha	>35 ha	Small plots
Plantation life	20–30 years	30 years	12–15 years	30 years
Yields (kg/ha)	190–300	500–1000	>1300	300–600

[a] Pre-harvest annually.
Sources: Boot (2002); Sorby (2002).

First, there is traditional extensive coffee farming as an integrated agro-forestry system with trees of different species for shade and sub-products. Coffee varieties like Tipica and Bourbon are used, while plant density, inputs and labour intensities, as well as yields, are all rather low. Smallholders with family labour apply this system in Africa, Asia and less-developed coffee regions in Latin America. Second is the coffee cultivation on cleared land with a range of technologically intermediate to more advanced shade-grown coffees. Basically medium-sized to large farms are applying higher plant densities with increasing use of pre-harvest contracted labour and agrochemicals. With increasing use of technology and plant densities, the number of shade trees is declining. The third category is the open sun-grown coffee system which is applied both on large plantations in Brazil and Colombia, but also on modernised smallholder plots in Costa Rica and other countries. About 69% of the coffee area in Colombia, 17% in Mexico, 20% in Guatemala and 35% in Honduras are planted with the sun-adapted and disease-resistant hybrids (Sorby, 2002: 1). Enhanced plant densities with technological and labour-intensive maintenance activities have brought very high yields. But the newly developed Caturra, Catuai and Catimorra varieties need abundant applications of agrochemicals to reach these results. One should remember that shade tree litter is a nutrient for the coffee plant and the canopy intercepts solar radiation, wind and rain. It prevents soil loss, reduces problems with insect pests and diminishes the competition for nutrients between weeds and coffee (Nestel, 1995: 174–176).

The high environmental costs in terms of soil erosion (sometimes deforestation), soil and water pollution, and health risks for land workers of sun-grown coffee have led to the (re)introduction of a series of environmentally friendly technologies. These belong to the fourth category which varies from organic to increased shade-grown coffee farming. The absence of any chemical input and therefore a return to low plant density with considerable labour inputs are the characteristics of the organic system. Yields will decline significantly, at least in the first years. The substitution of herbicides for manual weeding may increase soil erosion in the case of uprooted weed control. This and declining yields increase the pressure on land by organic coffee. Other environmentally friendly technologies are related to a revaluation of shade trees and corresponding positive effects. Because of the accumulated leaf litter, air humidity and organic matter, fewer chemical inputs are needed and yields may reach intermediate values. Sustainable technologies enjoy a price premium in the consumer markets, but lead to higher costs due to reduced yields and the required external certifications.

7.4.2   Coffee processing

This is the second segment in the chain which transforms the harvested berries into (internationally) tradable green coffee. This is done by either the dry method based on sun-drying as in Brazil and many West African countries, or by the wet one, where within 24 hours after harvesting the pulp should be removed by water, as in Central America, East African countries and Colombia. The wet procedure gives the better quality and higher priced Colombian and Other Milds because it requires more precise centralised quality control. The environmentally friendlier dry or natural method applies sun-drying on platforms by the farmers, which reduces the possibility of rigorous and uniform quality control. Environmental inputs in wet processing are water, energy and combustibles such as fuels and firewood. Outputs released to the environment are water pollution by organic and inorganic residues, eutrophication, solid waste (unripe, damaged and excessively ripe beans and pulp), bad odours and husks. The green coffee bean produced constitutes only about 18.5% of the berry, leaving more than 80% by volume as organic and liquid waste. Part of the sewage water and organic waste could sometimes be profitably recycled.

Most coffee mills make excessive use of the water from nearby rivers and discharge toxic components which are dangerous for both humans and aquatic life. Coffee mills may use up to 67 litres of water per kg of green coffee. The potential pollution extends also to the air, soil, fauna and human health diseases transmitted by insects caused by the accumulation of organic waste. At the end of the line the green beans are selected and packed in jute bags.

7.4.3   Coffee roasting

This includes the blending of different coffee varieties, roasting, grinding and packing to get finally the consumable coffee. The green coffee is roasted for 5–20 minutes at 350–550°C to get the right colour and flavour. Then it is quenched and cooled, waste materials are removed and the coffee is ground and packed. The main roasters are located in the developed countries near the great consumer markets. The roasting process brings a weight loss of 13–25% of the green coffee, but also an increase of volume of 50–80%. Inputs are water, energy, fossil fuels and packing material, while air pollution (CO_2, NO_x, SO_2 and smog), disposal of waste material and noise may be the outputs to the environment.

Coffee roasting is dominated by small numbers of multinational companies, which are sometimes vertically integrated to international traders and wholesalers or retailers (see Section 7.3). Their selling and buying markets are often very concentrated. These big roasters and increasingly also retail supermarket chains are controlling the whole coffee chain by establishing regulations and standards, including environmental ones, for upstream segments, deciding on the inclusion or exclusion of raw material providers from developing countries.

7.4.4   Coffee trading and consumption

In the roasting, wholesale and retail trade segments, extensive use is made of labelling or branding of the product. In this stage of the coffee chain the commodity is transformed into a very profitable differentiated product bought by the final consumers. It includes the gourmet, speciality and Indications of Origin coffees with high cup-values, and the environmental ones such as organic, bird-friendly and ECO-OK. In the US, price premiums of about 10% are paid to providers of organic and shade coffees. Those paid by retailers are higher than by roasters (Giovannucci, 2001: 10). Water and energy are the environmental inputs, mainly because of consumption, while air pollution (CO_2, NO_x, SO_2) and solid waste disposal (coffee and packing material) are the outputs. The value or price of the final consumer product and therefore the total value created in the chain may vary considerably depending on whether the coffee is consumed at home (in Western countries ±23% of the total) or outside in restaurants, bars or workplaces.

7.4.5   Coffee transportation and storage

Between each pair of the chain segments mentioned earlier, the coffee will be transported and stored by market parties. The coffee berries, or parchment coffee in the dry method, are normally transported in bulk by car or truck and frequently collected in warehouses (shortly) before processing. After processing the green coffee is transported by truck in bags of 45 or 60 kg to the exporters or harbours, and shipped by boat to be imported in the (developed) consumer countries.

Considerable stocks are kept in both producer and consumer countries, counting on average for respectively 3 months and 1½ months of world consumption (ICO data). Green coffee is transported by truck or train from the harbour to the roasters, then transported again as roasted ground and packed coffee to the wholesalers and retailers and bought by the consumers who may use their cars to transport it to their homes. In all these transport segments energy and lubricants are used and environmental outputs are air pollution (CO_2, NO_x, SO_2), CFCs, noise and others.

An overview of the environmental effects in the different chain segments gives an initial idea of their distribution and the problems which need to be tackled. It will be costly for upstream chain actors to reduce the corresponding effects. For most smallholders around the world it will be financially not viable to make the change to, e.g., organic technology, given the costs of the transformation and certification and the decline in yields. The same could be said of the investments needed to clean and reduce the use of water by the coffee processors. Actors in downstream segments have ample possibilities for improvement.

7.5   Identifying problem areas

The spread of the environmental effects of coffee production and trade is worldwide because of the fragmentation and geographical location of these activities. The diffusion does not have an even character and certain concentrations of environmental impacts may be observed, related to natural conditions, production relations and applied technologies. Due to the nature of the commodity, the impacts of coffee growing, initial processing and corresponding transport activities will also be located in the 52 tropical producer countries. Coffee roasting for their domestic markets will have negative environmental impacts, which may be stronger than in the developed countries because of the use of obsolete technologies. Brazil accounts for more than half of the roasting impacts of producing countries because of both its high number of urban consumers and its considerable consumption per head, which is only slightly below the average of the EU (see Table 7.2). Then come Ethiopia, Indonesia, Colombia, Mexico and India with a joint participation of 27% of the domestic consumption of producer countries. These are all big producers, but except in the case of Ethiopia the large urban populations are the important factor. The cradle of coffee, Ethiopia, has a relatively small urban population, but for cultural reasons it has a per capita consumption comparable to Colombia with an income per head of at least 20 times more.

As mentioned before, coffee cultivation on cleared land is increasing, which may imply increasing deforestation, specially in African and other countries where the coffee is moving to the agrarian frontier. In Ivory Coast small peasants have been driven towards this frontier by the gradual breakdown of traditional tenureship. Within 20 years they have cleared one million hectares of forest land to cultivate coffee (Seudieu, 1996). In mountainous countries with sun-grown coffees, soil erosion and depletion is a problem, as may be the case in Brazil, Colombia and Costa Rica. In these countries the presence of more technologically developed shaded and sun-grown coffees has brought the intensive use of agrochemicals as a consequence. The use of herbicides in weeding may reduce erosion but augments pollution. Large farms may use these substances more intensively to reduce labour costs for weed control. Extended coffee plantations and clustering of smallholders concentrate the pollution of (subsoil) water, rivers and soil in the rural coffee areas, mainly with nitrates from nitrogenous fertilisers.

With a reduced quantity and increased frequency of application, the absorptive efficiency of the plants may be increased and unused nitrates decreased. After World War II volumes and diversity of imported pesticides and fertilisers increased greatly to match the new high-yield varieties of technologically developed sun-grown coffees. Economic policies supported this process.

The effects of coffee processing depend on the method that is used and the scale of operation of the mills. The environmental impact of the dry method tends to be less and not concentrated, as is the case for most processing in Brazil, Ethiopia and robusta countries such as Ivory Coast, Indonesia and Vietnam. The three processing steps applied to the whole berry are cleaning, drying and hulling. In the dry method this is done by hand, using a sieve, and includes the selection process. Most of the moisture is eliminated by sun-drying for up to four weeks. Once dried, the beans are stored in silos and later sent to mills where hulling, sorting, grading and bagging take place. Organic waste and dust may be the environmental effects.

More than 35% of the global coffee harvest is processed by the wet method. Initial sorting and cleaning of the berries is done by washing with flowing water. After fermentation the pulp is removed from the berry by machines, passed through sieves to separate imperfect beans and waste, and to water channels for washing. After this it is sun-dried for about 10 days to reduce moisture, but in large mills hot air machines may be used. Then the parchment coffee is stored, and shortly before export it is cured to remove the parchment, cleaned, sorted and graded. This means that in many countries there is a considerable problem of water pollution by organic and inorganic residues, solid waste and water eutrophication. Regions in Colombia, Mexico, Central and South America, but also India (partly), Kenya and Tanzania have these problems. However, the seriousness may differ according to the degree of concentration of the mills. An example is Costa Rica where coffee processing is the most technologically advanced in Central America and very concentrated in a limited number of large mills. This has raised the coffee quality of this country which is considered as one of highest in the world. Guatemala, with hardly twice the Costa Rican harvest, has 20 times as many small and very decentralised mills. The concentrated water pollution in Costa Rica also gives a nasty odour around the mills which are often located near densely populated urban areas. In general there seems to be a negative correlation between price/quality of the variety of coffee (Colombian Milds, Other Milds, Brazilian Naturals and Robusta) and their environmental friendliness (see Section 7.4). The inter-firm land transportation causes CO_2 and NO_x air pollution in all producer countries.

About 75% of coffee production is exported to and roasted in the importing countries. Very large concentrated roasting factories are located in the 25 principal countries, as well as most of the final trade and consumption. Much of the CO_2, NO_x and SO_2 air pollution, solid waste (eventually) and water pollution from this stage are therefore concentrated in the principal importing countries such as the US, Germany, France, Italy and Japan. Together they may account for almost 60% of this pollution worldwide. Recent technological improvements due to regulations have probably reduced the environmental effects of the big roasters. Nevertheless, final consumption may add some pressure on the environment due to increasingly energy-intensive brewing methods and use of private cars to reach the more concentrated retail points in supermarkets.

The major share of the environmental burden in the coffee chain rests with the producer countries. The need to raise productivity and consumer quality has increased the use of environmentally unfriendly technologies. The international labour distribution and more stringent regulations in the developed countries have located the cleaner stages of coffee production in the main importing and consuming countries.

Article 35 of the International Coffee Agreement of 1994 between all growers and the main consumer countries states that due consideration should be given to the sustainable management of coffee resources and processing (see also article 39 of the renewed agreement ICO, 2000: 30). Because of the tradable nature of the commodity it is important to note that the 2001 Doha ministerial declaration of WTO considered that a non-discriminatory trading system could be consistent with the protection of the environment. Subsidies and other encouragements (incentives) for environment improvement programmes are allowed by WTO agreements on agriculture. Environmental clauses are considered as 'trade-facilitating' and described as WTO-mandated environmental regulations. Consumer countries apply rigorous food safety standards with maximum residue values for agrochemicals and control on microbial and ochratoxin contamination, which may appear because of excessive moisture of the coffee in storage. Developing countries are highly sensitive to sanitary and phytosanitary regulations and other norms, which may act as trade barriers. Compliance with these rules has led coffee farmers in India to recover their costs by unsustainable exploitation of biodiversity, e.g. by cutting trees. This is clearly a conflict between WTO standards and multilateral environmental concerns (Damodaran, 2002: 1123–1135). Coffee bags should also comply with environmental regulations for packaging.

Especially for coffee, governments do support the private eco-labels of organic, ECO-OK, bird-friendly and fair-trade coffees. Besides this there are the more general norms such as ISO 9000 and ISO 14000, which could be adopted by coffee production centres. These are all market access enhancing measures, while the sanitary ones are market access maintaining. It may be advisable to move certification into the public domain to enhance reliability, reduce costs and harmonise with WTO rules.[3] In the producer countries environmental regulations of governments are usually focused on the appropriate use of agrochemicals, prevention of erosion and deforestation, maintenance of biodiversity, prevention of soil and (river) water pollution and water saving (IDB 2002: 18–19). In most of these countries there are problems with implementation which may reduce the effectiveness of policies. Nevertheless, more severe enforcement may lead to excessive costs for local producers in the upstream segments of the chain. Declining coffee incomes and taxes have greatly restricted the financial space of grower countries' governments. Solutions to these environmental problems should be considered within a chain-wide perspective.

7.6 Sustainable coffee production

'Coffee, if grown right, can be one of the rare human industries that actually restore the Earth's health' (Halweil, cited in Varangis *et al.*, 2003: 55). But this will be conditioned by the right organisation of the global coffee chains. As we have seen in the preceding sections, this is all but a trivial undertaking. Solutions to the main environmental problems of deforestation, biodiversity reduction, soil depletion and erosion, water pollution and non-renewable energy use could not be isolated from the structural supply and demand mismatches and exclusion of

---

[3] The need for credible third-party intervention is also emphasised by Grolleau (2002: 345).

coffee growers. For a market-oriented approach it will be necessary to consider the future consumption trends of coffee. It seems that the crop fits well in the increasing preferences of affluent consumers for high value, exotic and differentiated goods, with a number of credence characteristics.[4] There is an ever-increasing variety of coffees where qualities (tastes), origins, environmental friendliness, etc., are explicitly distinguished and branded. As mentioned earlier, the growth rates of the markets for the environmentally differentiated coffees are much higher than for the generic ones. Practically all businesses in the US speciality coffee sector knew organic and shade coffee and more than half of them actually sold shade coffee and almost 80% organic coffee (Giovannucci, 2001: 7). It is important to note that quality as a sensory characteristic (taste) is considered as a kind of precondition for the acceptance of sustainability, which is mainly a credence one. The majority of the respondents saw certification as necessary and felt that price premiums of about 10% were reasonable for organic and shade coffees. Organic seemed to be better known in the coffee industry than shade-grown.

One of our conclusions is that the sustainable niche markets are an option for high-quality coffee producers only. The origin of sustainable coffee for the North American market may be an indication of this, in which Central and South America were by far the biggest suppliers and where Asia and Africa were not very significant (Giovannucci, 2001: 15–16). Leading supplier countries for both are Colombia, Costa Rica, Guatemala and Mexico, while Peru and Indonesia are also significant for organic coffee.

The diversion of low-quality coffee from the world market as proposed by the ICO looks like a correct step in the direction of better market orientation. Nevertheless, it may not be acceptable to exclude the African and Asian producer countries or the smallholders of the Latin American lowlands with their lower-quality coffees. The issue of certification costs may make things worse, since this will be a special disadvantage for low-income growers. It is questionable if this will be a necessary strategy, considering that mainstream coffee roasting is based on the pyramid of blends where wide ranges of qualities have their own place. Differentiation should be based not only on single origin or variety coffee, but also on the common practice of blending. There is no reason whatsoever to suppose that low-quality coffees could not adopt environmentally friendly technologies. It is quite the contrary. As we have seen in Section 7.4 these smallholders may very well be low-input farmers. The price fall may even have reduced the chemical input of poor farmers.

This brings us to the issue of what could be done in the upstream chain segments to match final developments in the consumer market. The approach that is currently most embraced in environmental circles is what could be called the micro-niche market one. This involves the persuasion of individual or groups of coffee growers to transform their farming system into a sustainable one, to

---

[4] Characteristics of goods that could not be tested by the senses, but are guaranteed by the providers (Grolleau, 2002: 343). Coffee 'with a story' is an example.

upgrade their produce and to look for access to niche-market buyers with price premiums (cases are given in Gobbi, 2000; DevNews, 2003; Ramírez, 2001: 4; Sorby, 2002). The small scale, long transition period and high costs make the approach unsuitable for mainstream coffee producers. The macro problem of overproduction will also remain.

For the mainstream producers two strategies may be possible. One is to promote high-quality and profitable coffee growing and to impose or stimulate the application of sustainable farming practices by the growers. This may reduce the oversupply, but the small share of growers in the generated value will not be changed. Examples are the earlier mentioned Global Quality Improvement Programme of ICO and the suggestions to phase out lowland low-quality coffee growing (as in Boot, 2002: 71–72). The asymmetric burden against numerous smallholders, the social costs and the obstacles and private costs of introducing sustainable techniques are among the main problems. This proposal also presupposes the disappearance of the actual blending customs of roasters.

We advocate the opposite strategy which goes from sustainability to quality enhancement and not the other way round as in the previous strategy. It should start with the identification within the chain of the actually most environmentally friendly producers, which will favour the traditional and low-input coffee-growing farmers, shade-grown mixed cultivation plots and dry-method processing. Eventually a broad classification from more to less environmentally sustainable green coffee production systems could be developed by mixed institutions of grower organisations, coffee companies and governments. The big buyers, roasters and trading houses in the chain should already pay the price premiums for environmental friendliness. This will immediately affect the income of the less wealthy smallholders favourably and make the current blending pyramid greener at the cheaper and bigger ends of the basis. In this option the upgrading of these smallholders should be dealt with through internal chain coordination. The strategy does not depend on the market niches of speciality coffees and may give advantages to Asian, African and many Latin American lowland growers. The economic improvement of these growers may be a more viable and less costly enterprise than drastic transformation of the established farm systems into sustainable ones. High-quality, technologically advanced and wet method producers should have the economic capacity to improve sustainability and pay the corresponding certification costs by receiving the price premiums for quality through the current systems. Steps should be taken initially in the direction of appropriate farm management in terms of the (reduced) use of agrochemicals for disease control and fertilisation, conservation of soil, water and biodiversity by shade management, and mixed farming (eventually, erosion control and waste reduction or recycling). For many smallholders shade-grown coffee of different densities may offer more realistic opportunities than organic coffee, despite the greater positive environmental and income impacts of the latter.

A second line of action is the use of water for first processing. Absolute reduction of the used volumes, the cleaning of water and the eventual recycling of wastes are technically feasible for the wet method (Obando and González, 1995; Wasser *et al.*, 1995). A cleaner upward agricultural segment will also benefit the cleaning up of first processing. There may be problems of the small scale of mills which make process transformation too expensive. For these cases horizontal as well as vertical integration may help.

A third line of action is the reduction of the use of non-renewable energy along the whole chain. This refers not only to the use within the different segments, but mainly to the long transport lines between these. Decentralisation of first processing to reduce the transport between farms and mills may have adverse effects on the energy use between mills and export points. However, this is an empirical matter because the latter may be more efficient. Energy use at the final consumer stage may also be reduced by propagating low-energy brewing methods. The trends towards concentrated retail points in supermarkets and the substitution for renewable energy are problems that could not be solved in the coffee chain. Finally, there remains the question of the financial coverage of sustainable strategies. Economic coordination and regulation is generally executed by the governance force of the commodity chain. As we have seen in Section 7.3, in the case of coffee these are the big multinational roasters and retailers in the consumer countries. It goes without saying that they may also be held responsible for the sustainability strategy in the coffee chain. Of all the actors in the chain they have at least the major economic power to do so (Pelupessy, 1999: 126–127), but responsibility should also be shared by the governments of the main importing countries, since coffee consumption and complementary spending are taxed considerably.

7.7 References

ACPC, 2002: Association of Coffee Producing Countries, coffee market report No. 23.
AKIYAMA, T., 2001: Coffee market liberalization since 1990, in: *Commodity Market Reforms*, World Bank.
ANACAFÉ, 2001: Investigaciones and Descubrimientos sobre Cultivo de Café, Guatemala.
BETTENDORF, L. and VERBOVEN, F., 2000: Incomplete transmission for coffee bean prices: evidence from The Netherlands. *European Review of Agricultural Economics*, 27(1): 1–16.
BOOT, W. J., 2002: *National policies to manage quality and quantity of coffee in Central America*, Report for the Inter-American Development Bank.
CEPAL, 2002: *Centroamérica: el impacto de la caída de los precios del café*. México D.F.
DAMODARAN, A., 2002: Conflict of trade-facilitating environmental regulations with biodiversity concerns: the case of coffee-farming units in India. *World Development*, 30(7): 1123–1135.
DE GRAAFF, J., 1986: *The Economics of Coffee*. Pudoc, Wageningen, The Netherlands.
DEVNEWS MEDIA CENTER, 2003: Cultivation of Eco-Friendly Coffee in El Salvador, http://web.worldbank.org.
FEUERSTEIN, S., 2002: Do coffee roasters benefit from high prices of green coffee? *International Journal of Industrial Organization*, 20: 89–118.

FITTER, R. and KAPLINKSKY, R., 2001: Who gains from product rents as the coffee market becomes more differentiated? *IDS Bulletin*, 32(3).

GILBERT, C. L. and ZANT, W., 2001: Restoring balance by diversion in the world coffee market. Paper, University of Amsterdam.

GIOVANNUCCI, D., 2001: *Sustainable Coffee Survey of the North American Specialty Coffee Industry*. World Bank.

GOBBI, J. A., 2000: Is biodiversity-friendly coffee financially viable? An analysis of five different coffee production systems in western El Salvador. Ecological *Economics*, 33: 267–281.

GROLLEAU, G., 2002: Proliferation and content diversity of environmental claims: an explanatory analysis applied to agro-food products. *Applied Economics Letters*, 9: 343–346.

ICO, 1995: Article 35 of the Agreement on Environmental Aspects. Executive Board, International Coffee Organization, London.

ICO, 2000: International Coffee Agreement 2001, London.

ICO, 2002: The Global Coffee Crisis: a Threat to Sustainable Development.

ICO, February 2003: Coffee Market Report.

ICO: http://www.ico.org/crisis/main.htm.

IDB, 2002: Managing the competitive transition of the coffee sector in Central America, Discussion document,World Bank.

LODDER, C. A., 1998: *Quality – the Key to Progress?* FO Licht GmbH, Köln.

MORISSET, J., 1998: Unfair trade? The increasing gap between world and domestic prices in commodity markets during the past 25 years. *World Bank Economic Review*, 12(3): 503–526.

NESTEL, D., 1995: Coffee in Mexico: international market, agricultural landscape and ecology. *Ecological Economics*, 15: 165–178.

OBANDO, S. and GONZÁLEZ, A., 1995: Incorporación en el Mercado del café. *Masoca*, 12.

OLEKALNS, N. and BARDSLEY, P., 1996: Rational addiction to caffeine: an analysis of coffee consumption. *Journal of Political Economy*, 104(5): 1100–1104.

PELUPESSY, W., 1999: Coffee in Côte d'Ivoire and Costa Rica, in: Dijkstra T., van Tilburg, A. and van der Laan L. (eds) *Agricultural marketing in Tropical Africa*, Ashgate/ Avebury, London.

PELUPESSY, W., 2001a: Smallholders and coffee markets, in: *Policy and Best Practice Document 8*, Ministry of Foreign Affairs, The Netherlands.

PELUPESSY, W., 2001b: *Market failures in global coffee chains*. Paper for the Conference on the Future of Perennial Crops, Yamoussoukro, Ivory Coast, 4–9 November, 2001.

PONTE, S., 2002: The 'latte revolution'? Regulation, markets and consumption in the global coffee chain. *World Development* 30(7): 1099–1122.

RAMÍREZ, L., 2001: Globalisation and livelihood diversification through non-traditional agricultural products: the Mexico case. *ODI Natural Resource Perspectives*, No. 67.

SELLEN, D. and GODDARD, E., 1997: Weak separability in coffee demand systems. *European Review of Agricultural Economics* 24: 133–144.

SEUDIEU D. O, 1996: *Impacts de la production du café sur l'environnement en Côte d'Ivoire*. Paper for the Seminar on Coffee and the Environment, 27 and 28 May 1996.

SORBY K., 2002: *What is sustainable coffee?* Background paper to World Bank Agricultural Technology Note 30, 'Toward more sustainable coffee'.

VARANGIS P., SIEGEL, P., GIOVANNUCCI, D. and LEWIN, B. 2003: *Dealing with the coffee crisis in Central America: impacts and strategies*. World Bank Policy Research Working Paper 2993.

WASSER, R., OROZCO, C., RODRÍGUEZ, J. F., GÁMEZ, S., MONGE, J. and DUARTE, R., 1995: Reducing water contamination caused by coffee industries with biogas technology, Biomass Technology Group, Enschede.

WILLIAMS, R. G., 1994: *States and Social Evolution: Coffee and the Rise of National Governments in Central America*. University of North Carolina Press, Chapel Hill and London.

8

Improving energy efficiency

H. Dalsgaard and A. W. Abbotts, COWI, Denmark

8.1 Introduction

Energy costs form a small proportion of the total costs within the food industry. As a result the resources available for investigating potential energy-saving measures are often quite limited. Consequently, the most relevant question to ask when working in the field of energy savings within this branch is not what kind of energy savings should be implemented but rather whether they should be implemented at all. Important prerequisites for success are therefore the use of the right arguments together with an understandable methodology and simplified and/or proved solutions.

Even though energy consumption and energy savings in the food industry have been in focus for the last 30 years, the tendency is still for energy consumption per produced unit to increase in some parts of production. This is due in part to comprehensive automation of production processes. However, the primary reason is the increasing demand for food safety. Increasingly stringent requirements from the veterinarian and the quality manager are, for example, dictating the need for higher levels of hygiene which subsequently leads to a larger consumption of cold and hot water as well as an increased number of cleaning cycles in production. These requirements are reinforced by the tendency of many food industries to adopt the same rule set as adopted in the medical and pharmaceutical industries (e.g. GMP and FDA, ISO 9000, etc.).

The rising levels of energy consumption can also be explained by a series of barriers that exist towards focusing on an optimisation of energy systems. Examples of these barriers include:

- Lack of time and money
- Lack of knowledge and methods (in industry as well as consultants)

- Lack of success stories
- Lack of motivation (size of saving potential).

The risk or rather fear for production losses that can result from the implementation of complex energy systems, together with a lack of interest and awareness of the potential for energy savings within top management, is prevalent within the food industry and a major barrier towards the achievement of more comprehensive energy savings. The gains are perceived as small compared with the effort and the risks that need to be taken, and in addition far more pressing issues seem to be piling up on the manager's desk.

It is seldom, either, that optimal energy systems are achieved within design and construction projects in industry. Such projects consist of a combination of many tasks, each with its own criteria of success. The best solution for the control engineer differs from those of the quality manager responsible, the operations staff, the maintenance personnel, the executive management, or the system designer. Some of the criteria can easily be combined but others are in direct conflict. Thus, the final design often ends up being a compromise between the different active partners. This compromise is often very difficult to predict because of its dependence on human relations, knowledge and the power structure within the company or project. The resultant energy systems are normally not 'energy-perfect' but practical and understandable, and easy to purchase.

8.2   Analysing energy use in food processing

In order to obtain an indication of how energy savings can be made, it is important to understand the nature of the energy consumption in the specific branch and the actual production plant. This understanding can only be achieved by conducting a mapping of the plant's energy distribution. This mapping must be carried out both for an existing plant and when building new and energy-improved production facilities. Tables 8.1 and 8.2 show the distribution of energy in the food and beverage industries in Denmark in 2002. In general, with the exception of domestic purposes, the majority of the thermal energy is used for water heating or drying, evaporation and frying purposes. Energy consumption is, however, closely related to the kind of food processed and the stage of the processing. Slaughterhouses, for example, often have a very high

Table 8.1   Energy consumption in Danish industry 2002 – processes

	Energy consumption (TJ/year)
Energy intensive processes	83 500
Energy light processes	37 300
Room heating	14 400
Total	135 200

Table 8.2   Energy consumption in Danish Industry 2002 – sources

	Energy consumption (TJ/year)
Power	30 700
Heat	104 500
Total	135 200

consumption of hot water and hardly any other heat demands. Heat consumption in processing plants where the food is refined, meanwhile, is not as easy to categorise because of its close interrelationship with the product. Usually, the processes include a mixture of boiling, pasteurisation, evaporation, cooking, baking, and frying. Electrical consumption (besides for lighting and ventilation) is primarily used for pumps, cyclones, milling, conveyers and other transport systems, and in compressed air and cooling systems. The cooling system often represents approximately 50% of the total electricity consumption.

8.2.1   Measuring energy use

When an industry is analysed for energy savings, a cost–benefit calculation should be made for every project. The time and resources used for the analysis in this context is regarded as a project, thus the initiation of the energy analysis often depends on the size of the saving potential and probability for implementation.

In order to evaluate the value of the solutions, it is important that the increased complexity that the solution can present is compared with the savings potential. Malfunctioning caused by bad operation or lack of maintenance can, for example, easily turn promising solutions into a poor investment. For a proportion of the industries, the equipment must be operated by a number of different people in the course of a day or week, and consequently the solutions need to be robust. The simplicity and the easy adoptability of the solutions in the industry must be the objective and the goal.

If the study is concentrated on locating a very precise optimum, exact data are needed. To establish exact data takes time, and in many places such data are more or less impossible to generate due to the nature of the production and/or the knowledge of the energy consumption in general (lack of meters). Thus, it is recommended to focus effort on the data collection in order to estimate energy consumption per year, the maximum load and, for heat, also the temperature levels, and to rule out data that are clearly unimportant (energy consumption too small or too difficult to improve).

As a start, it will be enough to evaluate fluctuations in the average energy consumption over a certain time period. This can be based on existing meters or meters set up during the study in order to classify consumers. The investigation of the fluctuation can be very interesting and lead to actions in changing 'energy behaviour'. Continued data evaluation is one of the most common ways to measure and evaluate the energy efficiency. However, it is important that the data are connected with the actual production, so the generated key figures are directly linked to the amount of produced and approved products in the given time period (energy consumption per produced unit). The production-related energy key figure will open up the scope for energy savings, since minimising of product defects and increasing production can be ways of reducing the energy consumption per produced unit.

The example shown in Fig. 8.1 illustrates that one of the most profitable ways to decrease the energy consumption per produced unit is to increase production amounts, in other words to ensure that the OEE (overall equipment efficiency) for the different equipment and the entire plant is maximised.

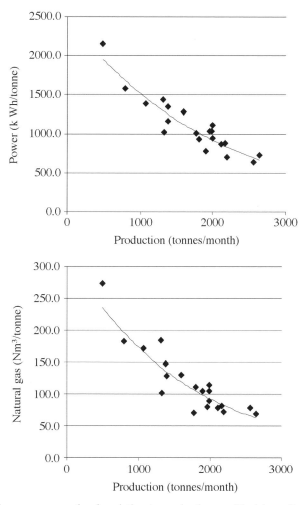

Fig. 8.1   Energy consumption in relation to production – a Danish poultry company, 2001 (masked figures).

8.3   Improving energy use

Energy consumption in the food industry, as in most other industries, consists of a mix of energy use necessary for the operation of the existing equipment and energy use as a consequence of idle-time operation or other irrational behaviour. The unnecessary use of energy caused by suboptimal operation of the production plant can be reduced by introducing energy management methods and systems. Experiences from Danish industry show that commitment from top management is critical if energy management is to be successful. In addition, it is essential that the responsibility for managing energy consumption is placed in the hands of one person who has sufficient time allocated for the necessary activities. Part of the task portfolio of the energy manager is to increase energy awareness on the production floor and follow up energy consumption data on a weekly basis. In addition, the manager is responsible for the creation of key figures that can be used by certain employees to measure performance in production and utility systems (compressed air and cooling plant, boiler house, etc.). It is recommended that energy management is implemented to a certain extent prior to the initiation of any investments in energy-saving equipment, in order to ensure that an energy conscious culture is, to a certain degree, in place. It is important to emphasise, however, that energy management is still strictly linked with the traditional ways of evaluating investment – it has to pay back.

Efforts to reduce energy consumption levels usually start with the appointment of an employee to work specifically with this field. The next step is then to attempt to improve the operation of the current equipment and the current behaviours and to get a solid understanding of where the energy is used (at both the utility level and the process level).

When the areas of primary interest are located, it is recommended to gather data from measurements and information provided by the suppliers of production and utility equipment. Suppliers can be used to bring in specific knowledge of feasible solutions and identify state-of-the-art energy-saving technology. The use of suppliers alone as the only specialist in the study is, however, not always recommended. Suppliers and the industry often agree to differ about what the 'optimal' solution actually is. If competence within the industry itself is considered to be inadequate, it is also recommended that a third party, without any interest in selling equipment, is involved.

In Denmark, the concept of 'Energy-Conscious Design' has recently been developed as a well-defined activity and as a supplement to traditional design. An energy consultant is placed in the project organisation, not as a competitor to the main consultant, but as a partner on the way to a 'better' solution. This concept has shown promising results, but as always, the outcome is very much related to the resources put into the task and the skills of and the motivation shown by the partners involved. The case studies in the following sections are all examples from energy-conscious design projects:

- Danish poultry processing plant
- Danish pig slaughterhouse.

8.4   Case study: improving energy use in poultry processing

During construction of a new processing factory (completed in December 1999) which substituted two small production facilities, an additional effort was made to reduce the energy consumption. The new plant used approximately 7200 MWh of electricity and 7250 MWh of natural gas in the second year after commissioning. The energy consumption corresponds to a cost of €500 000–550 000 per year. The results of the investigation and implementation are illustrated in Table 8.3.

During energy-conscious design at the Danish poultry processing plant, several improvements were recommended, but the major improvements were found in improvement of cooling systems and by increased heat recovery. During the first year of operation, the productivity of the new processing plant was increased and the additional speed of production resulted in remarkable improvements in energy key figures per produced product. As a follow-up, it has been recommended to improve the feeding of conveyor belts to increase production as much as possible.

8.4.1   Cooling systems

Instead of using liquid CO_2 for quick freezing of products after frying and cooking, it was decided to install a traditional low-temperature compressor-cooling plant. The influence on the product by changing the cooling method was only a slight increase in evaporation from the products. The cooling was entirely covered by the cooling compressor system. By using aluminium heat exchangers

Table 8.3   Energy-saving projects at poultry processing factory

Projects	Expected savings (euro/year)	Realised savings (euro/year)
Replacing existing oven	43 000	42 000
Buying new ovens with thermal oil supply	60 000	50 000
Reduction of steam exhaust from ovens	6 500	No measurement
Additional insulation on ovens and fryers	7 500	8 000
Compressor cooling instead of liquid CO_2	450 000	620 000
Waste heat recovery from cooling system	70 000	75 000
Export of waste heat to slaughterhouse	6 000	6 000
Additional cooling level (0°C)	7 500	15 000
Two-speed evaporator ventilators	6 000	6 000
Larger evaporators in spiral freezers (+2°C)	7 500	Larger production
Waste heat recovery on thermal oil boiler	8 000	4 000
Energy saving lighting armatures	7 000	7 000
Use of outside air for drying of rooms	3 500	3 500
Adjustable ventilation speed bag-cooling system	6 000	12 000
Total	690 000	850 000

and by increasing the heat exchanger area in the spiral freezers, it was possible to minimise the production losses without a too low evaporation temperature in the cooling system ($-40°C$). The annual electricity consumption for the $-40°C$ system was 2000 MWh. The savings from the reduced CO_2 purchase were almost €2 million. The optimisation was made possible only because the owner and the consultant were aware of the possibilities right from the beginning and because they involved the suppliers of the equipment early in the design process.

In addition to the two normal systems for cooling (the low-temperature cooling system for freezing and the $-10°C$ system for cool stores), an extra temperature level ($0°C$) was made for cooling of production rooms. The savings from using the $0°C$ system instead of using the $-10°C$ system amounted to around 300 MWh/year. The additional investment in the $0°C$ cooling system was €60 000, thus the payback period (PBP) was less than 4 years.

8.4.2   Heat recovery

By producing 155 m^3 of 65°C hot water by using the waste heat from the cooling plant, it was possible to save 2400 MWh of heat each year. The saving corresponded to approximately €75 000 year and the additional investment for the recovery system was €230 000, thus the PBP was about 3 years. Besides the financial benefits, the storage of 110 m^3 of hot water eliminated any problems with hot water shortages.

8.4.3   Cleaning

Most of the hot water was used for cleaning purposes. An extra effort was therefore made to develop better cleaning facilities and procedures. For instance, it was suggested only to clean freezers once a week instead of every day. The cleaning of a $-40°C$ freezer with hot water is very expensive and in many cases not necessary. Other procedures such as reducing hose pressure and measurement of water consumption, together with automation of cleaning procedures, were also discussed and partly implemented.

8.5   Case study: pig slaughterhouse

The Black Slaughtering Hall (BSH) is the place where the pigs are slaughtered but are still in one piece. When the pig enters the White Slaughtering Hall (WSH) it is automatically split in two and the bowels are removed. Table 8.4 lists the energy and water consumption costs of the various stages in the new slaughterhouse. The energy and water consumption reduction potential during the energy investigation was estimated at approximately 25% (€ 1.4 million/ year). Some of the savings potential was based on energy-conscious design, but the magnitude of the savings potential shows that a persistent and scrupulous effort can give a substantial reward.

Table 8.4 Heat, electricity, and water consumption in new slaughterhouse

Place	Energy and water costs (euro/year)
Pig reception	150 000
BSH (Black slaughtering hall)	670 000
WSH (White slaughtering hall)	320 000
Cooling	620 000
Blood treatment	170 000
Intestines workshop	420 000
Fat melting works	300 000
Ventilation and central heating	630 000
Washing machines	160 000
Cleaning	260 000
Pumps, air compressors, lighting	220 000
Cutting section	420 000
Total	4 340 000

The data collection and the task of getting to know the business of pig slaughtering was very time-consuming even though fast decisions were made and shortcuts taken. About 50% of the time available was used for establishing trustworthy data. The data is based on the existing slaughterhouses and key figures from suppliers and consulting engineers. The selection of focus areas was based on the established process overview of the estimated energy consumption. It was decided to look for improvements locally in the different workshops and in the equipment, before possibilities for integration between processes and optimisation of utility systems was investigated. It was agreed to do a closer examination of the following local areas:

- Optimisation of the singeing/scorching of pigs
- Optimisation of the decontamination process at the end of the white slaughtering hall
- Optimisation of the fat melting works
- Optimisation of tray washing procedures.

There was a need to make further investigations before the focus areas could be prioritised. As a result the optimisation of the fat melting works and the decontamination was stopped.

8.5.1 Singeing of pigs

In order to remove hair, improve the quality of the bacon rind, and kill bacteria, the pork is flamed with natural gas before it enters the White Slaughtering Hall – a sterile area before veterinarian control. It is standard practice to recover heat from the exhaust gas, which does not exceed 50%. It is not financially attractive to recover more because the temperature of the combustion products is lower (250°C) as a consequence of an open oven design. The oven is accessible to false air from the surroundings. The pork is exposed to the heat treatment for more than 60 seconds in order for the bacteria to be killed.

It was suggested to make a closed oven. The idea was to use not only the flames but also the exhaust gas to kill the bacteria. The closed design will improve the potential for efficient heat recovery (higher temperature of the combusted products) and at the same time reduce the amount of natural gas used in the oven. A development of a new and improved oven for the singeing of pigs was initiated and the outcome is still awaited. It is, however, expected that 15–25% of the natural gas consumption can be saved (300–500 kW). It is expected that additional development would lead to even larger savings potential.

8.5.2 Decontamination

The company had added (as a trial) an extra decontamination of the pork immediately prior to cold storage. The decontamination process had one purpose only, and that was to kill the remaining bacteria. For that purpose hot water at a constant temperature of 82°C had to be sprayed onto the pork. The water was returned for cleaning before recycling. At the same time as wastewater was removed from the system, fresh water was almost constantly added (because of the veterinarian demands it was not possible to use the 'almost' clean wastewater). Prior to the water being discharged to the sewerage system, however, a heat exchange between the outlet of wastewater and the inlet of fresh water was proposed.

The savings potential depended on the operation of the system. If the equipment for decontamination of the pigs was to be in operation for 16 hours per week, the potential for heat recovery was $15\,000$ m^3 of 70°C hot water per year or 900 MWh/year. Based on the above assumption, the annual saving potential was estimated at €17,000. The saving was not implemented because a decision was made to leave out the decontamination process for the time being – the additional investment and operation cost did not match the value of the process.

8.5.3 Fat smelting works

In the process of smelting fat from the bacon rinds and other remains from pork cuts, steam is injected directly into the process. The waste product after separation of fat and protein is the so-called 'glue water'. Glue water is water with a very high protein content. The protein from the smelting and separation process is called greaves. The greaves are cooled and stored. Some of the greaves are used for hydrolysis where protein power is extracted and others are dried to extend the shelf-life of different products. The dried greaves are sold to other industries for various purposes.

In the hydrolysis, the proteins from the greaves are transformed into small and refined particles mainly to be used in the food ingredients industry. The by-product or waste from the hydrolysis is fat and water. During the investigation in the fat works, it was suggested to use raw rind in the hydrolysis instead of greaves, and at the same time it was suggested to use indirect heating instead of direct injection of steam in the melting process. Both improvements would result in a decrease in the primary energy demand in the smelting works and would, at the same time, reduce the amount of glue water to less than 33% of the originally estimated amount. The remaining hot glue water (60°C) would be mixed with the rind before the hydrolysis process. The utilisation of the glue water for the mixing would eliminate the use of cold tap water and steam, and would at the same time add extra protein to the process. Finally, it was suggested to recover the heat from cooling the greaves and the products from the hydrolysis by preheating cold tap water. The maximum annual saving potential was estimated at €50 000. It was, however, a surprise that the melting of fat was consuming as much energy as it did – 15 000 MWh a year corresponding to an annual heating expense of €300 000. The result was that the fat works was downsized and the hydrolysis was left out. The additional investment and operation costs in the workshop did not correspond to the value of the increased production.

8.5.4   Washing of trays

During the need assessment, it was discovered that the water and energy consumption during the washing of bowels trays in the slaughter hall could be reduced if the veterinarians accepted the proposal. Based on the information from the suppliers, it was expected that approximately €150 000 each year was to be used for washing and sterilising the trays. By examining the need for washing, it was discovered that less than 10% of the trays actually needed to be washed. However, the tradition and the focus on hygiene make it difficult to approve such a simple and inexpensive suggestion, although it is still under consideration.

8.5.5   Heat-exchanger network

It was planned that the waste heat in the new slaughterhouse was to be used through the hot water system (30°C, 45°C, 60°C, 82°C) and through the central heating system (T_{return} = 60°C, T_{supply} = 90°C). However, before the design of system was finalised it was decided to reduce the temperature for 30°C and 45°C. These levels were not crucial and experience from other slaughterhouses indicated that temperatures of 20°C and 42°C repectively were fully acceptable. This decision reduced the energy needed for hot water production by approximately 3–4%.

The hot water systems pick up waste heat from the cooling system and the air compressor systems. The savings potential in the hot water production was estimated to be 4000–5000 kW or €400 000/year and for the central heating the savings potential was 2000 kW or €190 000/year. The guidelines and restrictions from the veterinarians prevent the storage of hot water in open

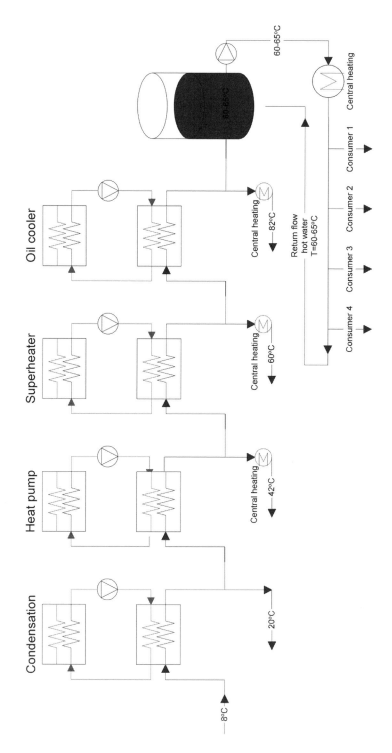

Fig. 8.2 Hot water production in slaughterhouse.

tanks except for the 60°C hot water used for cleaning of process equipment and facilities. The hot water used in processing has to be produced when it is needed or stored in closed water systems like normal hot water tanks. This circumstance has a large influence on the type of solutions that can be accepted and it narrows down the number of heat-exchanger structures that need to be investigated.

The suggested hot water system was supplied 100% by waste heat either by the direct heat recovery from the cooling plant and the air compressor stations or by the waste heat collected through the central heating system. The structure of the heat exchanger network was quite straightforward and was based on standard solutions that were fully backed up by the central boiler station. The structure of the suggested hot water system (Fig. 8.2) was very similar to that of the ordinary system where waste heat was not recovered. The major difference was the addition of plate heat exchangers connected to the source of waste heat.

8.5.6   Utility systems

Due to the Danish tax legislation, it was highly profitable to install a biomass boiler for the heating of the facilities. It was decided to use animal fat coming from a bonemeal factory. The savings were estimated at €240 000/year.

The cooling system and the cooling plant were closely examined. Based on research and experience, the cooling of the pork meat (half pigs) was changed in such a way that it was made possible to reduce the evaporation temperature. It is normal to operate with air temperature in the cooling tunnels at −25°C, but the cooling tunnel in the new slaughterhouse is designed to operate at a temperature of around −18°C. Besides the improved cooling for the tunnels, it was also decided to install an additional cooling system for cooling of production rooms. The cooling system is designed at an evaporation temperature between −3°C and 0°C. It is common in other slaughterhouses only to operate with −40°C and −10°C in evaporation temperatures, but in the new slaughterhouses three temperatures are installed −25°C, −10°C and −3°C. At the same time, optimisation of the compressor constellation, fans, evaporators and condensors made it possible to reduce the electricity consumption for cooling by 38%. The total estimated saving in connection with the cooling plant was €400 000/year.

8.5.7   Suppliers

During the energy-conscious design of the slaughterhouse, it has been very important to have close contact with the suppliers of equipment in order to discuss all the opportunities for energy savings at the process level early in the design project. The involvement of suppliers at this predesign stage has at the same time made it possible to foresee some of the problems and to make necessary changes in the layout of the slaughterhouse before the changes became too difficult to incorporate.

8.6   Future trends

Optimisation of production is basically achieved via manufacturing the right product and the right quality at the right time as cheaply as possible. If the product cannot be sold or produced when it is needed, there is no point in trying to minimise the energy costs. Thus, production development and production management will often be prioritised far more than energy savings.

The key to a better utilisation of resources will, however, always lie in a good overview and registration of the production key figures. One way of achieving this is via the introduction of ERP (Enterprise Resource Planning) systems such as SAP, BAAN, Navision, etc., a tendency we have seen in the last decade. The systems in short make handling, execution and the follow-up of orders easier and more transparent. What is particularly interesting is that the system can be extended so that there is direct communication with the production equipment – the extension is called MES (manufacturing execution system).

By creating a direct transmission of data between the production floor and the central computer system, it is possible to establish a greater control over the resources and thereby establish the key figures for the daily production, or item/ equipment-related key figures.

When the production and resource data are generated, the optimisation possibilities become more visible. A combination of the right organisation, an awareness of the potential rewards, and an increase in the utility expenses can then be the right ingredients for the introduction of process integration technology. As with most other technology, the research and the development of energy-correct systems needs, however, to start from the bottom and work its way up. Development must start with simple solutions. If the simple systems become a success, they can then form the foundation for increasingly more energy-efficient systems, even though these are often more complex.

Involvement of suppliers and increased research and development of standard solutions with focus on energy (cooperation between suppliers, trade organisations, universities and consultants) will be necessary in order to make the market for energy savings attractive.

As a consequence of a relatively low energy-saving potential, it is necessary to focus on the design and construction process as the way to achieve energy improvement. Retrofit solutions are much more difficult to sell, partly because the changes are not necessary, but primarily because the investment is much more difficult to support from a financial point of view. The time and resources for need assessment as part of the energy-conscious design must be allocated if substantial energy savings are to be achieved.

Standard solutions that have a future are standard heat recovery components fitted into standard heat recovery systems, high-reliability and low-cost heat and power units (e.g. fuel cells), and reuse of water technology.

Technology for water minimisation and reuse is becoming increasingly ingenious and can be adapted as one of the standard solutions. The development of equipment that can guarantee a certain quality of recovered water is, however, necessary. The risk of production losses because of contaminated 'dirty' water will prevent the implementation of water reuse for the time being, unless it can be used for secondary purposes such as for cooling towers.

8.7   Sources of further information and advice

American Council for an Energy Efficient Economy: Okos, M. R. *et al.*, *Energy Usage in Food Industry*, 1998 (www.aceee.org/pubs/ie981.htm)
COWI (www.cowi.com)
Energy Research Institute at the University of Cape Town: *How to Save Energy* – guide books 1–7 (www.eri.uct.ac.za)
European Integrated Pollution Prevention and Control Bureau (http://eippcb.jrc.es)
IEA Newsletter: CADDET (www.caddet-ee.org)

9

The environmental management of packaging: an overview

F. de Leo, University of Lecce, Italy

9.1 Introduction

Food packaging is an essential technique for preserving food quality, minimizing food wastage and reducing the use of additives. The food package serves the important function of containing the food, protecting it against chemical and physical damage and providing convenience in giving the product information to the consumers. The negative influence of packaging on the environment is often emphasized while its positive properties are overlooked. The basic role of packaging is to protect the product against stresses in the distribution chain and to facilitate handling of the product. Deficiencies in the packaging system lead to inefficiency from damaged and wasted products. The result is a considerably increased impact on the environment. However, food packaging is also perceived as a cause of environmental problems, because it contributes visibly to the solid waste stream.

This chapter will identify the key EU and national environmental laws affecting packaging and packaged goods in Europe, particularly the EU Directive on Packaging and Packaging Waste. The Directive is intended to reduce the overall impact of packaging and packaging waste on the environment by setting producer responsibility targets for recovery/recycling and establishing minimum design criteria (Essential Requirements). The Directive's broad aims are, as a first priority, to prevent the production of packaging waste and, as additional fundamental principles, to reuse packaging, to recycle and otherwise recover packaging waste to reduce the final disposal of such waste. Therefore this chapter:

- will describe solutions to reduce use of packaging, and will also show how these types of activities also can help companies to reduce energy and raw material consumption, cut waste and costs;

- will present and evaluate strategies for improving the environmental performance of packaging, such as use of recyclate in the context of limitations created for recyclate by food migration legislation, designing for recycling, designing for reuse and biodegradable/compostable packaging.

9.2   Packaging and the environment

Packaging performs a number of essential services in today's modern society. In fact packaging must be able to fulfil various functions such as:

- to protect, contain and preserve the product while at the same time allowing efficient manufacturing, handling and distribution methods;
- to provide commercial and consumer information;
- to present and market the product;
- to ensure tamper evidence and to facilitate product use (ergonomics);
- to ensure safe use and handling by consumers.

For food products packaging has an important role to preserve food quality. It has to act as a barrier to moisture and gas, microbial infection and insect infestation, other chemical and biological contamination, and light (where this can cause product degradation). In most cases packaging reduces the cost of goods, because its role as protector means reduced product damage. In the case of foodstuffs, processing costs are also reduced, as also are losses due to spoilage or vermin. In some cases packaging has a positive effect on reducing overall waste. In countries where packaging is not appropriate or does not exist, food wastes can exceed 50% compared to 2–4% in other countries (Pre, 1997). Moreover the trend to buy, in industrialized countries, more prepared meals which need sophisticated packaging to provide adequate protection may increase packaging waste, but as a result food waste in the home can often be reduced. In fact, all along the food chain losses occur or some form of waste is produced. It is better for food preparation waste to be generated at the beginning of the supply chain, because waste from food preparation in the processing plant can often be used as a by-product for use in other processes (Bickerstaffe, 1996). In Table 9.1 various definitions of packaging functions are shown.

Legal definitions for packaging are in the EC Directive on Packaging and Packaging Waste (94/62/EC) in which packaging is defined as 'all products made of any materials of any nature to be used for the containment, protection, handling, delivery and presentation of goods, from raw materials to processed goods'.

There are three main categories of packaging:

- Primary (sales) packaging, around a product at the point of purchase by the user/consumer, e.g. bottle, plastic bag
- Secondary (grouped) packaging, which groups a number of items together until the point of sale, e.g. a box or strapping around a number of items

Table 9.1   Definitions of packaging functions

Reference	Function
Kelsey (1979)	Protection, containment, information, utility of use
Paine and Paine (1983)	Contain, protect, preserve, inform, sell and provide convenience
Robertson (1993)	Containment, protection, convenience, communication
Soroka (1995)	Contain, protect/preserve, transport, inform/sell
Hanlon *et al.* (1998)	Protect, contain, carry, dispense, preserve, measure, communicate, display, motivate, promote, glamorize.

- Tertiary (transport) packaging, which allows handling and transport of a number of grouped items as a unit, e.g. a pallet or banding/shrink-wrap.

Despite all the positive functions fulfilled by packaging, in the last few years it has been singled out as a wasteful representative of our throw-away society. As consumption of consumer goods has increased, so too has the profile of packaging and packaging waste. The visibility of packaging in the domestic wastebin has contributed to the common perception that packaging is an important waste stream. However, packaging should be considered within the context of the total waste production in Europe. Recent studies have shown that, in Europe, packaging forms only around 16% of Municipal Solid Waste (MSW) and around 2% of Non-Municipal Solid Waste (Non-MSW) (Sturges, 2002).

Packaging can have numerous environmental impacts over its life cycle related to the use of resources, the emission of pollutants and the impact of amenity (see Table 9.2) (Envirowise, 2002). It is important to consider the whole range of impact, direct and indirect, across the packaging life cycle, from the extraction/harvesting of raw materials, through manufacture and use of the packaging, to its final disposal. It is possible to adopt a systematic approach to this type of packaging assessment using life cycle analysis (LCA) in its various

Table 9.2   Packaging actions with a potentially adverse effect on the environment

Environmental impacts of packaging over its life cycle

- Use of non-renewable and polluting energy resources (oil, gas, coal)
- Use of non-renewable materials (e.g. non-recycled plastics, steel)
- Unsustainable use of renewable resources (water, trees, etc.)
- Emission to water (e.g. hazardous substances, suspended solids, oxygen-reducing materials with a high chemical or biological oxygen demand (COD/BOD))
- Emission to air (e.g. particulates, acid gases, ozone depleters, global warming gases)
- Emission to land (e.g. solid waste and hazardous substances that contaminate land)

The resulting deterioration in amenity can be reflected in, for example, landscapes, access, habitats, noise, vibration, odour, traffic congestion and litter.

Source: Envirowise (2002).

forms. LCA can be used to evaluate the environmental performance of processes and products 'from cradle to grave' and identify potential cost savings. LCA identifies the material, energy and waste flows associated with a product over its entire life cycle so that the environmental impacts can be determined. LCA is an effective instrument for informing a company's management about the company's environmental performance, improving its understanding of the environmental impacts of company products, and identifying cost savings associated with manufacturing and waste disposal methods. While LCA is a valuable tool, it should be noted that it is a *decision-aiding tool* – it does not make decisions for you (Envirowise, 2000).

LCA can also be used by companies to help review environmental impacts as part of the implementation of an Environmental Management System (EMS) such as the International Standard ISO 14001 or the European-based Eco Management and Audit Scheme (EMAS). ISO 14001, Environmental Management System, is an international standard that specifies the requirements for an EMS to be certified by an accredited, independent third party which assesses whether a company meets the requirements to be awarded an ISO 14001 certificate. ISO 14001 contains five main principles:

- Establishing a company environmental policy
- Planning by setting objectives and targets to address company environmental impacts
- Implementing and operating the EMS
- Monitoring the system and taking corrective action, as appropriate
- Undertaking a management review to assess the effectiveness of the system.

EMAS is a voluntary European Union Initiative, introduced in 1995. It recognizes the achievements of companies which have taken positive steps to protect the environment. It requires a company to report its environmental performance publicly and have the report audited independently. The European Commission has agreed that ISO 14001 satisfied the management system requirements of EMAS: they are complementary rather than competitive.

9.3   The regulatory context

The EU's Packaging and Packaging Waste Directive (94/62/EC), which came into force in 1994, harmonizes national measures covering the management of packaging waste – to ensure that Member State restrictions on packaging do not have the effect of creating barriers to trade, and to reduce the overall impact of packaging and packaging waste on the environment. The principal objectives of the Directive are:

- to harmonize national measures concerning the management of packaging and packaging waste, whilst providing a high level of environmental protection;

- to ensure the functioning of the internal market and avoid obstacles to trade.

To this end this Directive lays down measures aimed, as a first priority, at preventing the production of packaging waste and, as additional fundamental principles, at reusing packaging, at recycling and other forms of recovering packaging waste and, hence, at reducing the final disposal of such waste. It also calls for life cycle assessment techniques to be used to justify a clear hierarchy between reusable, recyclable and recoverable packaging.

In the final version of the Directive adopted in 1994, targets were established to be achieved in each member state by June 2001. The rates for recovering and recycling packaging waste are shown as follows:

- 50% minimum to 65% maximum by weight of packaging waste will be recovered.
- Within the above, a minimum of 25% and a maximum of 45% by weight will be recycled (within this target there is a minimum target of 15% by weight for each packaging material).

The Council of the European Communities will review these targets every five years. More recently, the EU (Proposal, 2001) has proposed increased targets for packaging waste recovery and recycling to be achieved by 2006 (Table 9.3).

In order to meet the objectives of the Directive, Member States shall take measures in order to ensure that:

- packaging or packaging waste is collected and/or returned in order to direct it to the most appropriate waste management solution;
- packaging and packaging waste collected is reused or recovered, which includes recycling.

The legislation stipulates that Member States should take the necessary steps to set up systems capable of handling the return, collection, reuse or recovery of waste. Member States had already started to introduce legal measures to regulate packaging waste at the end of the 1980s and the beginning of the 1990s. The majority of Member States had implemented Packaging Regulations in 1997. Only Greece has not yet transposed the EU Packaging Directive into national

Table 9.3   Targets for packaging waste recovery and recycling to be achieved by 2006

Recovery	Recycling

60–75% by weight	- 55–70% by weight of packaging materials
	- Differentiated material specific recycling target of:
	  – 60% for glass:
	  – 55% for paper and board:
	  – 50% for metals:
	  – 20% for plastic
	  – No specific targets for wood, textiles or composites

Source: Proposal of the European Parliament and of the Council Amending Directive 94/62/EC.

law. Depending on national waste management traditions, the regulation of packaging waste recovery can be accompanied by voluntary agreements (Denmark, the Netherlands).

The Directive also stipulates that packaging placed in the market must meet certain essential requirements. There are three categories of essential requirements, plus a heavy metal content limit:

- Packaging must be minimized by weight and volume.
- Packaging must be produced so as to minimize the content of harmful substances in the packaging material, with regards to their presence in emission, ash or leachate generated from waste management operations.
- Packaging must be designed, produced and commercialized so as to permit its reuse or recovery, including recycling.
- Packaging must be produced so that the sum of concentration levels of lead, cadmium, mercury and hexavalent chromium does not exceed 100 ppm by weight.

Reusable packaging must principally be able to support several cycles or rotations and must satisfy the requirements for recoverable packaging at the time that it ceases to be reused and consequently becomes waste.

For recoverable packaging there are four considerations for the different forms of recovery:

- Packaging recoverable by material recycling must be produced in such a way that a certain percentage by weight of the materials used are recycled for the production of marketable goods.
- Packaging waste recoverable by energy generation must have a minimum calorific value allowing the recuperation of energy to be optimized.
- With regard to the recovery of waste by composting, packaging waste must be sufficiently biodegradable in order not to present an obstacle to the composting activity to which it is introduced.
- Finally, biodegradable packaging waste must be able to undergo physical, chemical, thermal or biological decomposition such that the largest part of the compost obtained decomposes into carbon dioxide, biomass and water.

A number of Member States (Belgium, Denmark, France, Portugal, United Kingdom) have transposed the EU Packaging Directive in regulating the recovery requirements and the environmental requirements in the design and manufacture of packaging ('essential requirements') in separate legal acts. Table 9.4 lists the legal basis for the transposition of the Packaging Directive.

In order that essential requirements are observed, the Directive encourages the development of European Standards intended to confer an assumption of compliance with the requirements for packaging placed on the market in accordance with these standards. As of today only two Member States – France and the UK – and the Czech Republic are actively enforcing the essential requirements and have adopted detailed regulations explaining what companies must do to comply.

Table 9.4   Legal basis for the transposition of the Packaging Directive

Country	Legal basis
Austria	Packaging Ordinance of 1992, amended 29 November 1996
	Target Ordinance (Federal Law Gazette No. 646/1992, as amended by 649/1996)
Belgium	The Ecotax-Act (ordinary Law of 16 July 1993 aiming at completing the federal structure of the State)
	Interregional Co-operation agreement Packaging Decree of 30 May 1996 (came into effect on 5 March 1997)
	Law of 21 December 1998 (essential requirements)
	Royal Decree of 25 March 1999 defining standards for packaging
Denmark	Statutory Order no. 298 of 30 April 1997 on certain requirements for packaging
	Statutory Order no. 299 of 30 April 1997 on waste
	Statutory Order no. 124 of 27 February 1989 on packaging for beer and soft drinks as amended by statutory order no. 540 of 1991, no. 583 of 1996 and no. 300 of 30 April 1997
Finland	Decision of Council of State on Packaging and Packaging waste 1997
	Law on Alcohol Excise, No. 1471 of 29 December 1994
	Law on Soft Drinks Excise, No. 1474 of 29 December 1994
France	Lalonde Decree No. 92-377 of 1 April 1992, in force since January 1993, setting out conditions for the collection and the recovery of packaging waste produced in households
	Decree No. 94-609 of 13 July 1994 on packaging waste for which the holders are not households
	Decree No. 96-1008 on the disposal of household waste which contain the quotas set by the European Packaging Directive
	Decree No. 98-638 of 20 July 1998 related to the environmental requirements in the design and manufacture of packaging
Germany	Packaging Ordinance of 1991, amended 21 August 1998
Greece	Draft Law 'Measures and conditions for the alternative management of packaging and other waste products. Foundation of the National Organisation for the Alternative Management of Packaging and Other Waste (NOAMPOW)'
Ireland	Waste Management (Packaging) Regulations 1997
	Waste Management (Farm Plastics) Regulations 1997
	Waste Management (Packaging Amendment) Regulations 1998
Italy	'Ronchi Decree', Law effective from 5 February 1997 implementing EC Directives (Directive on waste, hazardous waste and packaging waste) amended 28 November 1997
Luxembourg	Grand Ducal Regulation of 31 October 1998
Netherlands	Packaging and packaging Waste Decree of 4 July 1997
	Packaging Covenant II of 26 December 1997
Portugal	Decree Law No. 366-A/97 of 20 December 1997 (modified by Decree Law No. 162/2000 of 27 July 2000)
	Ordinance No. 29-B/98 of January 1998
	Decree-Law No. 407/98 of 21 December 1998 for essential requirements and maximal concentration of heavy metal

Table 9.4 Continued

Country	Legal basis
Spain	Packaging Law 11/1997 of 24 April 1997
	Royal decree 782/98 of 30 April 1998
	Law 10/1998 of 21 April 1998
	Order 50/1998 of 30 December 1998
Sweden	Decree (1997 – 185) on producer responsibility for packaging
UK	Producer Responsibility Obligations (Packaging Waste) Regulations 1997
	Packaging Regulation (1998)
	Packaging (Essential Requirements) Regulations 1998
	Producer Responsibility Obligations (Packaging Waste) Regulations (Northern Ireland) 1999

It is necessary to emphasize that without standardization essential requirements only have limited application. In fact they explicitly refer to a quantitative notion or to criteria or methods to be precisely determined. Thus, reusable packaging must be able to support *several* cycles or rotations (quantity not specified), a package that is recoverable through material recycling must be produced so as to allow *a certain* percentage by weight of the materials used to be recycled (percentage not specified), and the packaging waste treated for the purpose of energy recovery will have *a minimum calorific value* (combustion index not specified) (Demey *et al.*, 1996). In response, the EU Commission mandated CEN to draw up:

- a set of standards intended to provide an assumption of compliance with the essential requirements of the Directive;
- standards in support of the environmental objectives of the Directive;
- reports on special areas, including subsequent proposals for standards linked to these areas.

CEN standards in the field of packaging and the environment are shown in Table 9.5.

CEN submitted the complete package of texts to the Commission, with a request of the references in the *Official Journal*. Following an objection from two Member States and given that the Commission considered that these standards do not fully meet the essential requirements of the Directive, it was able to cite only two standards (EN 13428 and EN 13432) in the *Official Journal* with a note in which it is explained that these standards do not cover the essential requirements of the Directive. Individual Member States are free to accept compliance with them as evidence of conformity with the Essential Requirements (Commission Decision, 2001).

The Commission has called CEN to undertake work to revise the standards with a new mandate. The basic standard EN 13427 was not part of the original mandate but has been included in the second mandate.

Table 9.5   CEN standards in the field of packaging and environment

CEN ref no.	Title
EN 13427: 2000	Packaging-requirements for the use of European Standards in the field of packaging and packaging waste
EN 13428: 2000	Packaging-requirements specific to manufacturing and composition – Prevention by source reduction
EN 13429: 2000	Packaging-requirements for relevant materials and types of reusable packaging
EN 13430: 2000	Packaging-requirements for packaging recoverable by material recycling
EN 13431: 2000	Packaging-requirements for packaging recoverable in the form of energy recovery, including specification of minimum interior calorific value
EN 13432: 2000	Requirements for packaging recoverable through composting and biodegradation – Test scheme and evaluation criteria for the final acceptance of packaging

9.4   Packaging minimization

The first priority of packaging waste management options is prevention, i.e. the reduction of the quantity and of the harmfulness for the environment of:

- materials and substances contained in packaging and packaging waste;
- packaging and packaging waste at production process level and at the marketing, distribution, utilization and elimination stages.

Packaging waste prevention can be:

- eliminating packaging;
- using less material in each pack (known as light-weighting or down-gauging);
- substitution by a lighter material;
- distribution of product in minimal packaging to be transferred into reusable containers at home;
- product re-design (e.g. concentrated products).

Only light-weighting or down-gauging will invariably reduce waste, provided product wastage does not increase as a result. The elimination of packaging will certainly reduce packaging waste, but if the products are inadequately protected as a result, then overall waste will increase.

Progressive use of less material has taken place in most product sectors (some examples are shown in Table 9.6) (Bickerstaffe, 1994). As new technology emerges, this is continuing. There are many environmental advantages of light-weighting: lower raw materials usage, lower energy consumption in manufacture and transportation and low materials/costs, or lower tonnage going to landfill. There are also benefits for manufacturers: reductions in unit package costs, increased production capacity without significant capital investment, and

Table 9.6   Light-weighting in different sectors

	Weight	Year	Weight	Year	Weight	Year	Weight	Year
Glass milk bottle	538 g	1939	397 g	1950	245 g	1960		
Plastic yoghurt pot	12 g	1965	7 g	1984	5 g	1990		
Metal food can	90 g	1950	69 g	1970	58 g	1980	57 g	1990
1.5 l plastic fizzy								
drinks bottle	66 g	1983	57 g	1985	45 g	1987	42 g	1990
Plastic carrier bags[a]	47 μm	1970	37 μm	1980	25 μm	1990		
Cardboard box	559 g	1970	531 g	1990				

[a] Film thickness/weight reduction
Source: Bickerstaffe (1994).

more containers per tonne of raw material recovering overhead and capital expenditure with greater efficiency. Companies that are managing their packaging use carefully are saving money and increasing profits, as well as saving valuable resources and reducing disposal of waste. Light-weighting can be achieved by using less materials (see Case study 1) or eliminating one or more layers to reduce the weight of packaging used (see Case studies 2 and 3).

It is also possible to reduce voidspace by improving packaging design, e.g. when there is a cardboard pack around plastic inner packaging. This is fundamental for reducing distribution costs (see Case study 4).

Case study 1:   Reduce thickness of materials

A health food company, Country Harvest Natural Foods, has reduced the thickness of plastic packaging on 1 kg packets of its range, reducing the gauge of plastic film from 75 μm to 60 μm. A larger transit box of higher-grade cardboard was also introduced to improve transit efficiency, increasing the number of packets carried from six to ten per box. Savings and other benefits achieved:

- Overall savings achieved of £7000/year – a saving of 1.12 pence per 1 kg packet
- Increased line efficiency due to easier loading onto pallets.
- Overall reduction in weight of packaging, leading to a reduced obligation under the packaging waste regulations
- Benefits for the producer responsibility obligations of those companies supplied by Country Harvest.

The investment cost was zero because the change merely entailed purchasing thinner specification packaging materials.

Source: Envirowise, 1999.

Case study 2   Redesign of garlic bread packaging saves tonnes

A supermarket company, Sainsbury, has redesigned the packaging of one of its own-brand food products. Sainsbury's garlic bread packaging was a 500 μm white-lined cardboard carton weighing 34 g, made with 75% recycled material and a 20 μm plastic film (3.0 g). The product is now packaged in a single layer of polypropylene and polyester plastic film weighing 5.9 g. Savings and other benefits achieved:

- Net savings of 160 tonnes/year of packaging
- Improved transport optimization due to space saved on lorry loads
- Reduced obligation under the packaging waste regulations for supplier
- Reduction in packaging waste to the consumer.

Source: Envirowise (1999).

Case study 3:   Elimination of inner collation wraps saves packaging costs

A chocolate manufacturer, Nestlé Rowntree, has eliminated a layer of wrapping for bumper packs of one of its best-selling biscuits. Owing to the limitations of the packaging machinery previously used by Nestlé, the two layers in bumper packs of Kit Kat chocolate biscuits had to be wrapped individually before they could be collated into the single bumper pack. In accordance with Nestlé's ongoing policy of packaging minimization, Nestlé has developed a new wrapping machine capable of wrapping without the need for the inner collation wrap. This has achieved a reduction of up to 100 tonnes/year of polypropylene

Source: Envirowise (1999).

Case study 4:   Reduce pack volume

A supermarket company, Sainsbury, has reduced the volume of one of its own-brand cereal packs by 24% allowing more units per case and more cases per pallet. This has achieved a 2% reduction in material costs and an annual saving of 7.2 tonnes of cardboard, with associated reductions in the company's obligations under the Packaging Regulations.

Source: Envirowise (2002).

9.5 Packaging recycling

Recycling involves the use of materials at the end of their previously useful lives as the raw material for the manufacture of new products. If recycling is employed intelligently and carried out in conjunction with good design, many materials can be reclaimed after their first useful life is over. The two major objectives should be:

- to conserve limited natural resources;
- to reduce and rationalize the problems of managing municipal solid waste disposal (Lauzon and Wood, 1994).

Regulatory requirements have been a major factor in the development of food contact processing. Food safety is the paramount consideration for use of recycled materials in food contact applications. At the present time no European Commission Directive applies specifically to food contact materials and articles made from recycled materials. The regulatory situation in individual European countries varies, with some forbidding the use of recycled materials (e.g. Italy and Spain) and others leaving it up to the user to ensure full compliance with the European Directives and national regulations on food contact.

The basic principle of European Union regulations for materials and articles intended to come into contact with foodstuffs is laid down in the Council Directive 89/109/EE. This Directive covers the general requirements for all packaging for health and safety, for solely justifiable changes to foodstuff composition and for lack of deterioration of the organoleptic properties. In accordance with Directive 89/109/EE, 'materials and articles must be manufactured in compliance with good manufacturing practice so that, under the normal or foreseeable conditions of use, they do not transfer their constituents to foodstuff in quantities that could:

- endanger human health;
- bring about an unacceptable change in the composition of the foodstuffs or deterioration in the organoleptic characteristics thereof'.

Even if at the moment there is no clear legislation, several factors support application of using recycled materials for food contact packaging:

- Mandatory recycled content legislation in some countries
- The potential for reducing the packaging costs
- Advances in technology
- Market competitiveness.

Considering solely the plastic sector, which represents about 50% of total packaging for the food and drink sector (Pira International), industry has developed three routes for the use of post-consumer recycled (PCR) plastics in food contact application:

1. Chemical recycling, in which recovered polymers are broken down to chemical precursors, which may then be purified and repolymerized to give new polymers.

2. Multilayer extrusion, in which high-quality mechanically recycled food grade polymer, such as that derived from in-house scrap or drinks bottles, is sandwiched between two layers of virgin polymer. Extensive testing involving the deliberate introduction of a range of model contaminants into the PCR PET middle layer of a multilayer PET bottle, followed by testing under specified conditions, has demonstrated the action of PET as a functional barrier. The US FDA granted a 'Non-Object' for multilayer PET containers in 1993, and requires a 0.025 mm thick inner virgin PET layer to prevent food contact by the inner PCR PET layer.
3. Superclean recycling processes using mechanical and non-mechanical procedures to recycle high-quality post-consumer material, producing polymer suitable for use in mono-layer application i.e. use in direct contact with food. The processes are proprietary, but where details are known they generally involve a combination of standard mechanical recycling processes with non-mechanical procedures such as high-temperature washing, high-temperature and pressure treatments, use of pressure/catalysts, and filtration to remove polymer-entrained contaminants (Recoup, 2002; WRAP, 2002).

The topic of reuse of post-consumer plastics for food packaging is currently under discussion at the European Commission. In front of this background an expert group under the responsibility of ILSI Europe and with the participation of Fraunhofer IVV has proposed specific guidelines (Castle *et al*., 1998) on the basis of the results obtained from the European Project (EU Agro–AIR, 1997) dealing with the question of recyclability and reusability of post-consumer plastics for new food packaging applications.

In the meantime the Plastics Commission of the German BgVV published its recommendations for the quality of recycled PET for food contact (BgVV, 2000). The guidelines cover:

*   Quality control of incoming materials: food contact application calls for control of the input fraction; it should have a mass fraction of more than 99% PET. The first use of the PET bottle is also of interest; only bottles initially used for foodstuffs should be accepted as input fraction for recycling.
*   The challenge test on the recycled material to determine the cleaning efficiency of the recycling process proposed by the Fraunhofer IVV is a modified, more user-friendly version of the FDA challenge test (FDA, 1992).
*   Analytical quality assurance based on the results of a Europe-wide recyclate screening that Fraunhofer IVV has developed, which is a rapid analytical method for routine control of recyclate produced.

Countries such as Germany and Austria have adopted the BgVV guidelines. Nevertheless, whilst there may be guidelines for the use of recycled plastic, there is no organization that will give its approval to any specific material or process. The safety of the recycled material is the responsibility of the company putting the foodstuff products on the market (Welle, 2002).

Case study 5:   Bottle made by RPET

The commercial use of PCR PET in food contact applications was pioneered by Coca-Cola and Pepsi. Coca-Cola initially used bottles with chemically recycled PET content in the US in 1991. Coca-Cola has used multi-layer bottles supplied by Continental PET in Australia and New Zealand since 1993 and in Switzerland and Sweden since 1995. Coca-Cola introduced food contact RPET bottles to the UK in 1992, primarily to demonstrate that it was technically possible and to encourage European chemical industries to produce food contact recycled PET. Initially recycled PET was imported from the USA, but because of the high costs incurred multilayer bottles superseded this approach. Coca-Cola stopped using RPET bottles in UK in 1994 because of the falling price of virgin PET and difficulties encountered obtaining sufficient quantities of high-quality post-consumer PET from the UK. Multilayer bottles are currently used by Coca-Cola in other European countries including Switzerland and Belgium.

Pepsi has used PET bottles made from chemically recycled PET and mechanically recycled Supercycle PET. Pepsi-Cola has a goal of placing 10% recycled content in all its plastic bottles in the US by 2005.

Source: Recoup (2002).

9.5.1   Design for recycling
Packaging design for recycling should take the following into account.

- Make possible separation of materials for recycling. Recycling of materials requires the use of uniform materials. Packaging made with a single material is preferred; alternatively use compatible materials that are easy to deal with during sorting and reprocessing. For example, there are several combinations of materials that are best avoided on plastic bottles, such as PET and PVC (avoid PVC labels on PET bottles, or PVC seals in the closure of PET bottles and PET labels on PVC bottles), coloured coatings on PET bottles, thermoset and metal closure.
- Ensure labelling of materials conforming to standards. It is essential that materials can be unequivocally identified by means of standardized labelling. Products containing materials that have not been labelled correctly very often end up in a waste dump, although recycling of materials would be possible. There is currently no harmonized global set of standards. The European industry supports the material identification system developed by the European Committee for Standardization, CEN TC261 SC4/WG1, or the identification system adopted by the Association of Plastic Manufacturers in Europe (APME), even though the European Commission published its own

Table 9.7   Polymer identification codes

Polymer type	CEN Recommendation CEN WI 261 070	EU Commission Decision 97/129/EC	APME Position
Polyethylene Terephthalate	01 PET	1 PET	1 PET
High Density Polyethylene	02 PE-HD	2 HDPE	2 HDPE
Polyvinyl Chloride	03 PVC	3 PVC	3 PVC
Low Density Polyethylene	04 PE-LD	4 LDPE	4 LDPE
Polypropylene	05 PP	5 PP	5 PP
Polystyrene	06 PS	6 PS	6 PS
Unallocated References	07–20	07–19	07–19

proposed identification system in the Decision on Material Identification in 1997 (97/129/EC). Table 9.7 lists the European Commission, CEN and APME identification systems, which are currently in use throughout Europe. At present, any of the identification systems are suitable for use to identify plastic packaging.

*   Minimize contamination, The labels, inks and glues used on a pack all influence the recyclability. Designers should minimize contamination and make these easier to remove (for example, use individual blobs of hot melt adhesive on paper packaging rather than thin strips that can break up in the pulping process; consider using water/acrylic based emulsion and starch-based coating on paperboard instead of PE and wax laminates).
*   Provide for instruction for disposal. Usually, consumers are not prepared to make great efforts for the disposal of a product. An environmentally acceptable disposal option should therefore be simple and should not require facilities that regular end-users do not have. In addition, it is useful to provide the user with information concerning adequate disposal and to encourage correct user behaviour.

The CEN standard, EN 13430:2000, on 'Packaging – Requirements for Packaging Recoverable by Material Recycling', currently under revision, will define:

- Substances or materials that are likely to create problems in the recycling process
- Materials, combinations of materials or packaging designs that are likely to create problems during the collecting and sorting process prior to material recycling
- Substances or materials that are likely to have a negative influence on the quality of the recycled material.

Performance against targets for recycling established by Directive 94/62/EC is good. As reported by Member States to the European Commission (Proposal, 2001), many Member States had easily reached the 2001 targets as early as 1997/98 (the most recent available performance data are for 1998). All Member States had achieved the overall recycling target and all had achieved the minimum 15% material-specific target for glass and paper and board. Only Italy and Luxembourg had failed to achieve 15% recycling for metals, while eight Member States had not reached the target for plastic.

Table 9.8 shows critical factors, per material, identified for recycling. They are classified in factors that are reasonable reasons to limit recycling and factors that may be valid points but not really a reason to limit recycling (RDC, 2001).

Table 9.8 Summary of the recycling difficulties

Recycling difficulties	Glass	Plastic	Paper/ board	Metals	Composites
Capacity	x		x		
Output market/Market price	x	x			
Contamination	x	x	x	x	
Imbalance supply – demand	x		x		
Insufficient amount of waste		x		x	x
Recycling lifetime			x		
Nature of waste (too thin, etc.)		x	x	x	
Recycling costs		x			

Factors which are not really a reason to limit recycling

Noise	x				
Danger from breakage	x				
Insufficient maintenance	x				
Disposers' participation	x	x	x	x	
Colour, odour		x			
Resistance to recyclate use		x			

Source: RDC (2001).

9.5.2   Design for reuse

Reuse means any operation by which packaging, which has been conceived and designed to accomplish within its life cycle a minimum number of trips or rotations, is refilled or used for the same purpose for which it was conceived, with or without the support of auxiliary products present on the market enabling the packaging to be refilled; this reused packaging will become waste when no longer subject to reuse.

In accordance with CEN standard EN 13429:2000, currently under review, packaging is considered to be reusable if:

- it was conceived by the filler for multiple use;
- it is possible to empty the packaging without damage so as to process it for further use;
- it can be subject to some form of processing (e.g. cleaning, washing, repair);
- a multiple-use system is principally in place (open loop system, closed loop system, hybrid system).

Three systems for reuse are provided:

- Closed loop system: a system in which reusable packaging is circulated by a company or an organized group of companies
- Open loop system: a system in which reusable packaging circulates among unspecified companies
- Hybrid system: a system consists of two parts
    - reusable packaging, remaining with the end-users, for which there exists a redistribution system leading to any commercial refilling
    - one-way packaging, used as an auxiliary product to transport the contents to the reusable packaging.

Case study 6:   Packaging re-use

A major UK food retailer has replaced cardboard cartons with returnable plastic crates for deliveries from suppliers to stores via depots. Together with the use of returnable trolleys for fresh milk and pot plants, the company has reduced its packaging waste by 28 000 tonnes per year.

Source: APME (2001).

Reuse is a convenient and important way of saving resources, often offering an energy-free alternative to recycling (see Case study 6).

However, sometimes reusing a package creates an unacceptable environmental burden. For example, bottles for beer brewed, distributed and consumed locally can be reused many times, with minimal environmental impact. But for a specialist beer produced on the other side of the continent, the environmental impact of transporting used bottles over long distances back to the bottling plant would far outweigh any saving.

In 2001 the consulting firms RDC-Environment and Pira International completed a Cost Benefit Analysis (CBA)[1] study for the European Commission that investigated the cost and benefits of reusable primary packaging (RDC, 2001). This study compared 330 ml refillable glass bottles with one-way glass bottles of the same size, and 1.5 litre refillable PET bottles with one-way PET bottles of the same size. The study concluded that from a total social cost perspective[2] there is no case for universally encouraging the use of returnable packaging. The environmental and economic costs and benefits are dependent on a range of local and regional factors, including the frequency of reuse and the distance to market.

Critical factors for reusable packaging are the following:

- Economic constraints mainly concern the much higher initial capital investment of reusable packaging compared with that of disposal packaging and the costs of returning used packaging to its point of origin (e.g. transport costs). The latest constraint can be reduced by the development of European standards such as Europallets. In this case, reuse can occur in the same geographical area as initial use. Maintenance (e.g washing) and repairs can also be costly and time consuming.
- Consumer convenience can influence the reuse rate by the choice between one-way use and reuse and by the level of return (the end user may not return the packaging after use).
- Technical limits can relate mainly to the quality of reuse packaging and the necessity of an effective control.

9.5.3   Biodegradability and compostability

Biodegradable items are those in which the degradation results from the action of naturally occurring microorganisms such as bacteria, fungi and algae. The term biodegradable, by itself, is not useful. In fact degradation will not occur in an unfavourable environment, or the biodegradable material will not degrade within in a short time. It is, therefore, important to couple the term biodegradable with the specification of the particular environment where the biodegradation is expected to happen and of the time scale of the process.

Composting is the managed process that controls biological decomposition and transformation of biodegradable material into a humus-like substance called compost. The European Directive 94/62/EC has specified that composting of

---

[1] Life cycle cost–benefit analysis (CBA) is an economic evaluation tool used to compare the costs against the benefits of different activities. CBA analysis combines aspects of financial cost–benefit analysis with the economic valuation of the environmental impacts which are determined by life cycle assessment techniques.
[2] From a total social cost perspective the reuse system is preferable from short and medium transportation distances (under 175 to 280 km).

packaging waste is a form of recycling, owing to the fact that the original product, the package, is transformed into a new product, the compost.

The CEN standard EN 13432:2000, currently under review, specifies requirements and procedures to determine the compostability and anaerobic treatability of packaging and packaging materials by addressing four characteristics:

- Biodegradability
- Disintegration during biological treatment
- Effect on the biological treatment process
- Effect on the quality of the resulting compost.

If the packaging contains a non-compostable component it is considered as not in conformity with the standard unless the components can be easily separated by hand. The standard contains references to the relevant test methods.

Biodegradability is a property which all biopolymers are likely to have. Some materials based on a mixture of synthetic materials and biobased material can have this property, and some synthetic materials have been shown to be fully biodegradable and in compliance with the norms. Table 9.9 lists some materials currently on the market. It should be stressed that many of these materials may be available only in test quantities; only very few of them will have been used for direct food contact packaging (Marron *et al.*, 2000). Paper and board materials are not included, because normally composting of paper packaging does not make sense as the environmental benefit of material recycling to secondary paper fibre is economically and environmentally much more beneficial than composting. Composting of paper makes sense only if the paper has a high humidity and/or is contaminated with organic food residues.

The final treatment of the waste originated from the biobased products must be taken into consideration. The final system of waste treatment has an important role in the overall ecological balance of the biobased materials and can affect the final result. If a biobased material is recycled through composting, it will contribute to the formation of compost; if a biobased material is recovered by incineration with energy recovery, it will contribute in sparing some fossil fuel. On the other hand, a biobased material dumped in a landfill site could produce negative effects by uncontrolled evolution of methane.

9.6   Future trends

The environment will remain a major issue for the packaging industry. The principal driving force is legislation. The current revision of the Packaging and Packaging Waste Directive will set new higher targets for recycling of packaging (see Table 9.3). The effects of this proposal will be stronger in countries that so far have not yet achieved high recycling rates. The scale of the task faced by many EU Member States is illustrated in Table 9.10 which summarizes the percentage increases in recycling (using 1998 performance data

Table 9.9 Biobased packaging materials and biodegradable material currently available in the market

Material	Supplier	Trade name (if known)	Polymer linkage
Biodegradable materials based on natural renewable resources – Biopolymers			
PHB/PHV (Polyhydro – xyalkanoate)	Was Monsanto	Biopol	Ester
	Biomer	Biomer	Ester
Cellulose acetate	Courtaulds		Acetal
	Mazzuchelli	Bioceta	Acetal/ester
Polylactide /PLA	Cargill Dow Polymers	Nature Works PLA	Ester
	Mitsui	Lacea	Ester
	Hycail		Ester
	Galactic	Galactic	Ester
Starch	National Starch	Eco-foam paragon	Ester
	Avebe	Paragon	Ester
Biodegradable materials based on blends of biopolymers and synthetics			
Starch based	Novamont	Mater Bi	Acetal/Ester
	Biotec	Bioplast	Acetal/Ester
	Earth Shell	Earth Shell	Acetal/Ester
	Biop	Biopar	Acetal/Ester
Biodegradable materials based wholly on synthetics			
Copolyester	Basf	Ecoflex	Ester
	Eastman Chemical	Eastar Bio	Ester
Polycaprolactone	Union Carbide	Tone Polymer	Ester
	Solvay	CAPA	Ester
Polybutylene succinate	Showa Highpolymer	Bionolle	Ester
Polyesteramide	Bayer	BAK	Ester
Polyesterurethane	Bayer	MHP 9029	Ester
Polyester copolymer	Bayer	Degranil VPSP42002	Ester
Polyactic acid	Fortum		Ester
Polyester	Dupont	Biomax	Ester

Source: Marron *et al.* (2000).

as the baseline) required by each Member State in order to achieve the 2006 targets.

By 1998, both Germany and Belgium had already achieved or exceeded all proposed recycling targets for 2006. Austria and Sweden have performed well, but must secure a 32% and 61% increase in metals recycling, respectively.

Metal recycling represents a challenge for many Member States. Spain and the UK must more than double metals recycling performance, whilst Finland, Luxembourg and Italy must increase metals recycling by more than 200%, 300% and 600%, respectively.

Table 9.10 Percentage increases in performance required to achieve 2006 targets (baseline 1998 performance)

	Overall recycling performance	Paper and board recycling	Metals recycling	Glass recycling	Plastics recycling
Austria			32		
Belgium					
Denmark	10		25		186
Finland	22		213		100
France	30		11	33	150
Germany					
Italy	61	48	614	62	81
Luxembourg		Not available	355		122
Netherlands					43
Spain	61	6	127	62	122
Sweden			61		
UK	96	4	117	161	150

Source: Sturges (2002).

Plastic recycling is also a challenge. Denmark, Finland, France, Luxembourg and the UK must all more than double 1998 performance. Italy and the Netherlands must also make significant progress.

Considering 1998 performance data, Finland, France, Luxembourg, Italy, Spain and the UK are presented with major challenges if they are to achieve the 2006 targets. Italy and UK must improve performance in all categories.

Regarding heavy metals content it has been proposed that the concentration levels of heavy metals will be reduced to zero by end of 2006 (European Parlament, 2002). This has no scientific justification and is technically impossible to achieve. The consequence of a zero limit on heavy metals in packaging would be enormous for the European packaging industry and disproportionate to any possible environmental benefit.

It could be suitable to use 'environmental indicators' for various materials as a means of measuring progress. Progress would vary according to the different situations in the Member States and it would not be necessary to impose the same recycling figures for all Member States.

Changes in lifestyle, which are occurring in industrialized countries, are also affecting food packaging. As a result, time-saving, convenient and long-life products are increasingly in demand. The development of several innovative packaging techniques has allowed processes to push up the limits on shelf-life and consequently to optimize quality, minimize food waste and maximize sale and export. An example of this could be Modified Atmosphere Packaging (MAP) which has the potential to increase shelf-life and therefore reduce product wastage. However, more research is required to determine if this is in fact the case in practice (FAIR project 'Actipak', 2001).

9.7   Sources of further information and advice

EU institutions, www.europa.eu.int
International associations – general packaging:

- Pira International, www.piranet.com
- European Organisation for Packaging and the Environment (EUROPEN), www.europen.be
- Packaging Recovery Organisation Europe (PRO Europe), http://www.pro-e.org
- The Industrial Council for Packaging and the Environment, www.incpen.org
- Waste Resource Action Programme, www.wrap.org.uk
- Valpack, www.valpak.co.uk
- Remade, www.remade.org.uk
- Institute of Packaging, www.iop.co.uk
- Institute of Grocery Distribution, www.igd.org.uk

International associations – plastics:

- Association of Plastics Manufacturers in Europe (APME), www.apme.org
- European Plastics Converters (EuPC), www.eupc.org
- European Council of Vinyl Manufacturers (ECVM), www.ecvm.org
- European Federation for the Flexible Packaging Industry (FEDES), www.fedes.com

International associations – paper and board:

- Alliance for Beverage Cartons and the Environment (ACE), www.ace.be
- European Federation of Corrugated Board Manufacturers (FEFCO-Probox), www.fefco.org www.probox.com
- European Tissue Symposium (ETS), www.europeantissue.com
- Confederation of European Paper Industries (CEPI), www.cepi.org
- International Confederation of Paper and Board Convertors in Europe (CITPA), www.citpa-europe.org

International associations – glass:

- European Container Glass Federation (FEVE), http://www.glasschange.com

International associations – metals:

- Beverage Can Makers in Europe (BCME), www.bcme.org
- The Can is Cool, www.thecaniscool.com
- European Aluminium Association (EAA), www.aluminium.org
- Association of European Producers of Steel for Packaging (APEAL), www.apeal.org

International associations – food industry:

- Confederation of the Food and Drink Industries (CIAA), www.ciaa.be
- European Food Information Council (EUFIC), www.eufic.org
- Department of Environment, Food and Rural Affairs, www.defra.gov.uk

9.8   References

ASSOCIATION OF PLASTICS MANUFACTURERS IN EUROPE (APME) (2001), *Insight into consumption and recovery in Western Europe – Plastics: a material of choice for the packaging industry*, Brussels: APME.

BGVV (2000), *Gesundheitliche Beurteilung von Kunststoffen im tem Kunstoff aus Polyethylenterephthalat für die Herstellung von Bedarfsgegenstanden, Bundesinsitut fur gesundheitlichen Verbraucherschutz*, Berlin: Bundesgesundheitsblatt – Gesundheitsforschung Gesundheitsschutz, 43, 826–828.

BICKERSTAFFE J. (1994), *Minimising Packaging Waste*, Conference publication, Strategies for Meeting European Packaging Legislation, London, 17–18 March.

BICKERSTAFFE J. (1996), *Environmental Impact of Packaging in the UK Food Supply System*, London: The Industry Council for Packaging and the Environment (Incpen).

CASTLE L., DE KRUIJF N., FRANZ R., GILBERT J. ROSSI L. (1998), *Recycling of Plastic for Food Contact Use*, Brussels: Guidelines prepared under the responsibility of the International Life Science Institute (ILSI) Europe Packaging Material Task Force.

Commission Decision of 28 June 2001 relating to the publication of references for standards EN 13428:2000, EN 13429:2000, EN 13430:2000, EN 13431:2000 in the Official Journal of the European Communities in connection with Directive 94/62/EC on packaging and packaging waste, *Official Journal of the European Communities* L190, 12 July 2001.

DEMEY TH., HANNEQUART J.P., LAMBERT K. (1996), *Packaging Europe – A Directive Standing up to Transposition into 15 National Laws*, Brussels, Brussels Institute for Management of the Environment.

ENVIROWISE – ENVIRONMENTAL TECHNOLOGY BEST PRACTICE PROGRAMME (1999), *Reducing the Cost of Packaging in the Food and Drink Industry,* Harwell, Oxon: Envirowise.

ENVIROWISE – ENVIRONMENTAL TECHNOLOGY BEST PRACTICE PROGRAMME (2000), *Life Cycle Assessment – An Introduction for Industry*, Harwell, Oxon: Envirowise.

ENVIROWISE – PRACTICAL ENVIRONMENTAL ADVICE FOR BUSINESS (2002), *Packaging Design for the Environment: Reducing Costs and Quantities*, Harwell, Oxon: Envirowise.

EU AGRO – INDUSTRIAL RESEARCH PROGRAMME (AIR) (1997), *Program to establish criteria to ensure the quality and safety of recycled and re-used plastics for food packaging,* Brussels, Final report of EU funded AIR project AIR2 CT93 1014.

European Parliament and Council Directive 94/62/EC of 20 December 1994 on packaging and packaging waste, *Official Journal of the European Communities*, L365, 31 December 1994, 10–23.

EUROPEAN PARLIAMENT (2002), *Draft Report on amending Directive 94/62/EC on packaging and packaging waste*, Proposal for a directive (COM (2001) 729-C5-0664/2001/0291 (COD)), PE 314.355/19–123, 8 May 2002.

FAIR PROJECT 'ACTIPAK' (CT98–4170) (2001), *Evaluating safety, effectiveness, economic-environmental impact and consumers acceptance of active and intelligent packaging*, Brussels: European Union shared cost research project.

FDA (1992), *Points to consider for the use of recycled plastics in food packaging: chemistry consideration*. US Food and Drug Administration, Center for Food Safety & Applied Nutrition, Office of Premarket Approval, http://vm.cfsan.fda.gov/~dms/opa-cg3b.html.

HANLON J.F., ROBERT J.K., HALLIE E.F. (1998), *Handbook of Package Engineering*, 3rd edn, Lancaster, PA: Technomic Publishing Co.

KELSEY R.J. (1989), *Packaging in Today's Society*, Lancaster, PA: Technomic Publishing Co.

LAUZON C., WOOD G. (1994), *Environmentally Responsible Packaging – A guide to development, selection and design*, Leatherhead, Surrey, UK, Pira International Ltd.

MARRON V.J.J., BOLCK C., SAARI L., DEGLI INNOCENTI F. (2000), The market of biobased packaging materials, in Weber C. J, *Biobased Packaging Materials for the Food Industry Status and Perspective*, Copenhagen: Department of Dairy and Food Science, 125–131.

PAINE F.A., PAINE H.Y. (1983), *A Handbook of Food Packaging*, Glasgow: Blackie.

PIRA INTERNATIONAL, personal communication.

PRE G. (1997), 'Packaging of food products – its role and requirements', *Packaging India*, 30/1, 9–11.

Proposal of the European Parliament and of the Council amending Directive 94/62/EC on packaging and packaging waste, 2001/0291 (COD).

RDC-ENVIRONMENT AND PIRA INTERNATIONAL (2001), *Evaluation of costs and benefits fot the achievement of reuse and recycling targets for the different packaging in the frame of the packaging and packaging waste directive 94/62/EC*, Brussels, RDC and Pira International for DG Environment.

RECOUP (2002), *Fact Sheet: Use of recycled plastics in food grade applications*, www.recoup.org.

ROBERTSON G.L. (1993), *Food Packing: Principles and Practice*, New York: Marcel Dekker.

SOROKA W. (1995), *Fundamentals of Packaging Technology*, Herndon, VA: Institute of Packaging Professionals.

STURGES M. (2002), *Packaging and the Environment – Arguments for and against packaging and packaging waste legislation*, Leatherhead, UK: Pira International.

WASTE AND RESOURCES ACTION PROGRAMME (WRAP) (2002), *Plastic Bottle Recycling in the UK*, Banbury, Oxon: WRAP.

WELLE F. (2002), Bottle to bottle recycling: the legislation, *PETplanet* 3(6–7): 26–27.

9.9   Acknowledgement

I would like to thank the colleagues of the Environmental Packaging consultant group of Pira International and, in particular, Michael Sturges for suggestions on how to improve the paper.

10

Recycling of packaging materials

D. Dainelli, Sealed Air Corporation, Italy

10.1   Introduction

Before discussing recycling of packaging materials, the following definitions must be given:

- Recycling: defined as the reprocessing of the waste material in a production process either for the original purpose or for other purposes. The definition includes organic recycling but excludes energy recovery[1]
- Recycling process: consists of the physical and/or chemical process which converts collected and sorted packaging or scrap into secondary raw material or products[2]
- Secondary raw material: defined as the material recovered as a raw material from used products and from production scrap. However, scraps arising within a primary production process shall not be seen as secondary raw materials[2]
- Organic recycling: consists of aerobic or anaerobic treatment of the biodegradable part of the packaging waste to produce stabilized organic residues or methane[1]
- Energy recovery: consists of the combustion of packaging waste as a means to generate energy.[1]

The availability of packaging materials accurately sorted, washed and free from contaminants is a prerequisite for their recycling. Consequently the mere technical feasibility of the recycling is only one of the elements rendering the process economically worthwhile. Collection and sorting systems are of primary importance, since they affect the quality of the input stream, and therefore influence the quality and the value of the secondary raw materials. Costs related to logistics and transportation must be also taken into account in the economic balance, in relation to the volume of packaging to be recycled.

In addition, acceptance of recycled materials by consumers is also a sensitive issue: some materials have a better image than others for consumers, such as recycled paper and glass. Packaging consisting of these materials is considered more 'environmentally friendly', being more easily associated with the concept of 'recyclable packaging'. On the other hand, plastics have a poorer image in consumers' eyes, even if their recyclability is often technically feasible and is important from the economic and environmental viewpoints.

Sometimes the decision of the public authorities to promote packaging recycling schemes is driven by political reasons, associated with citizens' concerns over waste incineration plants and landfill sites, or by 'green marketing' operated by retail chains or packaging manufacturing companies. Recyclability of packaging, in summary, is not always driven by environmental reasons, but often economic, political and social factors play a major role. The presence of a well-established market for products containing recycled glass and paper makes the recycling processes of these materials largely independent of environmental considerations. The opposite situation arises for plastics, where the market for recycled products is not highly developed, with the exception of some specific materials, and therefore the environmental driver is predominant.

Waste minimization, best use of natural resources and limitation of environmental impact of post-use packaging are evident benefits of packaging recycling. Recycling, however, could not always represent the best environmental option: in fact, it should be weighed against other forms of minimization of the environmental impact, such as energy recovery, that in specific eco-balances might lead to more favourable scenarios.

In conclusion, three elements should be taken into consideration when dealing with recycling of packaging materials:

- Best environmental performance
- Economic balance of the whole processes, including collection, sorting and transportation
- Consumer acceptance of recycled materials, particularly in sensitive applications such as food packaging.

All of the above are of the utmost importance in determining whether recycling of a given packaging material is worthwhile, and each of them can kill off a recycling initiative if not properly addressed.

10.2   Regulation in the EU

The legislation on recycling (as well as other forms of recovery) of packaging materials is stipulated by the body of law of the European Union, through Council Directive 94/62 of 31 December 1994, also referred as the 'Packaging and Packaging Waste' (P&PW) Directive.[1] Prevention at source is the main object of this Directive. Besides prevention, recycling, recovery and reuse are indicated as means for minimizing packaging waste and consequently environmental impact. In this Directive, recycling is one of the options for recovery of packaging, together with energy recovery, and organic recycling.

10.2.1   The recycling and recovery targets

According to Directive 94/62/EC, targets for recovery and recycling in the European Community have been laid down. These targets are the following:

- Global recovery between 50 and 65%
- Global recycling between 25 and 45%
- Recovery by materials of 15%.

These targets are referred to packaging composed of plastics, paper, wood, glass, aluminium and steel. The objective of the recovery target imposed on single materials can be interpreted as an obligation to recycle a certain quantity of waste also for the materials for which incineration would be a more economically viable short-term solution, in particular plastics.

The Community Member States were expected to achieve these objectives within five years from their implementation in the national laws of the Directive itself, and not later than 30 June 2001.

Other notable provisions contained in the Directive are:

- To limit combined lead, mercury, cadmium and chromium (VI) content to 100 ppm
- To ensure that packaging materials are introduced in the marketplace only if they meet the so-called 'essential requirements', i.e. characteristics that include minimization of weight and volume and suitability for material recycling.

10.2.2   The 'essential requirements'

The Directive specifies the nature of the 'essential requirements' mentioned in the above paragraph, in its Annex II. Requirements have been placed relative to the manufacturing and composition of packaging, their reusable nature and their recoverable nature. The last one has to do with recycling: it stipulates that packaging materials recoverable in the form of recycling shall be manufactured in such a way as to enable the recycling of a certain percentage by weight of the raw materials used for their production.

Successively the European Parliament gave a mandate to the European Committee of Standardization, CEN, with the aim of developing standard norms which would allow packaging materials to meet the 'essential requirements', thereby complying with the P&PW Directive. As a result of the mandate, in 2000 CEN published five European Standard Norms,[3] namely:

- EN 13428:2000 – Prevention through reduction at source
- EN 13429:2000 – Reuse of packaging materials

- EN 13430:2000 – Requisites for packaging recoverable through materials recycling
- EN 13431:2000 – Requisites for packaging recoverable through energy recovery
- EN 13432:2000 – Requisites for packaging recoverable through composting and biodegradation.

After careful scrutiny the European Parliament decided to accept in full only EN 13432:2000. The norm EN 13428:2000 was only partially accepted. The norms EN 13429:2000, EN 13430: 2000 and EN 13431:2000 were not accepted because, in the Parliament's opinion, they did not correctly address the issues of reuse, materials recovery and energy recovery required by the Directive. As a consequence, the criteria by which packaging materials can be declared reusable or recoverable via material or energy recovery are still unclear, and somewhat left to common sense. In early 2002, CEN was given a new mandate with the aim of reviewing the norms concerned. New drafts are now available, and the completion of this new mandate is foreseen by the end of 2004.

10.2.3   Implementation in the Member Countries

The P&PW Directive required that the Community Member States would achieve the targets indicated in Paragraph 1.1.1 by 30 June 2001, with the exception of Portugal, Ireland and Greece for which the terms of compliance were delayed until 31 December 2005. More importantly, the Member States have been given the possibility of adopting economic instruments aimed at achieving the above-mentioned objectives.

The implementation of the P&PW Directive in the Community Member States gave origin to a rather complex legal framework, due to the overlap of the Directive itself with the national measures already existing in some countries such as Germany, the Netherlands, France and Austria. Germany, in particular, is the first European country that has introduced a packaging recovery scheme (Decree Toepfer, 1992), and can be considered a leading model in Europe. Its approach is based on the possibility of transferring the responsibility of recovery to a third party, authorized and controlled by the Regions. Such a third party is called DSD-Duales System Deutschland AG, and enables its Member Companies to use a special logo, the Green Dot, on the packaging introduced in the market. The Green Dot has become very popular also outside Germany, so that many other Community countries have adopted the same logo in order to avoid the formation of commercial barriers. These countries are Austria, Belgium, France, Spain, Portugal, Ireland, the UK, Sweden, Luxembourg and Greece. For the same reason, several other countries not belonging to the European Community have adopted the Green Dot logo, such as Norway, Poland, the Czech Republic, Hungary and Latvia. It must be highlighted, however, that the use of the same logo in different countries does not mean that the compliance schemes are the same: on the contrary, each country adopts its own scheme, and the function of the logo is limited to the mutual recognition of these schemes. In particular, the financing systems of the recovery and recycling schemes may differ significantly.[4] The countries that are outside the Green Dot area are Italy, the Netherlands, Denmark and Finland.

10.2.4   The future legal framework

Although the attainment of the recovery and recycling objectives is still controversial in several Member States, a proposal of amendment of the P&PW Directive was presented in December 2001. Actually, Article 6 of the P&PW Directive already stipulated the revision of the targets before the end of the five years in which the targets themselves should have been met. Due to the political pressures in the European Parliament, this proposal envisages a substantial increase of the targets. The latest proposal from the Council of Ministers of the Environment of the European countries, currently under discussion at the European Parliament, envisages an increase of the minimum recovery up to 60%, an increase of the overall recycling from a minimum of 55% to a maximum of 80%, and the raising of the specific targets to 60% for glass, 60% for paper, 50% for metals and 22.5% for plastics.[5]

Another part of the proposal deals with the introduction of 'Packaging Environmental Indicators', or PEIs, that should represent the technical characteristics of the packaging materials that would provide indications of their environmental impact. For example, the introduction of PEIs should be able to give information as to whether a packaging material has been designed and manufactured in accordance with practices and procedures that would minimize its environmental impact. However, because of the lack of more precise details on definitions and use of PEIs, further comments cannot be made at the present time.

The proposal of amendment is under discussion at the European Parliament and Council; therefore it has to be considered provisional, including the new targets, as it might still undergo significant modifications. It is clear, however, that the packaging industry should be prepared to face more challenging tasks to respond to the environmental requirements set forth by the Community law. As a consequence, environmental requisites are increasingly becoming an essential feature of packaging materials that should be taken into consideration at the very early stage of their development and become a built-in characteristic of new-generation packaging materials.

10.3   Recycling paper packaging: collection and separation

Recycled paper is an important source of raw materials for the paper industry: about 42 million tonnes of waste paper were recycled in Europe in 2001, confirming a substantial growth in the last ten years (26 million in 1991, with a growth of 38%) (Fig. 10.1). These 42 million tonnes represent 52.1% of the total

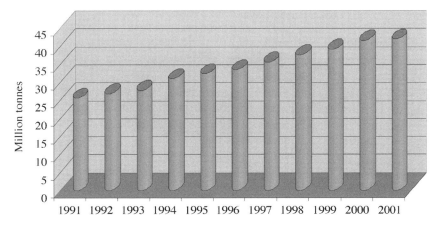

Fig. 10.1   Recovered paper utilization in Western Europe (Source: CEPI).

paper used in Europe. They also represent the largest majority of the collected paper, which was 44.7 million tonnes in 2001.[6] Recovery of paper is operated mainly through recycling, i.e. reprocessing of waste paper into conventional pulpers, with the use of appropriate technologies for cleaning the product. Only a small fraction, equal to 2.7 million tonnes or 6%, is either incinerated (with or without heat recovery) or landfilled.

The biggest volume of paper collected was in Germany (11.5 million tonnes) followed by France (5.5 million tonnes) and Italy (5.1 million tonnes); offices, commercial businesses and households generate large quantities of paper waste, including packaging materials. The highest potential for an increase in collection today in Europe lies with households, though the cheapest source of post-use paper is represented by commercial business, where large quantities of paper foils and cardboard boxes are wasted, thus decreasing the logistic costs of collection. In fact, often the paper packaging destined for recovery pulpers does not come from municipal collection and sorting schemes, but directly from commercial businesses, allowing savings because of the avoidance of trade operations, sorting costs and often volume reduction.

As already mentioned, recovery in Europe as a whole has progressed significantly over the last few years, such as to make realistic the target of 60% recycling of paper and board packaging contained in the proposed amendment of the P&PW Directive (see above). Actually, if the volume of non-collectable/non-recyclable paper and board is deducted from the total consumption, the real collection rate is already very close to that figure.

The declared objective of the European Recovered Paper Council, the European association of paper and board manufacturing and recycling industries, is to achieve a recycling rate of 56%, relative to all paper (not only packaging) by 2005.[7] In theory, 70% of the paper produced is recyclable, although an excessive use of recycled paper would result in degraded properties of the finished products. This could be compensated only by weight increase, or

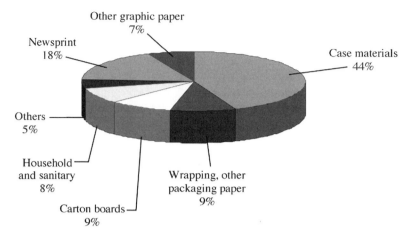

Fig. 10.2   Recovered paper usage by sector in Western Europe, 2001 (Source: CEPI).

less degradation, with questionable environmental impact and economic balance.

Packaging materials are the largest sector in which recycled paper is used: 26.2 million out of 42.0 million tonnes of recycled paper in 2001 were used for manufacturing new packaging, i.e. case materials, corrugated board, wrapping and others [6] (Fig. 10.2). This quantity represented 74.8% of the paper and board packaging produced in Europe in 2001.

10.3.1   Separation of contaminants

The collected paper has different technical and economical value depending on its degree of purity. Sorting operations obviously have a large influence on the final purity. Conventional sorting is via mechanical processes, while more accurate sorting can be achieved only manually.

Commercial separation machines are available, capable of separating paper from plastics and other packaging. These machines operate from 1.5 to 3 tonnes/ hour feedrate, generating a minimum product purity of 90%. In a typical separation process a mixed packaging waste stream is randomly fed in the machine through an acceleration conveyor; the separation criteria are based upon infrared sensors, which are capable of classifying cartons, paper and mixed plastics. Other types of separation can be also operated, such as punched screens, slot screens and centrifugal purifiers; non-paper components that are mechanically stable so as to remain in the form of large particles can be removed more easily through these systems, while materials of very small dimensions cannot be removed.

There are more than 60 types of fractions of waste paper that can be selected, stipulated by the European List of Standard Grades of Recovered Paper and Board, EN 643.[8] Sorting is done on the basis of the final use of the fraction, and is strongly influenced by the requirements and prices of the different fractions.

Both market requirements and prices may vary significantly; therefore the economic worth of the process resides in its ability to adapt the grades produced to the actual market request. Paper sorted in fractions according to EN 643 is no longer considered waste, but secondary raw material, as it is fully used in pulpers to produce new paper materials.

10.4   Recycling paper packaging: processing[9]

The standard paper production process includes the following steps: preparation of fibres, introduction of additives, preparation of sheets, drying and coating. The production of paper from recovered fibres does not differ significantly from the process that utilizes virgin cellulose fibres. The main difference resides in the preparation of the fibres; several processes can be used to clean up the recovered material, with the aim of removing contaminants prior to further processing. Fibres preparation in the case of virgin cellulose consists of a series of treatments that have the objective of dissolving the natural wood and reconstituting fibres with a desired consistency that would allow the production of a conventional paper sheet. When recovered fibres are introduced in the production process, the fibres preparation phase is associated with treatments capable of removing contaminants and inks. These treatments that fibres undergo when recovered material is present are briefly described in the following paragraphs.

10.4.1   Re-pulping

In the case of processing native wood, pulping is the treatment that removes the non-cellulose fraction, preparing material for paper production. It consists of a treatment to alkaline conditions at pH 8–10, operated in suitable reactors (pulpers) of different dimensions and sizes, with the aim of defibrizing and achieving a given pulp consistency. The term 're-pulping' is referred to recovered fibres, and consists of the water dissolution of recovered paper for further processing; re-pulping is the first step, and is associated with mechanical cleaning tools such as removal of inks. Often the desired efficiency of ink removal is a parameter of choice of the pulper to be used.[10] In many processes virgin and recovered fibres are treated together, to obtain paper products with selected technical characteristics.

10.4.2   Other mechanical cleaning processes

Mechanical cleaning is also carried out by washing, screening and suspension, i.e. treatments using differences in size, density and surface properties. These steps have the aim of removing extraneous contaminants from the fibres, such as metals, sand, glass and other agglomerated particles derived from inks, varnishes and adhesives. Being operated after the pulping step, the control of contaminants removed during pulping is the key parameter that would determine the efficiency of these mechanical cleaning techniques.

10.4.3   De-inking by flotation[11]

The most common technique for de-inking consists of flotation: the paper pulp formed is insufflated with airflow in the presence of surface-active substances. In these conditions 'flotation bubbles' are formed, on the surface of which inks and pigments of particle size 20–100 μm are suspended, together with fragments of fibres and other paper additives. The foam formed on the surface of the pulper is then removed. There are limitations in the ink particle size: inks containing components soluble in alkaline solutions break into particles of about 1 μm and less, which are unsuitable for flotation. In this case an option is represented by an additional washing cycle to remove them. On the other hand, coarse particles resulting from cross-linked inks or inks coated with cross-linked varnishes, e.g. through UV cross-linking technology, are too heavy to be floated. In this latter case there is still the option of dispersing these particles and submitting again to a floating process.

10.4.4   Washing

Washing can remove fillers, finely divided ink particles and other colloidal materials. This process can achieve high cleaning efficiency, but it implies the use of high volumes of water, with associated costs, and leads to significant loss of fibres removed by mechanical effects exerted by the water pressure.[9]

10.4.5   Thermal treatment

Also known as 'hot-dispersion', this treatment is operated on high-consistency material and consists of submitting fibres to high mechanical forces together with vapour or steam heating at 60–100°C, depending on the process. Sometimes the thermal treatment is associated with chemical treatment. Hot-dispersion has the purpose of reducing chemical and microbiological contamination, especially in the processes where recycled water is used in the various cleaning and de-inking steps.[9]

10.4.6   Chemical treatment

Known also as 'bleaching', this consists of treatment with biocides, slimicides, whitening agents, enzymes, ozone, hydrogen peroxide, oxygen and other chemicals. The purpose is to increase brightness and reduce contamination.[9]

10.5   Food packaging from recovered paper

As already mentioned, recovered fibres represent an important source of raw materials for the paper industry, and suitable physical–mechanical characteristics are obtained by mixing virgin and recovered fibres in different proportions. The paper packaging industry does not represent an exception: paper packaging is produced largely with the use of recovered fibres. One of the sensitive areas where the use of recovered fibres must be carefully taken into account is the production of paper destined for food contact. Packaging materials and articles destined for food contact use are submitted to the safety requirements of Council Directive 89/109/EC.[12] This Directive prescribes that the said materials and articles fulfil the basic safety requirements, i.e.

- They should not endanger human health
- They should not cause unacceptable organoleptic and nutritional modification of the food.

Materials and articles made of paper and board for food contact application fall within the scope of the above-mentioned directive. However, contrary to the case of other food contact materials, such as plastics, which are submitted to more specific legislation, paper and board for food contact do not have further legal requirements at the European level. They are submitted, however, to national provisions in some of the Community Member Countries, such as the following:

- The Ministerial Decree of 21 March 1973 and subsequent amendments in Italy[13]
- The Recommendation XXVI of BGVV (Federal Institute for Consumers' Health Protection and Veterinary Medicine) in Germany[14]
- The 'Materials in Contact with Food' Ordinance in the Netherlands.[15]

The above legislation is also used as a reference for recycled fibres. Another important text that lays down the technical specifications and safety requirements in this field is represented by the Resolution of the Council of Europe AP (2002) 1.[16] The Council of Europe, within the frame of the Partial Agreement in the Social and Public Health Field, works actively in the development of recommendations in the areas not fully covered by the Community legislation. Although these recommendations do not have a binding character, they are widely used by industry and by the public control authorities of the Council of Europe countries as reference norms. In the case of paper and board for food contact use, Resolution AP (2002) 1 establishes the purity criteria that shall be fulfilled by recovered cellulose fibres in order to be used safely in food contact applications. This Resolution must be seen in combination with the 'Council of Europe Policy Statements concerning Paper and Board Materials and Articles intended to come into contact with Foodstuffs',[9] consisting of the technical documentation that specifies the characteristics that these materials have to fulfil to be suitable for food contact application. According to the above-mentioned Resolution, three types of foodstuffs have been identified:

- Type I: aqueous and/or fatty
- Type II: dry, non-fatty
- Type III: food shelled, peeled or washed before consumption.

The use of fibres derived from recovered paper in contact with food is allowed provided that it originates from paper of suitable quality, and not from sources that may carry potential chemical or microbiological contaminants, such as waste paper packaging from households. In all cases an appropriate processing and cleaning treatment must be carried out as a function of the food type for which the final paper product will be used.

In practice, recycled fibres are suitable for manufacturing packaging for Type II and Type III foodstuffs, having limited capacity of extraction, and in these cases raise only little, or no, concern in terms of food safety or organoleptic damage. On the contrary, paper packaging for food of Type I, such as dairy products, chocolate and cardboard for pizza trays, has to be manufactured with the exclusive use of virgin fibres, or consist of multilayer materials, in combination, e.g. with plastics and aluminium, where paper is not in direct contact with the food.

10.6   Recycling plastic packaging

The total consumption of plastics in Europe was about 36.8 million tonnes in 2000, 13.7 million tonnes (37.3%) of which were used for manufacturing packaging materials (Fig. 10.3). Plastics accounted for 17% of the total packaging usage in Western Europe.[17] A wide variety of polymers are used in the production of plastic packaging, the most widespread being:

- Polyethylenes with different densities, i.e. high density polyethylene (HDPE), low density polyethylene (LDPE) and linear low density polyethylene (LLDPE)
- Polypropylene PP
- Ethylene copolymers with propylene, vinyl alcohol, vinyl acetate
- Polystyrene PS in sheets or films, or foamed
- Polyesters, mainly polyethylene terephthalate (PET)
- Polyamides, mainly nylon 6 and nylon 6/12
- Polyvinyl chloride (PVC) and vinylidene chloride copolymers.

Each of the above-mentioned polymers has special properties, such as being a barrier to oxygen or to water vapour, easy sealability, thermal and dimensional stability and others that may be desired in packaging applications. In other cases, specific properties are imparted by the process technology, such as orientation, cross-linking, etc. Since packaging applications may require a combination of some of these properties, polymers can be put together in composite structures to achieve the specified performances. This leads to multi-material products, such as multilayers, that pose considerable difficulties during recycling.

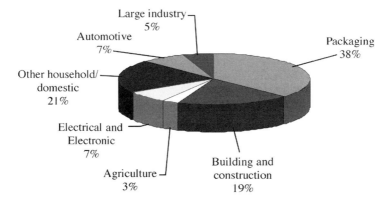

Fig. 10.3   Plastic consumption by industry sector in Western Europe, 2000 (Source: APME).

During 2000, 19.5 million tonnes of plastics were wasted,[17] 13 of which derived from Municipal Solid Waste (MSW), and 6.5 from commercial business and industries. Besides industrial scraps and secondary raw plastics collected from commercial businesses, MSW has been identified as a source of plastics suitable for recovery, since it contains about 7% of plastics. As explained in the Introduction, environmental factors play a major role in the recovery of plastics from MSW; however, MSW collection schemes always result in mixed polymer fractions that are not easily sorted and need extensive labour cost to obtain homogeneous materials that are suitable for recycling. In addition, composites and multilayers cannot be separated into their components, and the fractions containing these materials can be recovered only via energy recovery.

10.6.1   The market for recycled plastic packaging

Plastic recycling in the packaging industry is well established for three polymers, namely PET, HDPE and PP. PET recycling in Europe amounted to about 350 000 tonnes in 2002, and is expected to grow to 600 000 tonnes in 2005.[18] The main market is represented by the fibre industry but a significant amount is also used for production of soft drink bottles. For example, the Coca-Cola Co. used 2.5% recycled PET in bottles in 2001, and has a target of 10% usage by 2005.

The amount of recycled HDPE was about 100 000 tonnes in 2001, consisting of grinding and reprocessing of pallets and crates for fish and vegetables in a closed loop. PP crates are also recovered in the same way; recovery in this latter case was 50 000 tonnes in 2001.[19] One of the main obstacles to the development of the market for these products is the lack of harmonization in their use in food contact applications. Such use is currently based on approvals granted to specific processes by the single EC Member Countries, and in at least two of them, namely Italy and Spain, the use of recycled plastics for food contact applications is forbidden. The modern cleaning technologies, especially for PET (see Section

10.7), are able to achieve a high degree of purity, as has been demonstrated by a pan-European study carried out between 1999 and 2001.[20] However, the harmonization of the European legislation in this field does not seem a priority for the European Commission. Recently an industry proposal for the issuing of a European legislative instrument has been put forward, and will probably be discussed by the Community Health Authorities during the course of 2003. Harmonization of rules and duties in the food contact field would represent a great impulse for the plastic packaging recycling industry, as well as a guarantee for a high level of consumer protection.

10.6.2   Economic considerations

In order to make economic sense, the cost of collecting, sorting, cleaning and reprocessing plastics has to be competitive compared with the virgin material. To obtain a reasonable economic framework, a balance should be achieved between the quality of the recovered plastic fractions and the cost needed to obtain the desired degree of quality. The composition of the recovered fraction does have a major influence on the quality of the finished products obtained from the recycled plastics, while the cost that the industries are prepared to bear can be only determined as a function of the value of the secondary raw material or, in other words, of the market requirements of the different fractions. The two elements are, however, strictly interconnected: for example in the case of the mixed polymer fraction, recyclers do not pay excessive attention to obtaining highly refined secondary raw materials since an established market for this fraction does not exist. On the contrary, in the case of PET and, to a certain extent, HDPE, for which a well-established market does exist, the quality requirements of the secondary raw material are higher, and the cost of refinement is covered by the selling price.

It makes a big difference whether the plastic packaging is recovered from commercial businesses, or in closed loops or from MSW. While in the first two cases the recovered packaging is homogeneous in composition and needs relatively little cleaning, sorting and cleaning heavily influence the cost of packaging fractions derived from MSW. In this latter case, since sorting is often operated manually, the labour cost is often 90% of the cost of the final recyclate. For example, the cost for recycling LLDPE from MSW is around €0.7/kg, excluding collection cost and price paid for the waste polymer. Since the same base virgin resin is sold at around €0.8/kg (up to €1.1/kg, depending on the grade), it can be clearly understood that recycling LLDPE from MSW does not make economic sense. On the other hand, LLDPE industrial scraps cost €0.3–0.4/kg, therefore their competitiveness is extremely high, while quality, especially in terms of absence of contamination, is higher than for plastics recovered from MSW.[21]

PET is quite an interesting case: the recycled polymer for non-food application is sold at 50% of the virgin resin; for this grade of resin the primary application is in the fibre industry. On the other hand, recycled PET for the soft drink bottles industry, which undergoes special purification and processing to compete with the virgin resin in terms of quality and safety, is sold at 10–20% less than the virgin resin. In this latter case the high price volatility of virgin PET can explain the strategies of investment of intensive food contact grade PET users in recycling technologies, which are aimed at protecting against price increase through feedstock diversification. Moreover, despite the high cost of recycling, gaining access to lower-priced recycled food contact grade PET does result in an increase in profit per bottle in such a way as to exceed the recycling cost. However, with the exception of PET and, to a certain extent, HDPE, for all other polymers the spread between cost and market value is such that the industry of the post-use recycling of plastics needs to be supported through economic measures that would encourage its development.

The enforcement of the P&PW Directive in the Member Countries leads in all cases to the adoption of levies on plastic packaging; these economic measures have the aim of keeping the price of recycled plastics at a market competitive level, 20–50% less than the corresponding virgin resin depending on the applications and the grade of purity. Sometimes the cost of recovery may compete with landfilling, and the lowest quality fractions that are obtained in a separation plant may be sold at negative cost. In addition, because of the general low quality of the products, the market for recycled plastic packaging can be easily pulled down by factors such as variations in the quoted price of the virgin materials and exchange rates.

The above considerations lead to the conclusion that the plastics packaging recycling industry can be market driven only when sorting and cleaning costs are minimized. For this reason the treatment of MSW should be seen in the context of reduction of the environmental impact, rather than an economic opportunity.

10.7   Collection and separation of plastic packaging

10.7.1   Collection schemes for plastic packaging

Collection schemes for post-use plastic packaging, particularly PET plastic bottles, from MSW already existed in many EU Member States before the enforcement of the P&PW Directive. Post-use plastic scraps are generated at industrial, commercial and residential levels. Industrial scraps are the main source of secondary raw materials that are often reused without any treatment in the same production cycle, thus they cannot be defined as 'secondary raw materials'. Part of this stream, however, is introduced in the recovery chain, and therefore it constitutes a source of 'secondary raw materials'.

Commercial scraps represent a great potential for high recovery and recycling rates, as well as the main source of plastic packaging. These plastics consist of products that are generated by commercial business, such as shipping and receiving departments, warehouses, distribution centres, wholesalers, or distributors' return schemes for retail stores. Large quantities of commercial scrap minimize logistics and cost of transportation, maximize the efficiency of collection and ultimately result in a favourable economical balance of recovery programmes. Labels, staples and tape typically contaminate this type of plastics, while microbiological contamination is unlikely; this makes them an attractive source for the production of secondary raw materials.

Post-consumer plastics from MSW derive from households, and costs of collection can be significantly different depending on whether they are dispersed over large geographic areas or concentrated in highly populated zones. The cost associated with collection and transportation to the recycling facility might be prohibitive, and for this reason in the former case it becomes mandatory to maximize the yield of transportation by submitting the collected material to size reduction processes. These processes may consist of simple pressing and baling for materials that are to undergo sorting at a later stage, or cutting, comminuting and granulating for obtaining homogeneous-mix plastics.

Although business and public authorities now largely sponsor the use of separation bells and return schemes are increasingly used, these plastics are rather expensive to recover and have the highest potential for contamination. Non-plastic residues such as paper, metals, earth, powders, etc., are typical contaminants, and microbiological contamination is usually high.

10.7.2   Sorting of plastics packaging for recovery

Sorting is the critical step in plastics packaging recovery; this in fact determines the purity, and ultimately the value, of the secondary raw materials. Sorting is operated to separate high value fractions from mixed plastics and contaminants. In order to provide a better picture of plastics packaging sorting processes, in this paragraph we will describe the plant of Seriplast Srl, a major sorting plant located near Milan, Italy,[22] and we will use this plant as a model to describe conventional sorting processes in more general terms. Seriplast Srl processes 12 000 tonnes per annum of plastic packaging from MSW collected in the whole area of northern Italy; PET is the primarily recovered resin, though HDPE is also recovered, as well as some other lower quality fractions. PET flakes are sold to fibres production industries, while the recovered HDPE ends up in the production of shopping bags. The mixed plastics fraction is either sold to energy recovery plants or may find some application in the production of filling fluffs. The scheme of the plant is described in Fig. 10.4.

Receiving
The material contained in plastic collection bells is received by the separation plant either as collected or after volume reduction, being the two alternatives chosen on the basis of cost of transportation. This plastic stream consists mainly of films, foamed trays, thermoformed cups and sheets, bottles and other containers for liquid detergents. The large majority of bottles, as well as parts of foamed trays and thermoformed materials, consist of PET. Containers for detergents are composed primarily of HDPE, while LDPE and LLDPE are the main constituents of films. Other foamed trays, cups, sheets and thermoformed

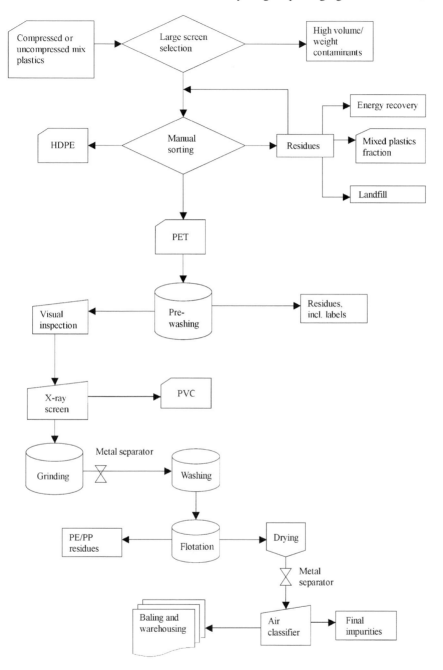

Fig. 10.4 Scheme of plastic packaging sorting plant of Seriplast.

products consist of PP, PS and PVC. PVC composes also a small fraction of soft drink bottles.

Sorting

After a preliminary separation to eliminate high volume or high weight contaminants, operated mechanically through a large screen, the main separation is operated manually. The material is conveyed through a belt in a separation area where sorting takes place and the fractions are put into different containers. Normally the separation is made in more than one transit in the separation area, to increase the yield. This step results in a PET fraction that may or may not consist of bottles of the same colour. HDPE for production of shopping bags is treated separately. Since this fraction is subjected to heavy odour contamination (the majority of HDPE comes from containers for detergents), thorough washing is needed prior to commercial application. Films, trays and other rigid and expanded components compose the 'mixed plastics' fraction, while non-plastic contaminants are removed to the highest possible extent.

In some plants, pre-screened fractions are treated via automatic air classifiers, where the separation is based upon detection of the composition via an IR detector and the airflow expulsion of the non-PET fraction. It must be said, however, that this process does not achieve the same level of accuracy as manual separation, and needs additional measures of purification at a later stage. An additional separation can be also operated between clear and coloured PET by visual inspection or with the use of a further IR detector.

The separation of light components such as films can also be operated; these products can be ground to produce flakes, but because they often contain multi-layer films and are composed of different polyolefins, nylon, polyesters and other polymers, their properties are rather poor and they can only be recovered via energy recovery. In some plants the separation of films through a pneumatic pump is an option that has been trialled, but in addition to the high cost not being justified by the value of the fraction obtained, the yield from separation is not always satisfactory.

This separation step results in the following fractions:

- Clear PET bottles
- Coloured PET bottles
- HDPE containers
- Mixed plastics/residual fraction for energy recovery
- Residual fraction to landfill.

In the plant that we have examined the residual fraction sent to landfill did not exceed 10% of the total processed material.

Washing

The resulting PET fraction is successively pre-washed. Washing is operated by a moderately alkaline hot solution, and results in the removal of paper labels, caps and other extraneous components. After a further visual inspection, and removal of extraneous materials that may have escaped the separation steps, PET undergoes X-ray scanning for the detection and elimination of residual PVC via, e.g., air classifiers.

Grinding

The resulting material is then ground. Dry grinding is the preferred option, capable of producing flakes 4–20 mm in size. Since grinding knives are particularly sensitive to the presence of metals, particular attention must be paid to removing metals prior to this step. To this aim, magnetic metal separators are placed at the entrance of grinders. Other magnetic separators are also used at various points of the process, for a more efficient metal removal. Other types of grinding may be used, such as wet grinding, which allows washing of the flakes while reducing knife wear.

Refinement

After grinding, the flakes are sieved and washed to remove the glue derived from labels. As a secondary effect of washing, the residual paper is also removed to a part-per-million level. The resulting material still contains a certain quantity of polyethylene and polypropylene deriving mainly from caps. The separation of such fraction is made on the basis of the different wettability of PE and PP with respect to PET in the presence of selected surfactants. The separation technique consists of flotation, where flakes are introduced into water together with surfactants and left there to float or sediment; flotation can be operated in more than one step, adjusting the surfactant's concentration to separate different components. Other practices may envisage stirring and gassing of the material, which would improve the separation efficiency.

   After drying and further control for the absence of metals, a final refinement is carried out via 'visual inspection' operated by a high-resolution video camera equipped with an image analyser, which removes micro-powders and other potential extraneous particles through an air classifier. Finally, the PET flakes obtained by the process are baled and warehoused.

10.8   Recycling techniques and uses of plastic packaging

Plastic packaging obtained from collection/separation schemes is addressed to recovery. Recovery can consist of either recycling or energy recovery; the two options are chosen on the basis of the quality and the market demand for secondary raw materials. Despite the legislator's tendency to raise the targets for plastics recycling, this is not always the best option from both the economic and the environmental viewpoints. Energy recovery remains the preferred option for mixed and highly contaminated plastics, often resulting as by-products in the production of secondary raw plastics, while recycling is preferred for refined and homogeneous secondary plastics. It has been demonstrated by accurate studies[23] that an integrated approach to plastic waste management, envisaging both recycling and energy recovery, results in the greatest level of environmental benefit.

The process of separation described in the above paragraph, as well as other analogous processes, results in quite a high purity of the PET flakes. It must be underlined, however, that the majority of the industries operating plastic separation are small and medium-sized enterprises, with limited investment capacity. Not all of them produce high-purity secondary raw materials, consequently a residual degree of contamination might be present, consisting of metals, paper, fibres, agglomerates, polymer gels and other components. In this case, filtering the recycled resins in the molten phase during the reprocessing step can attain a higher degree of purification. To this aim, porous or non-woven filters are inserted between the extruder and the extrusion dies. Several types of filters are available with different design and thickness, resulting ultimately in different degrees of purity.[24] The process of melt filtration, although very effective, is rather expensive in terms of both investment and yield reduction, and becomes justified by the value of the finished products when it is part of highly refined processes leading to food contact grade recycled PET.

As already mentioned, conventional recycled PET is used mainly by the fibre industry, which normally requires lower molecular weight resin than does bottle manufacturing, and therefore finds in the secondary raw material a cheap feedstock. Recycled PET can in fact be converted into fibrefill without further treatment; another application is the production of carpets, e.g. for the automotive industry.

However, when destined for the manufacture of new bottles or other types of food packaging, the recycled polymer needs specific treatments such as more accurate removal of contaminants, purification from acetaldehyde (a typical by-product of PET processing) through volatilization, and regradation, i.e. increase of molecular weight. To obtain food contact grade PET, specific processes have been developed (described in Section 10.8.4); such processes have the capacity of increasing greatly the quality and the value of the secondary polymer, and also have an influence on PET price, which becomes comparable with the virgin resin.

In more general terms, and in addition to the above-mentioned specific processes, recycling of plastics encompasses three types of technologies applicable at industrial level, namely mechanical, feedstock and chemical. Mechanical recycling is the most diffused technology for economic reasons, while feedstock recycling and chemical recycling are not widely applied because of their unfavourable economic balance, although they are capable of resulting in highly refined food contact grade secondary resins.

10.8.1   Mechanical recycling

Mechanical recycling consists of the reprocessing of plastic packaging deriving from collection and sorting systems by using technologies that are normally used for virgin resins, such as extrusion, co-extrusion, injection, blow-moulding, etc. These technologies offer a variety of possibilities for the use of secondary plastics, encompassing:

- Single polymer recycled articles, such as polyethylene films for agricultural application or production of shopping bags
- Embossing the secondary plastic into layers of virgin resin; this method is used for manufacturing, for example, PET bottles and other articles for food contact, where the secondary plastic cannot be in direct contact with the foodstuffs
- Mixing the secondary raw material with virgin resin, to balance the physical and mechanical properties with cost, for a variety of products, such as films, sheets and fibres.

Secondary raw plastic materials normally have lower molecular weight and lower thermal stability compared with virgin resins, because of the exposure to mechanical, thermal, oxidative and photochemical degradation conditions they underwent. Conventional resins normally contain stabilizers, such as antioxidants and light absorbers. During the processing and the service life of the packaging materials, the stabilizers are consumed to preserve the polymer matrix from degradation, therefore in the reprocessing phase it is common practice to add new stabilizers. The choice of stabilizer is easy in the case of homogeneous recycled materials (the same stabilizers as for the virgin resins are used), but less straightforward when mixed or contaminated plastics are processed: in this case the stabilizers must be compatible with the resin mix. Some antioxidants have been specifically developed for improving the thermal resistance of plastics to recycling, such as the Recyclostab® grades, Recyclossorb® and Recycloblend®, all from Ciba Specialty Chemicals. For the same reason, in the case of mixed or contaminated plastics, compatibilizing agents might be useful to avoid melt crack during extrusion due to incompatible melt phases.

Mechanical recycling rose from 1.2 to 2.2 million tonnes from 1995 to 2000 (Fig. 10.5), representing 11% of the total plastic waste collectable, and is expected to rise further to 2.7 million tonnes in 2006. The plastics recycling industry is quite important in countries like Italy, Spain and Sweden, where the rate of mechanical recycling has reached values between 15% and 20%.

10.8.2 Feedstock recycling

The term 'feedstock recycling' includes a series of processes that consist of the pyrolysis and gasification of plastics operated at high temperature in the absence or near-absence of oxygen. The processes lead to petroleum feedstock that is used for other purposes than producing the original material. Some specific types of feedstock recycling can generate synthesis gas (Texaco process), a mixture of carbon monoxide and hydrogen that finds application in oil refineries and the chemical industry to upgrade commercial products. Other processes lead to heavier petrochemical fractions, consisting of liquid hydrocarbons and waxes.

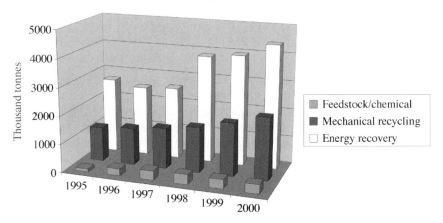

Fig. 10.5   Plastics recovery in Western Europe (Source: APME).

These products can find other applications, e.g. as fuel in the steel industry. Although several industrial processes have been developed for this type of recycling (by Texaco, BASF, Linde, SVZ, Akzo Nobel and others) the amount of plastic treated by feedstock recycling is rather small (see below). The rationale can be found in the unfavourable overall economics of feedstock recycling compared with more conventional petroleum refining, in particular high investments and limited yield derived from contaminated supply. All the above-mentioned technologies require, in fact, a significant financial support from public funds to break even, since the cost of feedstock recycling treatment has been estimated at higher than €1/kg, about double the cost associated with incineration (data referred to Germany, 2000).

10.8.3   Chemical recycling

The term 'chemical recycling' identifies the process of de-polymerization of PET leading back to the monomers terephthalic acid and ethylene glycol. These monomers, after purification, can be reused to produce to a new polymer. Chemical recycling can also be used for the de-polymerization of polyamides and polyurethane.

In the case of packaging materials, interest in chemical recycling concerns primarily the treatment of PET derived from soft drink bottles. The de-polymerization can be operated initially via glycolysis, consisting of a high temperature (240°C) treatment of PET in excess ethylene glycol, and resulting in the production of bis-hydroxyethyl terephthalate (BHET). After this step, BHET is further treated via

- hydrolysis, where the ester is transformed in the corresponding acid, or
- methanol with consequent production of the dimethyl ester of the terephthalic acid. This is a crystalline solid that can be easily purified.

Another process that has found commercial interest is methanolysis. It consists of the treatment of PET in the melt state (250C) with methanol in the presence of catalysts: the treatment results in ethylene glycol and dimethyl terephthalate, i.e. the constituting monomers. After purification, the monomers can be used for the production of new PET. Although this technology can result in very clean secondary raw materials, practically indistinguishable from the virgin polymer, its application in practice is extremely limited by the economics.

As a matter of fact, both feedstock and chemical recycling in Europe represent a very small fraction of recovered plastics. These two technologies accounted together for 329 000 tonnes in 2000, less than 2% of the total plastic waste collectable[17] (Fig. 10.5).

10.8.4   Recycling of PET for food contact

The economic importance of recycled PET in the production of food contact packaging, in particular soft drink bottles, has been explained in Section 10.6.2. Chemical recycling is able to provide secondary raw PET suitable for food contact application; another system for the food contact use of recycled PET, as explained in Section 10.8.1, is multi-layer processing, with the recycled polymer sandwiched inside the final product. This latter system is used for manufacturing PET bottles, and has been demonstrated to be safe through tests envisaging the introduction of model contaminants in the recycled PET layer and the study of a barrier to migration of these contaminants exerted by the virgin PET layers (defined as a 'functional barrier').

A third method for production of food contact grade PET consists of Superclean processing. This method encompasses a series of processes that are able to remove the volatile contaminants and increase the viscosity of the polymer to a grade suitable for injection-blow moulding (0.75–0.80 g dl^{-1}) and higher, for more demanding applications.

Superclean processing has also been demonstrated to lead to safe PET via the migration study of model contaminants. Several proprietary Superclean processes have been developed which combine high-temperature washing, melt filtration, pressure treatment, solid-state polycondensation and other treatments to achieve high purity/high quality recycled PET. The main advantage of a recyclate PET produced by Superclean is that it can be used in direct food contact, without the need of a 'functional barrier'. Some of the Superclean technologies available are reported below:[23]

- Ecoclear^{TM}, developed and used by Wellman in Holland, is one of the oldest processes (1997) with a capacity of 7500 tonnes per annum.
- Supercycle^{TM} by Schmalbach-Lubeca in France, operating since 1998 with a capacity of 6000 tonnes per annum.
- VacuRema, a process developed by Erema and adopted by Texplast in Germany, with an output of 6000 tonnes in 2002.
- Buehler AG also developed a process claimed to lead to a polymer

competitive with the virgin resin in terms of quality, in use at Schmalbach-Lubeca.

- PKR in Germany has adopted the process developed by OHL Stehning GmbH, with a production capacity of 10 000 tonnes per annum.
- URRC, United Resource Recovery Corporation developed a technology in collaboration with the Coca-Cola Company, which is in use at RecyPET in Switzerland since 2000 with an output of 15 000 tonnes per annum, and at Cleanway in Germany since 2002, with the same capacity.

Bottles manufactured by both the multi-layer and the Superclean processes have received 'non-objection' letters for food contact application by the health authorities of numerous countries in Europe and by the Food and Drug Administration in the USA. In Europe the use of PET feedstock originated from food contact applications is preferred as raw material for the recycling processes. In the USA the FDA has issued letters of 'non-objection' also for non-food-grade PET feedstock.

10.9   Conclusions and future trends

Packaging materials represent a fundamental part of modern distribution chains. They are of the utmost importance in the case of food, pharmaceutical products and fragile goods. Not only do they protect products from contamination and mechanical shock, extend the shelf-life of food and provide information about the product, but they also provide an increasing variety of opportunities such as the possibility of being used in microwave ovens, the capacity to increase shelf-life of food, drugs and nutritive products through both conventional and active barriers, and others. The environmental impact of packaging materials has been discussed for a long time, and a number of provisions have been issued by the public authorities to minimize packaging waste. The P&PW Directive is today the framework under which the national recovery schemes operate, even though such schemes are significantly different in the various countries.

Large investments in research and development have been made in the last ten years to develop new technologies for recovery and recycling of packaging; also, numerous studies have been undertaken in order to identify the best option to balance environmental needs with process economies, providing both authorities and industry with tools to maximize efficiency while minimizing the environmental impact. It is clear that there is no option that is valid for all packaging materials in all situations, but in each case the choice of the option would be influenced by boundary conditions. For example, there will be a big difference in considering the ecological/economic balance of a recycling operation depending on whether this operation is carried out in a closed loop or in an open loop. Also, the closed loop could fail in competition with reuse, for certain types of packaging, and could also be different depending on whether or not loss would occur during the loop.

In other words, the only correct way to decide the worth of a given packaging material for a defined application is to carry out an eco-balance study. In general, it is not true that one material is better than another in absolute terms, even though often the perception of environmental friendliness is linked to the material itself and overlooks all other factors. Within this framework, Life Cycle Analysis (LCA) is a powerful tool that could be of help in determining whether or not a packaging material in a defined application does represent the best option. LCA provides a systematic approach for the examination of all parameters that contribute to the environmental impact of the packaging material during its entire life cycle, comprising energy consumption, resource-intensity of production processes, air, water and soil emissions, generation of waste and macro-economics. LCA deals with systems rather than products; the same packaging material, for example, can have a totally different environmental impact in terms of its distribution system or the availability of recovery or disposal facilities at the end of its life cycle. Equally, the availability of adequate collection and sorting schemes, as well as consumers' attitudes towards final disposal, can influence to a great extent the analysis results. Therefore LCA can be extremely useful in understanding the impact of the whole industrial system associated with packaging materials, and in identifying the critical points that need to be addressed in order to maximize the environmental yield. On the other hand, it should be used carefully, avoiding comparing products to each other if their global industrial systems are not comparable; if not properly used the results may be strongly misleading. For a comprehensive introductory dissertation on LCA, references 26 and 27 should be consulted.

It has been demonstrated that recycling of paper and board packaging is an extremely attractive option from the economic point of view, leading ultimately to a great extent to the production of new packaging materials for both food and non-food applications. This industry is primarily market-driven, it has been developed independently from the legislation constraints introduced by the P&PW Directive, and there are reasonable expectations for further growth in the near future.

On the other hand, recycling is more difficult in the case of plastic packaging, due to the difficulty in consistency of supply in terms of both quality and costs. The industry, the public authorities and the consumers all have a role to play in steering plastic recycling:

- Industry should further invest in R&D aimed at developing alternative processes to mechanical recycling that are not extensively exploited today, as well as implementing 'design for recycling' practices, i.e. considering recyclability as one of the key parameters to be achieved by new packaging materials since the very beginning of their development.
- The public authorities should adopt further economic measures such as increasing landfill costs and economic facilitation of start-ups dealing with recycling. Another important provision that would bolster plastic recycling is

the harmonization and simplification of rules for the use of secondary raw plastic in manufacturing food contact packaging materials.

* Consumers should help in increasing the quantity and improving the quality of the collected packaging.

All the above provisions are likely to open new markets and new applications to secondary raw plastics from packaging, combining environmental and economic benefits.

10.10   References

1.   COUNCIL DIRECTIVE 94/62/EC, *Official Journal of the European Communities*, 31 December, 1994.
2.   Preliminary European Norm, Final draft prEN 13437:2002 (E.), 'Packaging and Material Recycling – Criteria for Recycling Methods – Description of Recycling Processes and Flow Chart',
3.   Available at the National Standardization Bodies, e.g. http://catalogo.uni.com/EN/home.html, or at the CEN website http://www.cenorm.be/sectors/sw_res/transport/packaging.htm
4.   S. POGUTZ, A. TENCATI, in: S. POGUTZ, A. TENCATI, *'Dal rifiuto al prodotto'*, ed. Egea, 2002, Ch. 13.
5.   M. CLEMENTS: 'The packaging and Packaging Waste Directive and its implication for flexible packaging business', Proceedings of the PIRA International Conference on Latest Innovations in Flexible Packaging, Birmingham, November 2002.
6.   CEPI – CONFEDERATION OF EUROPEAN PAPER INDUSTRY: *Special Recycling 2001 Statistics* (October 2002)
7.   EUROPEAN RECOVERED PAPER COUNCIL: 'The European Declaration on Paper Recovery', annual report 2001.
8.   European Norm EN 643:2001, available at the European National Standardization Bodies, e.g. http://catalogo.uni.com/EN/home.html.
9.   Council of Europe Policy Statements concerning paper and board materials and articles intended to come into contact with foodstuffs. Technical Document No 3: *Guidelines on paper and board materials and articles made from recycled fibers, intended to come into contact with foodstuffs*. 21 May 2002.
10.   CEPI – CONFEDERATION OF EUROPEAN PAPER INDUSTRY, *Guide for Good Manufacturing Practice for Paper and Board for Food Contact*, 2002.
11.   CEPE – EUROPEAN COUNCIL OF THE INDUSTRY OF PAINTING, PRINTING INKS AND ART COLORS: *Guide to Optimum Recyclability of Printed Graphic Paper*, March 2002.
12.   COUNCIL DIRECTIVE 89/109/EC, *Official Journal of the European Communities*, N L 40/38 of 11 February 1989.
13.   *Gazzetta Ufficiale della Repubblica Italiana* no. 104, 20 April, 1973.
14.   *Kunststoffe im Lebensmittelverkehr*, Carl Heymanns Verlag KG, edition January 2002.
15.   Verpakingen- en Gebruiksartiklenbesluit, VGB, Koninklije Vermande/SDU Uitgevers, December 2002.
16.   Resolution AP(2002) 1 on 'Paper and Board Materials and Articles intended to

come in contact with Foodstuffs', available at the Council of Europe website http://cm.coe.int/stat/E/public/2002/adopted_texts/ResAP/2002xap1.htm.

17.   APME, ASSOCIATION OF PLASTICS MANUFACTURERS IN EUROPE, *An analysis of plastics consumption and recovery in Western Europe 2000*, Spring 2002.

18.   J. Brueder (IK – Industrieverband Kunststoffeverpackungen), private communication (source: Shoenwald Consulting).

19.   M. Neal, presentation at the European Commission 'Food Contact Experts Group' meeting, 27 November, 2002.

20.   FAIR Project CT 98-4318 (Recyclability), available at *http://www.ivv.fhg.dde/fair*

21.   UK Deptartment of Trade and Industry, Survey 2002, Plastic Recycling Report Recoup, May 2002.

22.   Courtesy of Dr Cesare Anzivino, COREPLA – National Consortium of Collection, Recycling and Recovery of Plastic Waste Packaging, Italy, and the Board of Seriplast Srl, Novate Milanese, Italy.

23.   APME, ASSOCIATION OF PLASTICS MANUFACTURERS IN EUROPE: *Assessing the Eco-efficiency of Plastics Packaging Waste Recovery*.

24.   F. HENSEN, Filtration systems for recyclate processing, in J. Brandrup *et al. Recycling and Recovery of Plastics*, Hanser, 1995.

25.   K. FRITSCH and F. WELLE, *Kunstsoffe*, 92 (2002) 10, 111–114.

26.   I. BOUSTEAD, Theory and definitions in ecobalances, in J. Brandrup *et al. Recycling and Recovery of Plastics*, Hanser, 1995.

27.   P. FINK, Methods and approaches for the evaluation of ecobalances, in J. Brandrup *et al. Recycling and Recovery of Plastics*, Hanser, 1995.

11

Biobased food packaging

V. K. Haugaard, The Royal Veterinary and Agricultural University, Denmark, and G. Mortensen, Arla Foods, Denmark

11.1 Introduction

Traditionally, packaging materials for foods, with the exception of paper and board, have been based on non-renewable materials. However, due to an increasing focus on sustainability, major efforts have gone into development of renewable materials that have been applied successfully for production of packaging materials. At present, the market is based on biopolymers such as starch, cellulose, proteins, and monomers produced from fermented organic materials. So far, most biopolymer-based plastics on the market are used for different purposes such as composting (waste bags), catering (cutlery, drinking cups, plates), agriculture (mulch films, pots), and hygiene utensils (diapers, napkins) (Chandra and Rustgi, 1998; Guilbert, 2000; Bastioli, 2001). However, few references exist relative to application of biobased food packaging materials.

The performance of renewable packaging materials is undergoing progressive improvements as a result of extensive developments at industry and academic levels (Weber, 2000). Application of these novel materials for packaging of foods is challenging, since the demands made upon the packaging materials by the foods are often complex due to specific requirements in terms of oxygen and water vapour permeability, mechanical properties, and safety issues. Several materials have demonstrated favourable properties that may prove useful for packaging of specific foods. Some of the materials are already competitive alternatives to conventional food packaging, polylactate (PLA) being one, whereas other materials such as many starch-based materials still need further optimisation in order to be suitable for packaging of foods. As a result, the commercial use of these materials for packaging of foods is still in its infancy, and few commercially available food products are packaged in these novel materials. However, along with the increasing efforts put into research and development as well as growing environmental consumer awareness, this situation is changing.

This chapter provides an overview of biobased packaging materials for foods and presents the demands made upon biobased packaging materials by the foods. The wide range of biobased materials and properties, scientific and commercial state of the art, suggestions for future use of biobased packaging materials for foods, and future challenges are reviewed from a food packaging point of view. Biobased packaging materials are here defined as materials derived from primarily annually renewable sources. These materials may also be compostable. However, compostability in itself is not a focal point at this stage, since general waste management of compostable materials leaves a lot to be desired. Furthermore, a global standard on labelling is required to include origin of the materials and their disposal. Many of the biobased packaging solutions still contain non-renewable substances due to inferior properties of the materials and for cost saving reasons. However, on-going research and development will entail increased usage of packaging solutions based on fully renewable materials.

The definition used here also includes edible films and coatings and to some extent cellulose-based materials. However, these topics are not discussed in this chapter, because paper-based materials for foods have been used for decades. Furthermore, the use of edible films and coatings for foods has already been described in detail (Kester and Fennema, 1986; Krochta and De Mulder-Johnston, 1997; Miller and Krochta, 1997; Haugaard et al., 2001a). It is interesting to note that the first step in the study and development of biobased packaging materials for food applications was in the field of edible films and coatings (Haugaard et al., 2001a). However, despite the intensive R&D efforts, commercial application of edible films and coatings for foods is still in its infancy.

11.2 Biobased packaging materials

Traditionally, biobased packaging materials have been divided into three types, which illustrate their historical development. First-generation materials consist of synthetic polymers and 5–20% starch fillers. These materials do not biodegrade after use, but will biofragment, i.e. they will break into smaller molecules. Second-generation materials consist of a mixture of synthetic polymers and 40–75% starch. Some of these materials are fully biodegradable. Finally, third-generation materials consist of fully biobased and biodegradable materials (Gontard and Guilbert, 1994).

Polymers derived from renewable resources can be classified into three main categories according to method of production:

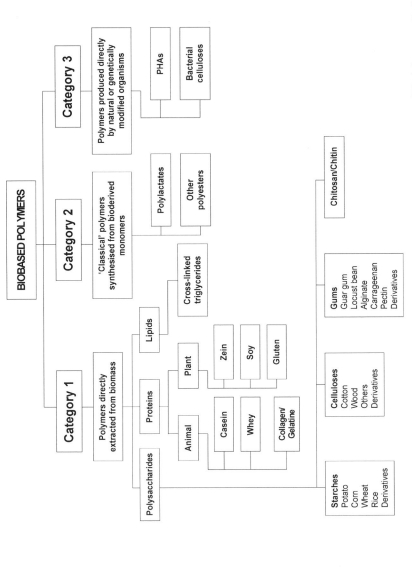

Fig. 11.1 Schematic presentation of biobased polymers based on their origin and method of production (modified from van Tuil *et al.*, 2000).

- Category 1: Polymers directly extracted from biomass
- Category 2: Polymers produced from classical chemical synthesis derived from biomass monomers
- Category 3: Polymers produced directly by natural or genetically modified organisms.

Materials from all three categories either are already used for packaging or have considerable potential for this application. An overview of the three categories is given in Fig. 11.1.

The most commonly available Category 1 polymers are extracted from agricultural or forest plants and trees. Examples include cellulose, starch, pectins, and proteins. These are cell wall, plant storage (starch), or structural polymers. Another example of a Category 1 polymer is chitosan (Fig. 11.2) derived from chitin from the exoskeleton of invertebrates. An interesting property of chitosan in relation to food packaging is its antimicrobial properties (Chandra and Rustgi, 1998). All polysaccharide-based Category 1 polymers are by nature hydrophilic and are somewhat crystalline, both factors causing processing and performance problems, especially in relation to packaging of moist foods. On the other hand, these polymers constitute materials with medium to excellent gas barriers (Petersen et al., 1999; van Tuil et al., 2000). Today, packaging materials based on Category 1 polymers are produced commercially, (see the list of manufacturers in Table 11.1). Due to the hydrophilic nature of many of the Category 1 polymers, many commercially available packaging materials are not fully biobased, but are blended with more hydrophobic polymers (first and second generation materials). An example is the starch-based packaging materials, which for many applications are mixed with the synthetic, but biodegradable, polymer polycaprolactone (PCL). So far, protein- and lipid-based materials have been the subject of limited research efforts with minimal commercial interest. The materials have proved promising within the area of edible films and coatings and may be used to improve selected barrier characteristics.

To date, PLA (Fig. 11.3) is the Category 2 polymer with the highest commercial potential for large-scale production of renewable packaging materials. PLA is a biopolyester polymerised from lactic acid monomers. The monomers may be produced by fermentation of carbohydrate feedstock. The carbohydrate feedstock may stem from agricultural products such as corn, wheat, or waste products, e.g. whey, molasses, or green juice (Garde et al., 2000; Södergård, 2000). PLA has a high potential for packaging applications and

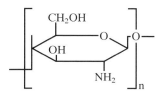

Fig. 11.2   Chemical structure of chitosan.

Table 11.1 Selected manufacturers of (partly) biobased packaging materials (see Section 11.7)

Materials	Suppliers	Trade names (if known)
Based on Category 1 polymers		
Starch and starch-based materials	Novamont (Italy)	Mater-Bi
	EarthShell (USA)	EarthShell
	National Starch (numerous countries)	Eco-FOAM
	Avebe (The Netherlands)	PARAGON
	BIOP Biopolymer (Germany)	BIOPar®
	Biopac (Austria)	
	Biotec® (Germany)	BIOPLAST®
	Plastiroll (Finland)	Bioska
	Apack (Germany)	
	Biologische Verpackung- systeme (Austria)	
	Cabot Plastics International (Belgium)	
	Rodenburg Biopolymers (The Netherlands)	Solanyl
Based on Category 2 polymers		
Polylactate and polylactate- based materials	Cargill Dow (USA)	NatureWorks PLA
	Hycail bv (The Netherlands)	
	Galactic (Belgium)	Galactid
	Trespaphan (Belgium)	Biophan
	Mitsui Chemicals (Japan)	Lacea® PLA
	Biomer (Germany)	
Based on Category 3 polymers		
Poly(hydroxybutyrate)-based materials	Biomer (Germany)	
	Metabolix (USA)	Biopol™
	Zeneca Bio Products (USA)	
	Procter & Gamble (USA)	NODAX™

Fig. 11.3 Chemical structure of polylactate (PLA).

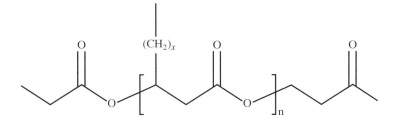

Fig. 11.4   Chemical structure of polyhydroxyalkanoate.

offers numerous opportunities for tailoring the properties of the final packaging solution. Furthermore, PLA packaging materials can be produced without blending with other polymers (third-generation materials). These properties may explain why PLA is the first biobased material produced on a larger scale, as evidenced by the production capacity of 180 000 tons PLA per year in 2002 by Cargill Dow (Rábago, 2002). See Table 11.1 for a compilation of producers.

To date, the Category 3 polymers consist mainly of the poly(hydroxy-alkanoates) (PHAs) (Fig. 11.4), of which poly(hydroxybutyrate) (PHB) is most common. PHAs are accumulated by a large number of bacteria as energy and carbon reserves. PHAs have a promising potential for food packaging applications, since they possess properties close to those of conventional packaging materials (see Section 11.3). However, due to exorbitant production costs, few suppliers exist in the market (Table 11.1). From 1996 to 1999, PHB was produced by Monsanto under the trade name Biopol[TM]. However, the production folded, as prices were prohibitive (ten times higher than those of conventional plastics) (IATP News, 2002). Production technology and the Biopol[TM] trade name were subsequently acquired by Metabolix (USA).

11.3   Requirements for biobased packaging materials

Protection, which is twofold, is often regarded as the primary function of the package. The package must first protect its contents from outside environmental effects such as water, water vapour, gases, odours, microorganisms, dust, shock, vibrations, compressive forces, etc., and second must impede product spill into the environment (Robertson, 1993). Thus, packaging requirements of foods are complex. Unlike inert packaged commodities, foods often constitute dynamic systems with limited shelf-life and specific packaging requirements. In addition, since foods are consumed to sustain life, the need to guarantee safety is a critical dimension of food packaging requirements. Biobased packaging materials must meet the food criteria applying to conventional packaging materials. Consequently, these materials must remain unchanged and function properly until time of disposal. These criteria relate to mechanical and barrier properties (oxygen, carbon dioxide, water, light, aroma). Furthermore, safety aspects (migration), resistance properties (temperature and chemical resistance), production requirements (welding and moulding properties), convenience, optical properties (transparency and haze), and marketing requirements (communication, marking, and printing properties) are important criteria of food packaging materials (Robertson, 1993; Haugaard *et al.*, 2001a).

Contrary to packaging materials based on non-renewable materials, future requirements for biobased materials should include full post-use compostability. Therefore, environmental conditions conducive to biodegradation must be avoided during product shelf-life, whereas optimised conditions for bio-degradation are desirable after discarding. The most important parameters for controlling stability of the biobased packaging materials include appropriate water activity, pH, nutrients, oxygen, storage time, and temperature. Another important aspect is the potential risk of undesirable mould and bacteria growth on the packaging, since some organisms may utilise the biobased materials as an energy source. Furthermore, should the biobased materials be degraded during product shelf-life, exterior microbial migration may occur, thereby con-taminating the food. Such aspects have been studied by Bergenholtz and Nielsen (2002), who evaluated the growth of 17 different food-related fungi on starch/PCL-based films and cups based on starch/PCL, PLA, and PHB. Results showed no significant growth on PLA and PHB to indicate the potential of these materials for food packaging. Growth did occur on the starch-based materials, and therefore the authors suggested a modification, i.e. incorporation of antimicrobial compounds into the packaging material.

Before using biobased materials for food packaging, the aspects listed should be evaluated in order to meet quality and safety requirements of foods and to comply with legal standards. In this context, testing of shelf-life under realistic storage conditions and evaluation of durability and migration aspects of the packaging materials are of vital importance (Haugaard *et al.*, 2001a).

11.3.1   Barrier properties

The proper gas barriers depend on the foods to be packaged. For instance, a high gas barrier is required for packaging of foods in modified atmospheres such as meat products, whereas a relatively low gas barrier is essential when packaging fruits and vegetables, since these produce carbon dioxide during respiration. This carbon dioxide must subsequently exit from the packages to avoid internal anaerobic conditions. Additionally, a high water vapour barrier is required when packaging dry products in order to avoid water absorption by the foods, and in the case of high-moisture products to prevent moisture loss.

The literature provides a vast amount of information on the barrier properties of biobased materials (Gontard *et al.*, 1996; Sinclair, 1996; Psomiadou *et al.*, 1997; Arvanitoyannis *et al.*, 1997, 1998; Kittur *et al.*, 1998; Barron *et al.*, 2001; Lehermeier *et al.*, 2001; Petersen *et al.*, 2001). However, comparisons between the different materials are complicated and often impossible due to the use of different types of processing equipment, variations in raw material parameters, and dissimilar measuring conditions.

Biobased materials mimic quite well the oxygen permeabilities of a wide range of conventional mineral oil-based materials. It is possible to choose from a range of barriers among the biobased materials, and it should be noted that improvements continue to take place (van Tuil *et al.*, 2000).

Packaging films based on PLA, blends of wheat starch and PCL, as well as blends of cornstarch and PCL, possess oxygen transmission rates in the range of 630–2640 ml/m^2/24 h, which is lower than those of low-density polyethylene (LDPE) and high-density polyethylene (HDPE) films measured at similar conditions and thicknesses, being 8800–9200 ml/m^2/24 h (Petersen *et al.*, 2001). Thus, these biobased packaging films provide a better protection against oxygen than do LDPE and HDPE. Chitosan films also offer better protection against oxygen than do LDPE and wheat gluten films. Kittur *et al.* (1998) found that chitosan was comparable to methyl cellulose/glycol and cellulose films, but inferior to polyester and polyamide (PA) films, whereas van Tuil *et al.* (2000) reported oxygen barriers of chitosan in the range of that of ethylene vinyl alcohol (EVOH). Generally, starch, protein, and chitosan-based materials have high oxygen barriers and may provide less expensive alternatives to the gas barrier materials presently applied, such as EVOH and PA. For comparison, an equivalent biobased laminate could consist of an outer layer of plasticised chitosan, protein or starch-derived film combined with PLA or PHA (van Tuil *et al.*, 2000).

Oxygen permeabilities of cups based on PLA and PHB are lower than those of polypropylene (PP), polystyrene (PS), and polyethylene (PE) (Petersen *et al.*, 2001). In general, the PLA properties resemble those of PS (Sinclair, 1996; Lehermeier *et al.*, 2001). Furthermore, PLA can be plasticised in many ways, resulting in properties which will match those of polymers such as LDPE, linear low-density polyethylene (LLDPE), and PS (Sinclair, 1996). Thus, PLA seems to be a very promising food packaging polymer.

In general, oxygen permeability and permeability of other specific gases are closely interrelated and, as a rule of thumb, mineral oil-based polymers have a fixed carbon dioxide:oxygen ratio of approximately 6:1 (Robertson, 1993). Studies have shown that this ratio does not apply to all biobased materials (see Table 11.2). Permeability ratios ranging from 4:1 to 14:1 for starch/PCL films, 7:1 for PLA films (Petersen *et al.*, 2001) and 15:1 for wheat gluten films have been reported in the literature (Barron *et al.*, 2001). An increase in the ratio at higher relative humidities implies that the films gradually become more

Table 11.2 Permeability ratios between carbon dioxide and oxygen of mineral oil and biobased films

Films	CO$_2$:O$_2$ ratios	References
Mineral oil-based films	Approximately 6:1	Robertson (1993)
Polylactate	7:1	Petersen *et al.* (2001)
Starch–polycaprolactone	4:1 to 14:1	Petersen *et al.* (2001)
Wheat gluten	15:1	Barron *et al.* (2001)

permeable to carbon dioxide as compared to oxygen (Gontard *et al.*, 1996), which may be desirable for some foods.

A major challenge is the hydrophilic behaviour of many biobased polymers, as many foods demand materials resistant to moist conditions. However, when comparing the water vapour permeability of various biobased packaging materials to materials based on mineral oil, it becomes evident that it is indeed possible to produce biobased materials with water vapour permeabilities comparable to those of certain conventional plastics. However, if a high water vapour barrier is required, few biobased materials apply. Development efforts are currently directed at this challenge, and hence, future biobased materials must mimic the water vapour barriers of conventional materials used today (van Tuil *et al.*, 2000).

The water vapour barriers of especially starch, protein, and chitosan-based materials need to be upgraded in order to offer alternatives to conventional food packaging materials (Petersen *et al.*, 1999; Guilbert, 2000; van Tuil *et al.*, 2000; Kantola and Helén, 2001). Increasing the amounts of starch in the packaging materials will result in increased water vapour permeability (Kim and Pometto III, 1994; Arvanitoyannis *et al.*, 1997; Psomiadou *et al.*, 1997). The water vapour permeabilities of blends of starch and oil-based films and cups have been compared to those of LDPE and HDPE films as well as PP, PE, and PS cups of identical thickness. Results showed that the water vapour permeabilities of the starch-based films were 4–6 times greater than those of conventional films. The starch-based cups had water vapour permeabilities of 100–300 times greater than those of PP and PE cups, and 5–9 times greater than those of PS cups. PLA provided a better water vapour barrier than did the conventional materials evaluated, i.e. four times higher than the conventional films, two times higher than PS cups, and 40–60 times higher than PP and PE cups (Petersen *et al.*, 2001). A very interesting property of PHAs with respect to food packaging applications is their low water vapour permeability, which resembles that of LDPE (van Tuil *et al.*, 2000). However, Petersen *et al.* (2001) found water vapour permeabilities of PHB cups in the range of starch-based cups.

Due to the hydrophilic nature of many of the biobased materials, their gas barrier properties depend on the humidity conditions during measuring, and the gas permeability of hydrophilic biobased materials may increase many times along with the increase in humidity (Krochta and De Mulder-Johnston, 1996; Despond *et al.*, 2001). This phenomenon is also seen with respect to conventional polymers such as PA and EVOH. Therefore, PA and EVOH are usually sandwiched between outer layers of e.g. LDPE, creating an effective water vapour barrier, which ensures that moisture does not interfere with the properties of PA or EVOH. In the same fashion, PLA and PHA may also protect the moisture-sensitive gas barrier provided by chitosan, polysaccharide, or protein. Developments have made it possible to improve water vapour and gas barrier properties of biobased materials substantially by plasma deposition of glass-like silicium oxide coatings or by applying nanocomposites from natural polymers and modified clay (Fischer *et al.*, 2000; Johannson, 2000). The gas barriers of PLA and PHB are not expected to be dependent on humidity (van Tuil *et al.*, 2000).

Aroma barrier is an important food packaging parameter. However, data is lacking on the aroma permeability of most packaging materials (Robertson, 1993). Furthermore, light barrier properties are important, but for most polymers these properties can be modified to match the requirements of the foods. With respect to the resistance of the packaging materials towards UV light, a study has shown that the resulting decrease in physical integrity and degradation of the polymer were much lower for PLA than for PE (Ho and Pometto III, 1999).

11.3.2   Mechanical properties

Most biobased polymer materials have mechanical properties similar to conventional polymers. Thus, mechanical properties can be altered by the processing method and selection of raw materials. This indicates that PS and polyethylene terephthalate (PET)-like polymers (relatively stiff materials) and PE-like polymers (relatively flexible polymers) can be found among the available biobased polymers (van Tuil *et al.*, 2000). The literature lists different comparisons, e.g. the tensile strength of blends of synthetic polymers (PE) and starch was found to be lower than that of PE (Holton *et al.*, 1994; Psomiadou *et al.*, 1997; Arvanitoyannis *et al.*, 1997, 1998). Furthermore, tensile strength, percentage elongation, and tear strength of starch/PCL films and PLA films were found to be lower than for LDPE and HDPE films (Petersen *et al.*, 2001). For cups based on starch, PLA, and PHB, compression was observed to be in the range of that of PP, PS, and PE cups (Krochta and De Mulder-Johnston, 1996; Petersen *et al.*, 2001). In general, PHB resembles isotactic PP with respect to mechanical behaviour (van Tuil *et al.*, 2000). Moreover, Ahvenainen *et al.* (1997) found that native starch alone does not provide sufficient mechanical strength and stability for food packaging applications, and a negative effect on the mechanical properties was pinpointed when the starch content was increased in LDPE/starch blends (Psomiadou *et al.*, 1997; Arvanitoyannis *et al.*, 1998).

Eventually, mechanical properties of most biobased and oil-based polymers can be tailored to meet the desired mechanical properties by means of plasticising, blending with other polymers or fillers, cross-linking, or addition of fibres (van Tuil *et al.*, 2000). Thus, a positive correlation can be obtained between the amount of plasticiser in the polymer and the tensile strength (Parris *et al.*, 1995). Furthermore, with respect to PLA, improved mechanical strength and heat stability can be obtained by polymer orientation during processing (Sinclair, 1996). Hence, the mechanical properties of PLA can be adjusted to a large extent, ranging from soft and elastic to stiff and high strength, e.g. by varying crystallinity and the molecular weight of the polymer (Södergård and Stolz, 2002). However, the mechanical properties of the PLA polymers are sensitive to heat and humidity. Studies have shown that PLA films lost mechanical properties faster when stored at high temperatures and humidities than when stored at low temperatures and humidities (Ho *et al.*, 1999a,b).

11.3.3   Thermal resistance

The usage temperature range of biobased materials with respect to food packaging is limited. For instance, stability of PLA is poor at elevated temperatures (Södergård and Stolz, 2002), and PLA cups will remain stable only up to a temperature of 55°C (Anon, 1997, 1998). A supplier of the two commercial products, PLA and PHB, states that the usage temperatures range from −10°C to 50°C and from −30°C to 120°C, respectively (Biomer, 2002). Furthermore, Rasmussen and Olsen (2001) found that deformation of cups and trays based on starch/PCL blends occurred between 60°C and 90°C. However, it is generally difficult to obtain information on the usage temperatures of biobased packaging materials for foods.

11.3.4   Migration

Migration is an important aspect to consider when employing packaging materials for foods, and the overall migration may not exceed the 10 mg/dm^2 limit. In general, because of the natural origin of the polymers and the fact that some of the raw materials are present in the foods, i.e. starch and lactic acid, many of the biobased packaging materials are considered safe for food packaging purposes. Thus, the categories lactic acid, edible and hydrolysed starch, and PHB, are approved for use in the manufacturing of materials and articles (Commission Directive, 1990). However, it is imperative that the final food packaging material is safe.

Conn et al. (1995) observed a limited overall migration of PLA into aqueous, acidic, and fatty food simulants. Therefore, the authors concluded that PLA is 'Generally Recognized as Safe' (GRAS) for its intended purpose as a polymer for manufacturing articles used for containment or as food packaging material. Likewise, Selin (1997) found that migration from PLA film was less than 1 mg/dm^2 (water, 3% acetic acid, 15% ethanol, and olive oil).

11.4   Using biobased packaging with particular foods

To date, considerable resources have been allocated to research, development, and pilot-scale studies, but usage of biobased packaging materials within the food industry remains limited. Technical packaging considerations as well as marketing aspects are important criteria when selecting a given packaging material or technique. These criteria are illustrated by numerous feasibility studies carried out for small and large food companies encompassing both technical and market-oriented aspects. However, the studies are confidential and are not available to the public (Haugaard et al., 2001a).

The potential for using biobased materials for food packaging has been evaluated both from a packaging material point of view, i.e. by conclusions drawn from the properties of the packaging materials including biodegradability, and from a food quality and shelf-life point of view. Examples of scientific studies are listed in Table 11.3 and commented on below.

Table 11.3   Scientific studies on packaging of foods in (partly) biobased materials

Packaging materials	Products	References
Polylactate (PLA)	Orange juice	Haugaard *et al.* (2001b, 2002, 2003)
	Mushrooms	Haugaard and Festersen (2000)
	Salad dressing	Haugaard *et al.* (2003)
	Yoghurt	Frederiksen *et al.* (2003), Junkkarinen (2002)
PLA-coated paperboard trays overwrapped with perforated starch bags	Organic tomatoes	Kantola and Helén (2001)
Polyhydroxybutyrate (PHB)	Orange juice	Haugaard *et al.* (2001b)
	Salad dressing	Haugaard *et al.* (2003)
PHB-coated paperboard trays overwrapped with perforated starch bags	Organic tomatoes	Kantola and Helén (2001)
Starch/polyethylene films	Ground beef	Holton *et al.* (1994), Kim and Pometto III (1994)
	Lean beef and bologna	Strantz and Zottola (1992)
	Bread	Holton *et al.* (1994)
	Broccoli	Holton *et al.* (1994)
Starch-based bags	Fruits and vegetables, bread	Bastioli (2001)
Starch laminate	Cheese	van Tuil (2000)
	Cut vegetables	van Tuil (2000)
	Muesli	van Tuil (2000)
	Dry products	Bastioli (2001), van Tuil (2000)
Laminate of chitosan, cellulose, and poly-caprolactone	Fresh produce	Makino and Hirata (1997)
Chitosan film and cellulose-based box	Mangoes	Srinivasa *et al.* (2002)
Zein	Frozen products	Padua *et al.* (2000)
Wheat gluten film	Mushrooms	Barron *et al.* (2001)

Makino and Hirata (1997) studied the applicability of a laminate film based on chitosan, cellulose, and PCL for modified atmosphere packaging of horticultural commodities. The suitability of the laminate for modified atmosphere packaging of head lettuce, cut broccoli, whole broccoli, tomatoes, and sweet corn was evaluated by measurements and calculations of gas permeabilities. The results encouraged use of the laminate for the studied products within a 10–25°C temperature range, since the permeability values of the laminate were equal to those of LDPE films preferred for fresh produce. Use of chitosan has also been examined in a storage study of mangoes packaged in cellulose-based boxes with top surfaces being covered with either chitosan film or LDPE as a reference material (Srinivasa *et al.*, 2002). The results showed that mangoes stored in

LDPE films developed off-flavours due to fermentation and fungal growth on the stalk and around the fruit. The mangoes were partially spoiled when packaged in LDPE. Mangoes stored in chitosan-covered boxes showed an extension of shelf-life of up to 18 days, without microbial growth or off-flavour formation. Thus, the authors concluded that chitosan films are a useful alternative to synthetic films for storage of mangoes (Srinivasa et al., 2002). However, the results also showed that weight loss of mangoes stored in chitosan films was considerably higher than when stored in HDPE due to poor water vapour characteristics.

Use of PLA for packaging of fresh, unpasteurised orange juice with a shelf-life of 10 days was studied in a storage experiment at 4°C under fluorescent light, to determine colour changes, loss of ascorbic acid, and limonene scalping. For comparison, HDPE and PS reference cups were used. Results showed that PLA provided a better protection against all the measured quality changes than did HDPE and at least as good a protection as did PS (Haugaard et al., 2002). These results are in accordance with the findings by Haugaard et al. (2001b, 2003) indicating that the quality of pasteurised orange juice and orange juice simulant packaged in PLA and PHB cups was equal to that of orange juice packaged in conventional materials (HDPE and PS cups). Thus, the fact that PLA properties resemble those of PS (Sinclair, 1996; Lehermeier et al., 2001) furthermore resulted in equal protection against quality changes.

The potential of using PLA for yoghurt has been studied by Frederiksen et al. (2003), who concluded that PLA cups offered a better protection against lipid oxidation and loss of β-carotene and riboflavin than did the conventionally used PS cups.

Use of PLA and PHB for packaging of high-fat foods has been demonstrated in a packaging experiment, in which salad dressing was packaged in PLA and PHB cups and as a reference in HDPE cups (Haugaard et al., 2003). The biobased cups were as effective as were the HDPE cups for protecting the salad dressing from colour changes and loss of α-tocopherols, whereas the biobased cups provided a better protection against lipid oxidation than did the HDPE cups (measured by formation of lipid hydroperoxides and secondary lipid oxidation products).

Quality changes in organic tomatoes packaged in PLA- and PHB-coated paperboard trays overwrapped with perforated starch-based bags were studied for 22 days and compared to LDPE packages (Kantola and Helén, 2001). The biobased materials offered the same protection against quality changes as did the conventional ones. However, the tomatoes lost more of their weight in the biobased packages than in the LDPE packages, which was identified as the limiting factor. Similar results were obtained in a study on packaging of mushrooms in a wheat starch/PCL film and a PLA film (Haugaard and Festersen, 2000). For comparison, a conventional packaging material was used. Results indicated that PLA provided the same protection against colour changes as did the reference. Mushrooms packaged in the wheat starch-based film lost lightness (change in L^* values) to a higher degree than when applying the reference material. However, mushrooms packaged in the biobased materials lost more weight than the reference samples, indicating that optimisation of the materials must take place before they can be used for packaging of mushrooms and other high-respiring produce (Haugaard and Festersen, 2000). Recently, Barron *et al.* (2001) have demonstrated that the use of wheat gluten film may actually be advantageous for storage of mushrooms. The wheat gluten film more efficiently removed carbon dioxide during storage of modified atmosphere packaged products than did a conventional film due to a higher carbon dioxide permeability of the wheat gluten film. However, the oxygen level in the headspace decreased to very low levels, which could result in detrimental anaerobic conditions. Quality assessments (cap opening and lightness of the mushrooms) showed that the gluten film delayed cap opening compared to the reference film and that colour was not affected by type of packaging.

Cornstarch-containing PE films have been studied for food packaging purposes. Holton *et al.* (1994) evaluated the suitability of a PE film containing 6% cornstarch (first-generation material, see Section 11.2) compared to a conventional PE film for packaging of broccoli, bread, and ground beef kept under normal storage conditions. The type of packaging film did not affect the evaluated parameters, i.e. bread staling, broccoli colour, and lipid oxidation of ground beef. However, a significant loss of elongation occurred in cornstarch-containing PE film, which could be ascribed to interactions between the film and free radicals developed during oxidation of ground beef during frozen storage. Therefore, the authors discouraged the use of the film for high-fat foods. Inconsistent results were found when packaging broccoli and bread in the cornstarch film. Therefore, the authors recommended that cornstarch-containing PE films be used only for packaging of moist and low-fat foods.

From a microbiological point of view, cornstarch-containing PE films have been suggested for use as primary containers for beef and bologna. Strantz and Zottola (1992) evaluated the effect of cornstarch in PE films on bacterial survival in various culture media under food storage conditions. They found that survival of *Salmonella typhimurium*, *Staphylococcus aureus*, and *Bacillus cereus* generally was not enhanced by the presence of cornstarch (6%) in lean beef and bologna. They also examined the migration of the same bacteria by inoculating the exterior of the packaging material with bacteria but found no migration of bacteria through either the cornstarch-containing PE film or the PE film itself. These results are consistent with those of Kim and Pometto III (1994), who found that starch addition (0–28%) did not accelerate microbial growth in ground beef. Both studies established that cornstarch-containing PE films, from a microbiological safety standpoint, could be used successfully for food packaging insofar as microbiological safety was concerned. From a packaging material and food quality point of view, Kim and Pometto III (1994) also found that the mechanical properties of the films applied and the colour stability of the ground beef did not change significantly after refrigerated or frozen storage.

As can be seen from the scientific state of the art, the numbers of research papers are few. However, papers do show that potential applications of biobased food packaging are many.

Table 11.4 Examples of commercial use of (partly) biobased packaging materials for foods

Packaging materials	Products	Example of producers	Example of end-users
Polylactate (PLA) cups	Organic yoghurt Fresh salads		Danone (Germany) McDonald's (Sweden)
PLA-coated paper cups and lids	Hot and cold beverages	Huhtamaki Van Leer (Finland)	Coca-Cola Co.: Olympic Games, Salt Lake City (USA)
Clear PLA cups and lids	Cold beverages	Autobar (France) Bartling GmbH (Germany) Fabri-Kal Corporation (USA) Fardis (Belgium) Rexam Thin Wall Plastics (Sweden) Termoplast S.r.L. (Italy) Biocorp (USA) Autobar (Spain)	
PLA punnets PLA trays and films PLA trays	Fruit and produce Fresh foods and pasta Mushrooms		Supermarket chain: IPER (Italy) Modellprojekt Kassel, e.g. Tegut supermarket chain (Germany)
PLA in general PLA films	Plates, cutlery and straws Bell peppers and cookies	Biocorp (USA) Trespaphan (Belgium)	Modellprojekt Kassel, e.g. Tegut supermarket chain (Germany)
Starch and starch-based materials	Hamburgers and sandwiches Dry products	Apack (Germany) Biologische Verpackungsysteme (Austria) EarthShell (USA)	McDonald's (USA)
Starch, limestone, and fibre mixtures	Plates, bowls, food service wraps and salad containers Dry pasta		Used in Italy
Starch and paper/pulp mixtures	Meat products French fries and chips	Apack (Germany)	Used in Belgium McDonald's (Sweden and Austria) Fast food restaurants
Starch films laminated on paper	Cutlery, cups and plates		Canteens
Starch and fibre-based trays Reinforced nets of starch blends	Organic fruits and vegetables Onions	Apack (Germany) Novamont (Italy)	Sainsbury supermarket chain (UK) Modellprojekt Kassel, e.g. Tegut supermarket chain (Germany)

11.5   Current commercial applications

Examples of commercial use of biobased materials for foods are given in Table 11.4 and commented on below. For many years, coated cellophane and cellulose acetate have been commercially utilised for food packaging. Coated cellophane is used for, e.g., baked goods, fresh produce, processed meat, cheese, and candy. Cellulose acetate is used mainly for baked goods and fresh produce (Krochta and De Mulder-Johnston, 1997). The moisture and gas barrier properties of cellulose acetate are not optimal with respect to food packaging. However, the film is excellent for high-moisture products as it allows for respiration and reduces fogging (Hanlon, 1992).

PLA-based packaging materials are now commercially available (see Table 11.1). One of the first commercial applications of PLA for foods was packaging of organic yoghurt in thermoformed PLA cups. This product was introduced on the German market by Danone (Anon., 1998) with technical success (Bastioli, 2001). However, the market introduction failed, and the product was subsequently withdrawn from the market. Use of PLA for packaging of organic yoghurt as well as milk has also been positively evaluated by the Finnish dairy company Valio Ltd, who will consider the use of PLA for like products, when prices are reduced to a competitive level (Junkkarinen, 2002). The fast food chain, McDonald's, is also using PLA cups for packaging of their fresh salads. Furthermore, PLA trays and films are used for packaging of fresh foods and pasta by the Italian supermarket chain, IPER. Other commercial applications of PLA include disposable bags (e.g. for bakery products) and food service items including hot and cold drink cups based on either pure PLA or PLA-coated paper. A commercial example of this is a corn-based PLA cup for cold drinks produced by Biocorp (USA), a manufacturer of compostable and biodegradable materials. Biocorp also supplied biodegradable and compostable cutlery for the Summer Olympic Games in Sydney, 2000.

The properties of PHB make the material interesting for food packaging, which is easily seen from the numerous scientific studies. However, lack of commercial use of PHB within the food industry is evident and is ascribable to exorbitant costs. Therefore, unless prices reach an acceptable level, PHB is not expected to appear on the food packaging market.

Starch-based materials within the food market segment are mainly used for bags for bread, fruits, and vegetables. The main advantage of using starch-based materials compared to traditional materials for packaging of fruits and vegetables is the breathability of the materials, which improves storage conditions (Bastioli, 2001). An example is the use of potato starch and fibre-based trays for packaging of organic fruit and vegetables, e.g. by the British supermarket chain Sainsbury. Starch-based materials are used for fast food applications, since few requirements are established for, e.g., the barrier properties of the packaging materials. In Belgium, starch-containing packaging is used commercially for french fries. Furthermore, EarthShell (USA) has developed foams based on organic/inorganic composite material containing starch, limestone, and reinforced wood pulp fibre. The materials are used to package dry and moist foods (Andersen *et al.*, 1999). Commercial examples include sandwich and salad containers and food service wraps. Other examples of use of starch-based materials for foods include disposable food service items and paper coatings (Krochta and De Mulder-Johnston, 1997). Furthermore, cups, plates, cutlery, and other containers, laminated or extrusion-coated with starch and PLA-based coatings, are available on the market for hot and cold liquids. An example is paper cups coated with PLA, which were used by the Coca-Cola Company at the Winter Olympic Games in Salt Lake City, 2002 (Bastioli, 2001).

In Kassel, Germany, a test market evaluation on compostable and biodegradable packaging has run since May 2001. A range of products packaged in materials such as PLA and starch have been distributed by the retailers. Applications include fresh produce packaged in compostable trays and flexible pouches, compostable bags, organic waste bags, single service cutlery and dishes, as well as cornstarch-based packaging materials for flowers (Modellprojekt Kassel, 2002a). Establishment of additional test markets are in the pipeline.

As is evident from the commercial state of the art, actual commercial applications are few, among other reasons due to high costs. Furthermore, actual applications are mainly used for foods with few packaging requirements, and for niche products such as disposable tableware, where environmental considerations are important (Guilbert, 2000).

11.6   Future trends

The market for biobased food packaging materials is expected to incorporate niche products, where the unique properties of the biobased materials match the food product concept (Weber *et al.*, 2002). Packaging of high-quality products such as organic products, where extra material costs can be justified, may be a starting point. Biobased materials are not expected to replace conventional materials on a short-term basis. However, due to their renewable origin, they are indeed the materials of the future (Weber, 2000). In the long term, the biobased materials are expected to compete with the conventional ones, when the properties and costs resemble conventional materials.

In the following, potential markets are pinpointed. Applications not yet thought of are likely to surface along with further development and optimisation of the materials. In the short term, biobased materials will most likely find usage for foods requiring short-term chill storage due to the biodegradability of the materials. Examples include fruits and vegetables, since biobased polymers facilitate production of films with variable carbon dioxide, oxygen, and moisture permeabilities (Haugaard *et al.*, 2001a). The high carbon dioxide to oxygen ratio of selected biobased packaging materials is interesting in relation to respiring foods. Cheeses, fruits, and vegetables produce carbon dioxide, which accumulates in the package headspace, resulting in inflation of the package. Therefore, using biobased materials for such products may result in improved quality and longer shelf-life. Furthermore, short shelf-life products with few gas and water barrier requirements may successfully be packaged in biobased materials (Weber et al., 2002).

High-value foods such as mushrooms, minimally processed salads, unpasteurised orange juice, organic foods with short shelf-life, etc. constitute a potential food group for biopackaging, since these foods typically are expensive and targeted at consumers who are willing to pay a premium for such products.

With biodegradability being a main issue, potential applications include packaging for short shelf-life foods such as fast food packaging or egg boxes, fresh or minimally processed fruits and vegetables, dairy products, organically grown products, etc. (Guilbert, 2000). Biodegradability is also an interesting property of food packaging materials, e.g. for military use with applications including utensils, trash bags, drink cups, and lids (Yang, 1995).

Biobased packaging for foods may also provide new possibilities for modified atmosphere packaging. Biobased packaging may create new atmosphere conditions, which improve the quality of specific food products during shelf-life, e.g. vegetables (Petersen et al., 1999). However, barrier properties of existing biobased packaging materials remain insufficient for modified atmosphere packaging, and developments are called for.

The above scientific and commercial overviews indicate great potential for using PLA for food packaging. It is expected that PLA will be suitable for dairy products other than yoghurt, to include milk, sour cream, and fresh and ripened cheeses. PLA, PHB, and paperboard cartons coated with PLA or PHB have been suggested for dairy products with short shelf-life (Haugaard et al., 2001a). Cargill Dow recommends the use of PLA for packaging of candy, fast food products, fresh foods, ice cream, cereals, coffee, and snack foods. Furthermore, market tests of PLA thermoformed trays and bi-oriented films for food packaging are in progress, and such applications may prove to be promising (Bastioli, 2001).

The novel hydrophilic, starch-based materials would be best suited for packaging of dry or frozen foods, since the materials do not come into contact with moisture. The materials may also be used as lipid barriers in multi-layer materials (Ahvenainen et al., 1997). Other suggestions include starch-based trays for packaging of fresh meats and trays and films for fruits and vegetables (Haugaard et al., 2001a). Furthermore, it is now possible to produce starch-based foams, which for some applications could replace expanded PS (Bastioli, 2001).

Extensive research and development on biopolymers is required before the materials become competitive alternatives with respect to food packaging (Ahvenainen et al., 1997). Pending issues are pinpointed below.

The materials must be tailored to specific food applications in order to fulfil the requirements of the foods. However, few packaging materials on the market today have been used for direct food contact, and assessment of package–product compatibility during realistic food storage experiments are called for. Consequently, tailoring of the biobased materials to the individual foods may become necessary. Furthermore, in many cases, it is necessary to laminate materials in order to improve the overall performance with respect to food packaging. Thus, materials based on starch, proteins, and chitosan may prove successful for packaging of foods, if they become less susceptible to water uptake. Modification of the polymers is discussed in Section 11.3.1 and includes silicium oxide coatings and incorporation of nanocomposites produced from natural polymers and modified clay (Fischer *et al.*, 2000; Johannson, 2000). Modification will obviously increase the cost of the end-product.

The cost of the packaging material also depends on the production scale, i.e. a lower cost is anticipated in the case of large-scale production. Today, the costs of many biopolymers almost equal or exceed those of PET and PA, with the exception of PHAs, which cost ten times as much as conventional plastics. It is not fair to compare costs to those of conventional polymers, since the materials are produced on very different scales. However, the cost of PLA is now €1.50–4.00 per kg compared to €1.00–2.00 for PE and PS, the reason being that Cargill Dow now produces PLA on a 180 000 tonne scale. Thus, when PLA is produced on an even larger scale, the costs are expected to drop further, thereby approaching the costs of PE and PS. Additionally, the issues of high-performance, biobased adhesives, inks, and additives still require undivided attention.

In order to ensure consumer acceptance, it is important to study consumer reactions to foods packaged in biobased materials. Little information is available in this area. However, in the Modellprojekt Kassel (see Section 11.5), consumer acceptance is being evaluated. So far, results indicate a high level of consumer acceptance. Almost 90% of the citizens of Kassel rate the idea as good or excellent, and the vast majority would even accept increased product prices (Modellprojekt Kassel, 2002b); 80% of the consumers who purchased the new products rated the quality as high and stated that they intended to purchase the products again. Furthermore, the survey showed that 33% of the consumers would pay a premium for the products. For instance, a yoghurt container may be a maximum of €0.05 more expensive; a compostable plastic bag would cost €0.15 instead of €0.10. Unfortunately, further increases in prices are expected to reduce the demand for biobased materials (East and Europe, 2002).

The biodegradability of most biobased packaging materials represents a challenge to retailers, since the packages should remain stable during the product's shelf-life and subsequently biodegrade efficiently at the time of disposal. Therefore, it is important to identify the shelf-life stability of biobased packaging materials. Temperature stability is another important aspect in order to ensure stability and high quality during processing and storage. As discussed in Section 11.3.3, many of the biobased materials degrade at relatively low temperatures, and therefore it is important to use these new materials for products not requiring, e.g., hot-filling or for products to be heated directly in the package. However, new developments may result in increased temperature resistance, and thus a broad spectrum of use will manifest itself.

Biobased packaging materials must comply with food and packaging legislation, and interactions between food and packaging material cannot compromise food quality or safety. In principle, biobased and conventional materials are treated equally in the European food contact material legislation. However, the biobased and the conventional materials may in fact interact differently with the surroundings, an example being the potential interaction between living organisms and the biobased material itself (see Section 11.3). Therefore, preparation of new test standards should be assessed on an international level. Additionally, issues of food safety and interactions between foods and biobased packaging materials must be addressed.

The Modellprojekt Kassel indicated that consumers are interested in these novel materials, and that the mere presence in the market of such materials will trigger increased curiosity at consumer, retailer, producer and legislator levels. This again will boost research and development activities in the industry and academic environments. The International Biodegradable Polymers Association and Working Groups (IBAW) predicts that if overall developments remain positive, the capacity in 2010 will be approximately 1 million tonnes of biodegradable materials, of which over 60% will stem from renewable resources (Käb, 2002). Novamont recently invested in a larger production facility, and the company foresees a 30% market growth in the years to come (Bastioli, 2002). Other suppliers are also making major capital investments. Thus, biobased packaging is picking up momentum, and suitable food packaging applications seem only steps away.

11.7   Sources of further information and advice

For a comprehensive overview of potential food applications of biobased packaging materials, see Haugaard *et al.* (2001a).

Selke (1996) is an excellent reference in the field. This book provides the groundwork for further elaboration.

The report 'Biobased Packaging Materials for the Food Industry. Status and Perspectives' (Weber, 2000) and the 'Conference Proceedings, The Food Biopack Conference' cited in this chapter can be downloaded from http://www.mli.kvl.dk/foodchem/special/biopack.

Information on EU financed food packaging projects can be located on the website http://cpf.jrc.it/webpack/projects.htm.

IENICA, the Interactive European Network for Industrial Crops and their Applications, is an EU project aimed at linking independent organisations and initiatives involved in the development of renewable materials. Further information is available on http://www.ienica.net.

The website http://biopolymer.net provides information on materials, products, news, etc. in relation to biopolymers.

A database provided by the Dutch Department of Agrotechnology and Food Science, http://www.ftns.wau.nl/agridata, lists information on processing, modification, etc. of biodegradable plastics.

A list of vendors of degradable polymers appears on the website http://www.moea.state.mn.us/berc/dfe/polyvend.pdf.

Information about Modellprojekt Kassel is provided on the website http://www.modellprojekt-kassel.de.

Packaging Network.com (http://www.packagingnetwork.com) provides information for the packaging industry, including products, suppliers, news, etc.

Pira (http://piranet.com) is a commercial consultancy business specialising within the packaging, paper, printing, and publishing industries. The company also provides information on biobased and biodegradable packaging materials.

Information on chitin and chitosan and related links can be found on the websites http://user.chollian.net/chitin/intro and http://primex.no/htm/links.

IBAW, the International Biodegradable Polymers Association and Working Groups, focuses on catalysing development of the biopolymer market. Further information on their activities can be obtained from http://www.ibaw.org.

Suppliers and converters of biobased materials:

- Apack: http://www.apack-ag.de
- Autobar Packaging: http://www.autobar.com
- Avebe: http://www.avebe.com
- Biocorp: http://www.biocorpna.com
- Biomer: http://www.biomer.de
- Biopac: http://www.biopac.com
- BIOP Biopolymer GmbH: http://www.tu-dresden.de
- Biotec®: http://www.biotec.de
- Cargill Dow: http://www.cargilldow.com
- EarthShell: http://www.earthshell.com
- Metabolix: http://www.metabolix.com
- National Starch: http://www.nationalstarch.com
- Natura Verpackungs GmbH: http://www.innovation-in-packaging.com
- NODAX™: http://www.nodax.com
- Novamont: http://www.materbi.com
- Plastiroll: http://www.plastiroll.fi
- Trespaphan: http://www.trespaphan.com
- Zeneca Bio Products: http://www.zeneca.com

11.8   References

AHVENAINEN R, MYLLÄRINEN P and POUTANEN K (1997), 'Prospects of using edible and biodegradable protective films for foods', *The European Food & Drink Review*, Summer, 73–80.

ANDERSEN P J, KUMAR A and HODSON S K (1999), 'Inorganically filled starch based fiber reinforced composite foam materials for food packaging', *Mat Res Innovat*, 3, 2–8.

ANON. (1997), 'Danone setzt auf Öko-Verpackung', *Welt der Milch*, 51, 882.

ANON. (1998), 'Focus on natural fibre', *Food Process*, 58(2), 38.

ARVANITOYANNIS I, PSOMIADOU E, BILLADERIS C G, OGAWA H, KAWASAKI N and NAKAYAMA A
O (1997), 'Biodegradable films made from low density polyethylene (LDPE),
ethylene acrylic acid (EAA), polycaprolactone (PCL) and wheat starch for food
packaging applications: Part 3', *Starch/Stärke*, 49(7/8), 306–322.

ARVANITOYANNIS I, BILLADERIS C G, OGAWA H and KAWASAKI N (1998), 'Biodegradable
films made from low-density polyethylene (LDPE), rice starch and potato starch for
food packaging applications: Part 1', *Carbohydr Polym*, 36, 89–104.

BARRON C, VAROQUAUX P, GUILBERT S, GONTARD N and GOUBLE B (2001), 'Modified
atmosphere packaging of cultivated mushroom (*Agaricus bisporus* L.) with
hydrophilic films', *J Food Sci*, 66 (8), 251–255.

BASTIOLI C (2001), 'Global status of the production of biobased packaging materials',
Starch/Stärke, 53, 351–355.

BASTIOLI C (2002), 'Materials from renewable resources: the future of the biodegradable
plastics industry'. In *Conference Proceedings Biodegradable Plastics 2002*,
Frankfurt 4–5 December.

BERGENHOLTZ K P and NIELSEN P V (2002), 'New improved method for evaluation of growth
by food related fungi on biologically derived materials', *J Food Sci*, 67(7), 2745–
2749.

BIOMER (2002), http://www.biomer.de/MechDatE.html (accessed November 2002).

CHANDRA R and RUSTGI R (1998), 'Biodegradable polymers', *Prog Polym Sci*, 23, 1273–
1335.

Commission Directive of 23 February 1990 relating to plastics materials and articles
intended to come into contact with foodstuffs (98/128/EEC). Amended by
Commission Directive 1999/91/EC of 23 November 1999.

CONN R E, KOLSTAD J J, BORZELLECA J F, DIXLER D S, FILER JR L J, LADU B N and PARIZA M W
(1995), 'Safety assessment of polylactide (PLA) for use as a food-contact polymer',
Fd Chem Toxic, 33(4), 273–283.

DESPOND S, ESPUCHE E and DOMARD A (2001), 'Water sorption and permeation in chitosan
films: relation between gas permeability and relative humidity', *J Polym Sci*, 39,
3114–3127.

EAST AND EUROPE (2002), http://www.estandardeurope.com/angol/baw.htm (accessed
November, 2002).

FISCHER S, DE VLIEGER J, KOCK T, GILBERTS J, FISCHER H and BATENBURG L (2000), 'Green
composites the materials of the future a combination of natural polymers and
inorganic particles', in Weber C J, *Conference Proceedings, The Food Biopack
Conference*, Copenhagen, 27–29 August, 109.

FREDERIKSEN C S, HAUGAARD V K, POLL L and MIQUEL BECKER E (2003), 'Light-induced
quality changes in plain yoghurt packed in polylactate and polystyrene', *Eur Food
Res Technol* DOI 10.1007/s 00217-003-0722-3.

GARDE A, SCHMIDT A S, JONSSON G, ANDERSEN M, THOMSEN A B, AHRING B K and KIEL P (2000),
'Agricultural crops and residuals as a basis for polylactate production in Denmark',
in Weber C J, *Conference Proceedings, The Food Biopack Conference*,
Copenhagen, 27–29 August, 45–51.

GONTARD N and GUILBERT S (1994), 'Bio-packaging: technology and properties of edible
and/or biodegradable materials of agricultural origin', in Mathlouthi M, *Food
Packaging and Preservation*, Glasgow, Blackie Academic & Professional, 159–
195.

GONTARD N, THIBAULT R, CUQ B and GUILBERT S (1996), 'Influence of relative humidity and film composition on oxygen and carbon dioxide permeabilities of edible films', *J Agric Food Chem*, 44, 1064–1069.

GUILBERT S (2000), 'Edible films and coatings and biodegradable packaging', *Bull Int Dairy Fed*, 346, 10–16.

HANLON J F (1992), 'Plastics' in Hanlon J F, *Handbook of Package Engineering*, 2nd edn, Lancaster, Technomic, 8.12–8.14.

HAUGAARD V K and FESTERSEN R M (2000), 'Biobased packaging materials for foods', in Weber C J, *Conference Proceedings, The Food Biopack Conference*, Copenhagen, 27–29 August, 119–120.

HAUGAARD V K, UDSEN A-M, MORTENSEN G, HØEGH L, PETERSEN K and MONAHAN F (2001a), 'Potential food applications of biobased materials. An EU-concerted action project', *Starch/Stärke*, 53, 189–200.

HAUGAARD V K, FESTERSEN R M and BERTELSEN G (2001b), 'Light induced changes in orange juice', in Oestergaard S, *Conference Proceedings. 2nd Nordic Foodpack Conference*, Stavanger, September 5–7. Taastrup, Danish Technological Institute.

HAUGAARD V K, WEBER C J, DANIELSEN B and BERTELSEN G (2002), 'Quality changes in orange juice packed in materials based on polylactate', *Eur Food Res Technol*, 214, 423–428.

HAUGAARD V K, DANIELSEN B and BERTELSEN G (2003), 'Impact of polylactate and poly(hydroxybutyrate) on food quality', *Eur Food Res Technol*, 216, 233–240.

HO K-L G and POMETTO III A L (1999), 'Effects of electron-beam irradiation and ultraviolet light (365 nm) on polylactic acid plastic films', *J Environ Polym Degrad*, 7(2), 93–100.

HO K-L G, POMETTO III A L and HINZ P N (1999a), 'Effects of temperature and relative humidity on polylactic acid plastic degradation', *J Environ Polym Degrad*, 7(2), 83–92.

HO K-L G, POMETTO III A L, HINZ P N, GADEA-RIVAS A, BRICENO J A and ROJAS A (1999b), 'Field exposure study of polylactic acid (PLA) plastic films in the banana fields of Costa Rica', *J Environ Polym Degrad*, 7(4), 167–172.

HOLTON E E, ASP E H and ZOTTOLA E A (1994), 'Corn-starch-containing polyethylene film used as food packaging', *Cereal Foods World*, 39, 237–241.

IATP NEWS (2002), http://www.iatp.org/iatp/News/news.cfm?News_ID=89 (accessed November 2002).

JOHANNSON K S (2000), 'Improved barrier properties of renewable and biodegradable polymers by means of plasma deposition of glass-like SiOx coatings', in Weber C J, *Conference Proceedings, The Food Biopack Conference*, Copenhagen 27–29 August, 110.

JUNKKARINEN L, Valio Ltd., Helsinki, Finland (2002), Personal communication.

KÄB, H (2002), 'Sustainable market development: assessing the role of EU governments'. In *Conference Proceedings Biodegradable Plastics 2002*, Frankfurt, 4–5 December.

KANTOLA M and HELÉN H (2001), 'Quality changes in organic tomatoes packaged in biodegradable plastic films', *J Food Quality*, 24, 167–176.

KESTER J J and FENNEMA O R (1986), 'Edible films and coatings: a review', *Food Technol*, 40(12), 47–59.

KIM M and POMETTO III A-L (1994), 'Food packaging potential of some novel degradable starch–polyethylene plastics', *J Food Prot*, 57(11), 1007–1012.

KITTUR F, KUMAR K R and THARANATHAN N (1998), 'Functional packaging properties of chitosan films', *Z Lebensm Unters Forsch A*, 206, 44–47.

KROCHTA J M and DE MULDER-JOHNSTON C (1996), 'Biodegradable polymers from agricultural products', in Fuller G, McKeon T A and Bills D D, *ACS Symp Ser*, Washington, DC: American Chemical Society, 121–140.

KROCHTA J M and DE MULDER-JOHNSTON C (1997), 'Edible and biodegradable polymer films: challenges and opportunities', *Food Technol*, 51(2), 61–74.

LEHERMEIER H J, DORGAN J R and WAY J D (2001), 'Gas permeation properties of poly(lactic acid)', *J Membr Sci*, 190(2), 243–251.

MAKINO Y and HIRATA T (1997), 'Modified atmosphere packaging of fresh produce with a biodegradable laminate of chitosan–cellulose and polycaprolactone', *Postharvest Biol Technol*, 10, 247–254.

MILLER K S and KROCHTA J M (1997), 'Oxygen and aroma barrier properties of edible films: a review', *Trends Food Sci Technol*, 8(7), 228–237.

MODELLPROJEKT KASSEL (2002a), http://www.modellprojekt-kassel.de/eng/seiten/news_press_releasea.html (accessed November 2002).

MODELLPROJEKT KASSEL (2002b), http://www.modellprojekt-kassel.de/eng/seiten/project_allgemein.html (accessed November 2002).

PADUA G W, RAKORONIRAINI A and WANG Q (2000), 'Zein-based biodegradable packaging for frozen foods', in Weber C J, *Conference Proceedings, The Food Biopack Conference*, Copenhagen, 27–29 August, 84–88.

PARRIS N, COFFIN D R, JOUBRAN R F and PESSEN H (1995), 'Composition factors affecting the water vapour permeability and tensile properties of hydrophilic films', *J Agric Food Chem*, 43, 1432–1435.

PETERSEN K, NIELSEN P V, BERTELSEN G, LAWTHER M, OLSEN M B and MORTENSEN G (1999), 'Potential of biobased materials for food packaging', *Trends Food Sci Technol*, 10, 52–68.

PETERSEN K, NIELSEN P V and OLSEN M B (2001), 'Physical and mechanical properties of biobased materials', *Starch/Stärke*, 53, 356–361.

PSOMIADOU E, ARVANITOYANNIS I, BILLADERIS C G, OGAWA H and KAWASAKI N (1997), 'Biodegradable films made from low density polyethylene (LDPE), wheat starch and soluble starch for food packaging applications: Part 2', *Carbohydr Polym*, 33, 227–242.

RÁBAGO, K (2002), 'Materials from renewable resources: the future of the biodegradable plastics industry?' In *Conference Proceedings Biodegradable Plastics 2002*, Frankfurt 4–5 December.

RASMUSSEN T and OLSEN M B (2001), The Danish Technological Institute, Taastrup, Denmark, and Togeskov P, Danisco Flexible, Lyngby, Denmark, Unpublished work.

ROBERTSON G L (1993) *Food Packaging. Principles and Practice* New York: Marcel Dekker.

SELIN J F (1997), 'Polylactides and their applications', in *Technology Programme Report 13/97*, Helsinki: Technology Development Centre Tekes, 111–127.

SELKE S (1996) *Biodegradation and Packaging*, 2nd edn, Leatherhead: Pira International.

SINCLAIR R G (1996), 'The case for polylactic acid as a commodity packaging plastic', *JMS Pure Appl Chem*, A33 (5), 585–597.

SÖDERGÅRD A (2000), 'Lactic acid based polymers for packaging materials for the food industry', in Weber C J, *Conference Proceedings, The Food Biopack Conference*, Copenhagen, 27–29 August, 14–19.

SÖDERGÅRD A and STOLZ M (2002), 'Properties of lactic acid based polymers and their correlation with composition', *Prog Polym Sci*, 27, 1123–1163.

SRINIVASA P C, BASKARAN R, RAMESH M N, HARISH PRASHANTH K V and THARANATHAN R N (2002), 'Storage studies of mango packed using biodegradable chitosan film', *Eur Food Res Technol*, 215, 504–508.

STRANTZ A A and ZOTTOLA E A (1992), 'Bacterial survival on cornstarch-containing polyethylene film held under food storage conditions', *J Food Prot*, 55, 782–786.

VAN TUIL R (2000), 'Converting biobased polymers into food packagings', in Weber C J, *Conference Proceedings, The Food Biopack Conference*, Copenhagen, 27–29 August, 28–30.

VAN TUIL R, FOWLER P, LAWTHER M and WEBER C J (2000), 'Properties of biobased packaging materials', in Weber C J, *Biobased Packaging Materials for the Food Industry. Status and Perspectives*, Frederiksberg, KVL Department of Dairy and Food Science, 13–44.

WEBER C J (2000), *Biobased Packaging Materials for the Food Industry. Status and Perspectives*, Frederiksberg: KVL Department of Dairy and Food Science.

WEBER C J, HAUGAARD V, FESTERSEN R and BERTELSEN G (2002), 'Production and applications of biobased packaging materials for the food industry', *Food Addit Contam*, 19, Supplement, 172–177.

YANG T C S (1995), 'The use of films as suitable packaging materials for minimally processed foods a review' in Barbosa-Cánovas G V and Welti-Chanes J, *Food Preservation by Moisture Control. Fundamentals and Applications*, Lancaster, PA: Technomic Publishing, 831–848.

11.9   Acknowledgements

This research was carried out as a part of the 'Biopack' project: Proactive Biobased Cheese Packaging (QLK5-CT-2000-00799), funded by the European Commission. The research does not necessarily reflect the Commission's views and in no way anticipates their future policy in this area. We thank all the participants for their co-operation and enthusiasm.

12

Recycling food processing wastes

M. Song and S. Hwang, POSTECH, South Korea

12.1 Introduction

Waste recycling has advanced as a method for preventing environmental decay and to help satisfy the increasing demands for raw materials. The potential benefits from successful recycling of food processing wastes are enormous (Allan, 1979). The characteristics of a particular waste will affect how readily a new recycling technology can show a return on investment, and will affect the variability in internal operating conditions. Also, waste disposal costs, which are a key driver for recycling technologies, will vary based on a given food processor's location and pertinent regulatory requirements, which will vary by region or city (Green and Kramer, 1979).

Bioconversion, or bio-recycling, can be defined as the reuse of organic waste to form new products by use of microbial activities, which means the organics in waste can be used as a substrate for microorganisms (Martin, 1998). Such technology would represent a way of meeting the increasing world demand for resource recovery and energy with waste materials. In this chapter, various bio-recycling technologies and a case study of mushroom mycelia production from food processing wastewater are discussed.

12.2 Bio-recycling technologies

This section discusses two options for recycling food processing wastes:

- Using waste to produce fuel or bioenergy
- Conversion of waste into new raw material.

12.2.1   Production of bioenergy

Due to dwindling fossil fuel resources, microbial production of biofuel from organic byproducts has acquired significant attention in recent years. Recent advances in biotechnology and bioengineering have resulted in remarkable success with fermentation routes using renewable resources for bioenergy production (Huang *et al.*, 2002). Bioenergy usually includes methane, hydrogen, and ethanol.

Anaerobic digestion for methane production is accomplished in a bioreactor, in which oxygen is completely excluded and all other parameters that govern anaerobic digestion are carefully controlled. Anaerobic digestion of organic waste can be thought of as a three-stage process in which the complex organic components of the waste are solubilized, broken down, and fermented into intermediate products that are subsequently reduced into methane and carbon dioxide (Stronach *et al.*, 1986). The stages of anaerobic digestion are referred to as hydrolysis or liquefaction, acidogenesis, and methanogenesis. These processes are carried out by many different species of symbiotic microorganisms, which are divided into two broad groups: acidogenic and methanogenic bacteria. Methane holds the most potential as a valuable fuel for producing electricity and heat. It releases energy at a rate of 12 000 kcal/kg (50 000 kJ/kg), and is a clean-burning natural gas (Ferry, 1993).

Biological hydrogen production using organic waste as a substrate has drawn considerable attention, because of the ability to produce an environmentally friendly energy source while simultaneously stabilizing the waste. This can be accomplished using a wide variety of bacteria through the actions of well-studied anaerobic metabolic pathways and hydrogenase enzymes. Generally, fermentation of both glucose and sucrose in food processing wastes under slightly acidic conditions in the absence of oxygen produces high concentrations of hydrogen gas (Haast *et al.*, 1986). Hydrogen has 2.4 times the energy content of methane (i.e., on a mass basis) and its reaction with oxygen in fuel cells produces only water, a harmless by-product. Hydrogen gas has valuable potential for producing clean and economical energy in the near future (Yang, 2002).

Another biological process useful for the conversion of biomass to fuel is ethanol fermentation by yeast or bacteria. The microorganisms carry out fermentation, such as the conversion of sugar to carbon dioxide, in the absence of free oxygen. Ethanol is an important product in the food, agricultural, and fuel industries, and has been labelled an alternative fuel for the future (Gong, 2001). Operation of ethanol fermentation on a full scale, however, usually requires a large amount of waste, which leads to transportation and storage problems.

12.2.2   Bioconversion of food processing wastes into higher value organic acids

Short-chain organic acids include acetic acids, propionic acid, and butyric acid, which are among the top-produced organic chemicals. They are used for various applications, and are produced mainly by petrochemical processes that cause pollution and use non-renewable raw materials (Huang *et al.*, 2002). Therefore, it would be ideal to produce organic acids using biological means, since it is generally considered that biological processes are environmentally friendly and use renewable resources.

Acetic acid is an important, widely used chemical with an annual production of approximately 2.12 million metric tons in the US in 1995, with a market price of $0.84/kg (Huang *et al.*, 2002). Acetic acid can be used to produce an environmentally benign and non-corrosive deicing agent to be used for airport runways, bridges, and strategic roads (Fritzsche, 1992). Acetate deicer is much more expensive, with a current market price of greater than $1.0/kg due to the high price of acetic acid produced through petrochemical processes. Because of the high cost of acetate, there have been many studies to produce low-cost acetic acid from renewable resources (Parekh and Cheryan, 1991).

Propionic acid and its calcium, sodium, and potassium salts, are widely used as food and feed preservatives. Due to the increase in consumer demands for natural food ingredients, commercial interests in producing propionic acid by fermentation are high (Crespo *et al.*, 1990). The price for propionic acid obtained by the fermentation route can be as high as $4.4/kg, compared with only $1.0/kg for the petrochemically derived acid. Butyric acid can also be generated by microbial fermentation of dietary substrates. Various drugs have been derived from butyric acid, which has therapeutic effects for treating colorectal cancer and hemoglobinopathies (Pouillart, 1998). It is desirable to use naturally derived butyric acid for these applications. Currently, butyric acid is produced mainly through petrochemical methods and has a market price of $1.21/kg (Huang *et al.*, 2002).

The possibility of using food processing waste as an inexpensive source of substrate for a variety of microorganisms has been investigated. For example, vast amounts of cellulose wastes, which could be measured in the billions of tonnes worldwide, are produced as residues from industrial food processing. Consequently, the use of microorganisms to remove and ameliorate these potential polluting materials is a real environmental challenge, but could be solved by a focused study concerning efficient methods applied in biological degradation processes. In this respect, the biodegradation of cellulose waste by continuous enzymatic activities of immobilized bacterial and fungal cells, as improved biotechnological tools, can be considered (Ropars *et al.*, 1992). In addition, there are new methods to immobilize microorganisms using polymeric hydrogels such as polyacrylamide (PAA), collagen-polyacrylamide (CPAA), elastin-polyacrylamide (EPAA), and poly-hydroxyl-ethylmethacrylate (PHEMA), which were achieved by gamma polymerization techniques. Unlike other biodegradation processes, these methods have been performed to preserve the viability of fungal and bacterial cells during long-term bioprocesses and their metabolic activity efficiency (Petre *et al.*, 1999).

12.3   Case study: recycling cheese whey

Humans have consumed many different types of mushrooms since early recorded history. Recent increase in mushroom consumption is due mainly to consumers' perception that mushrooms are low in calories and rich in vegetable proteins, chitin, vitamins, and minerals (Kang *et al.*, 2002; Wasser and Weis, 1999). The lifecycle of a mushroom proceeds as follows: First, the spores of the fruiting body germinate and form a network of threads called the mycelium, which develops into a pinhead under favourable conditions. The pinhead gradually grows and turns into the full mushroom fruiting body (Eyal, 1991). Figure 12.1 shows the general life cycle of a mushroom. Mycelium is a health food, considered to have outstanding medicinal qualities, including anti-tumour activity and the ability to lower cholesterol. Mycelia are presently grown in bioreactors that use a relatively expensive carbon source.

Cheese whey is a by-product of cheese production that remains when casein and butter fat are separated as curd from milk. Depending on the type of cheese being made, up to 9 litres of whey is generated for every kilogram of cheese produced (Hwang and Hansen, 1998). Worldwide whey production was approximately 150 million tonnes in 2001 (FAOSTAT, 2001). This particular

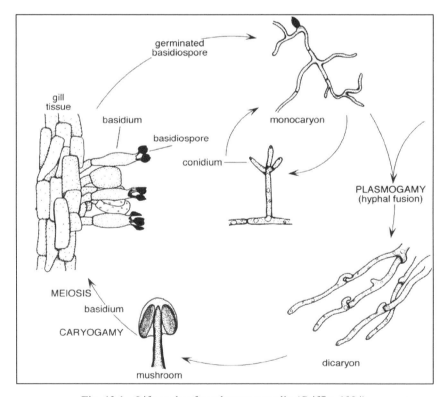

Fig. 12.1   Life cycle of mushroom mycelia (Griffin, 1994).

dairy waste should be viewed as an inexpensive potential source of raw material from which valuable products can be produced, since cheese whey includes approximately half of the original nutrients of milk; 4% lactose, nitrogenous compounds, trace minerals and vitamins make it nutritionally valuable (Haast *et al.*, 1986). Therefore, whey can be used as an alternative substrate for cultivating mushroom mycelium, thus providing a unique solution to the wastewater management of cheese whey. Growth of mycelia in whey provides a means of producing a value-added product from under-utilized media that remain after the three highest valued products from a cheese-making operation (cream, cheese and whey protein concentrate) are obtained. At the same time, the oxygen demand load carried by those low value waste streams is reduced so that discharge standards are met.

Control of environmental conditions, as well as the modification of media composition, has been vital in order to enhance the production efficiency in mycelial culture (Kang *et al.*, 2002; Fang and Zhong, 2002). Despite this effort, little information is available regarding the optimization of environmental factors affecting the growth of mycelium for various substrates in submerged culture. Furthermore, the biokinetics of mycelial growth when using whey, essential to control and to predict the production efficiency, is lacking in the literature.

The following case study describes how to find the optimal conditions, with respect to the simultaneous effects of environmental conditions, where mycelial production is maximized using cheese whey. The biokinetics of mycelium at the optimal growth conditions were also considered.

The treatability test was initially performed with five species of mushrooms: *Agaricus bisporus, Ganoderma lucidum, Pleurotus ostreatus, Lentinus edodes* and *Phellinus linteus*. The inoculums were stored on potato dextrose agar (PDA) medium at 4°C and were transferred to fresh PDA media every 3 months. The seed cultures of all organisms were transferred to petri dishes containing PDA media at 25°C. Mycelial agar discs (5 mm in diameter) were obtained by a round cutter and used as inoculums for subsequent experiments. Mycelial growth on the whey medium was monitored periodically to evaluate the radial growth rate, which was compared to those on six different types of commercial media. These were glucose peptone yeast (GPY), yeast malt (YM), Czapek Dox (CD), glucose ammonium chloride (GAC), malt (M), and potato dextrose agar (PDA). The results are shown in Table 12.1. All species had higher or similar radial growth rate on whey medium compared to those on the commercial media. Therefore, it can be concluded that it is feasible to use cheese whey as an alternative substrate for mycelial cultiviation and the cost of using a by-product such as whey permeate is less than 1% of the cost of presently used commercially prepared media.

Among the five species of mycelia, *G. lucidum* was selected for the maximal production and biokinetics study in submerged culture because it has gained wide popularity as an ingredient in many health foods and therapeutic medicines because of its perceived health benefits. The wild type of *G. lucidum* grows on

Table 12.1   Radial growth rate (mm/d) of mushroom mycelia on whey and commercial media

	G. lucidum	L. edodes	P. ostreatus	P. linteus	A. bisporus
Glucose peptone yeast	8.7	2.9	3.7	1.7	3.0
Yeast malt	7.2	3.8	3.9	2.0	3.3
Czapek Dox	4.3	3.1	1.8	2.1	1.6
Glucose ammonium chloride	3.8	3.0	1.6	1.2	0.8
Malt	8.4	2.0	2.6	1.8	1.7
Potato dextrose agar	8.8	4.5	4.0	2.0	5.0
Whey	16.7	4.2	3.3	2.5	4.9

old logs and stumps, especially maple and oak. For centuries, this mushroom has been regarded in northern Asia as a popular folk medicine used to treat various human diseases such as hypertension, arthritis, and bronchitis. Recent studies of this fungus have postulated that the polysaccharide 1,3-β-D-glucan inhibits a variety of cancers by enhancing the hosts' immune functions (Fang and Zhong, 2002). It has also been suggested that this mushroom has anti-inflammatory effects and cytotoxicity to hepatoma cells.

The conditions for maximal growth of G. lucidum associated with simultaneous changes in pH and temperature were investigated because these two parameters have been key variables to maximize the mycelial production of various edible mushrooms in submerged cultures (Cheung, 1997). A response surface method (RSM) was applied to optimize the factors affecting the growth. A sequential procedure of collecting data, estimating polynomials (Eq. [12.1]), and checking the adequacy of the model was used in this study.

$$\eta = c_0 + \sum_{i=1}^{n} \alpha_i x_i + \sum_{i=1}^{n} \alpha_{ii} x_i^2 + \sum_{\substack{i \\ i<j}} \sum_j \alpha_{ij} x_i x_j \qquad [12.1]$$

where

η is the experimental value of mycelial concentration (g/l)
x_i is the independent variable i (i = pH and temperature in order)
c_o is the regression constant
α_i are the regression coefficients of the independent variable i (i = pH and temperature in order)

The method of least squares was used to estimate the parameters in the approximating polynomials. The central composite in cube (CCC) design, which consists of an orthogonal 2^2 factorial design augmented by a centre and 2×2 axial points (Table 12.2), was employed in this research.

Batch data during fermentation at the estimated optimum conditions were used to evaluate the system performance and growth kinetics of G. lucidum. The substrate inhibition biokinetic expressions for microbial growth and substrate utilization (Grady et al., 1999) were used, and the numerical approximations are given as follows:

Table 12.2   Experimental conditions and results of the central composite design

| | Conditions of variables | | Responses | |
| | pH | Temperature (°C) | Mycelial dry weight (g/l) | Residual SCOD concentration (g/l) |
Trials				
1	3.5	25	14.3	6.8
2	4.5	25	16.5	4.1
3	3.5	35	13.4	10.3
4	4.5	35	15.1	7.8
5[a]	4.0	30	17.0 (0.5)[b]	4.5 (0.8)[b]
6	4.0	37.1	14.0	8.5
7	4.0	22.9	15.8	6.2
8	4.7	30.0	16.3	4.6
9	3.3	30.0	14.3	8.9

[a] Centre point. Experiment was replicated five times, with the response value representing the calculated average.

[b] Standard deviation in brackets.

$$X_{t+1} = \left\{ 1 + \left(\frac{\mu_m S_t}{K_s + S_t + \dfrac{S_t^2}{K_{si}}} - k_d \right) \Delta t \right\} X_t \qquad [12.2]$$

$$S_{t+1} = \left(1 - \frac{\mu_m}{Y} \frac{X_t}{K_s + S_t + \dfrac{S_t^2}{K_{si}}} \right) S_t \qquad [12.3]$$

where

X_t is the microbial concentration in the bioreactor at time t (microbial mass/volume)
μ_m is the maximum specific growth rate (time^{-1})
S_t is the residual substrate concentration at time t (mass substrate/volume)
K_s is the half saturation coefficient; numerically equal to the substrate-concentration at which specific growth rate is half of its maximum value (mass substrate/volume)
K_{si} is the substrate inhibition coefficient (mass substrate/volume)
k_d is the specific decay rate of the microorganism (time^{-1})
Δt is the time increment (time)
Y is the microbial yield coefficient (microbial mass/mass substrate utilized).

A fourth order Runge–Kutta approximation (Lim *et al.*, 2001) along with a multiresponse, nonlinear least squares (NLLS) method was employed to approximate kinetic coefficients with 95% confidence intervals (C.I.).

The quadratic model:

$$\eta = -9.1 \times 10^4 + 3.3 \times 10^4 x_1 + 2.7 \times 10^3 x_2 - 52 x_1 x_2 - 3.7$$
$$\times 10^4 x_1^2 - 44 x_2^2 \qquad\qquad [12.4]$$

was selected to describe the response surface within the experimental region (Table 12.2) since the residual plots of this model showed consistently less variance compared to other polynomial models. The optimal conditions for mycelial production were calculated by setting the partial derivatives of the function to zero with respect to the corresponding independent variables of pH and temperature, which were pH 4.2 and 28.3°C, respectively. The calculated model output at the optimal conditions was 18.1 ± 0.9 g dry weight per litre. Two- and three-dimensional response surfaces of the quadratic model for mycelial production, with estimated optima, are shown in Fig. 12.2. Adequacy of the model prediction was verified by comparing the maximum model output with duplicated experimental values at optimal conditions; residual plots for all observed values were then examined for any weakness in the models (Hwang *et al.*, 2001). Excellent prediction of maximum response values, along with constant variance in residual plots, indicated adequacy of the model. This meant the quadratic model allowed one to locate the conditions for maximum mycelial production of *G. lucidum* using cheese whey within the investigated experimental region.

The changes of microbial and residual substrate concentrations in batch culture of the mycelium at the optimal conditions are shown in Fig. 12.3. The mass of *G. lucidum* gradually increased to 20.1 ± 0.8 g dry weight per litre during 172 h of incubation. Residual substrate concentration decreased to 3.5 ± 0.1 g soluble chemical oxygen demand (SCOD) per litre for the same period, which was a 93.4% reduction of the initial wastewater strength. The concentrations of mycelial cells and SCOD were simultaneously used to estimate the biokinetic coefficients in Eqs [12.2] and [12.3]. The substrate inhibition model was used to fit the data. Table 12.3 summarizes the values of the biokinetic coefficients, which were verified by comparing the simulated response values of SCOD and cell concentrations with observed values. The model output values were very similar to the experimental values. Therefore, it could be concluded that the concept of bioconversion of the whey to mycelial cell mass was shown to be effective in pollution reduction, and that the kinetic coefficients estimated in this study could be used to design and to control a system for cultivating mycelium of *G. lucidum* with cheese whey as a substrate. This design includes determination of the mycelial production period and prediction of process performance, such as a degree of pollution reduction, as well as the amount of mycelium produced per unit time period for a scaled-up process.

12.4   Future trends in bio-recycling technology

Food-processing waste disposal and treatment are an important part of the overall food production and processing network. Among the variety of wastes,

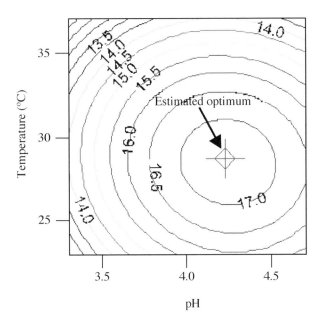

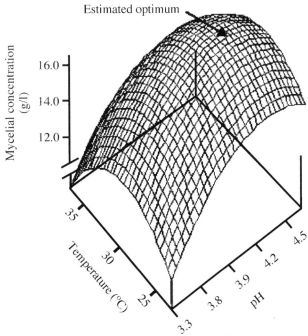

Fig. 12.2   Two- and three-dimensional contour plots of the quadratic model for the mycelial production with respect to pH and temperature.

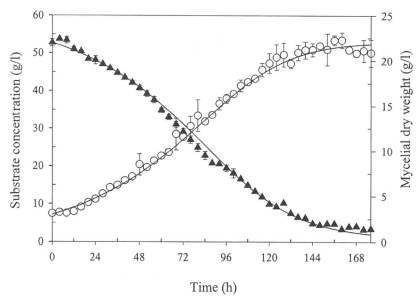

Fig. 12.3 Observed and predicted concentrations of substrates and *G. lucidum* in batch culture. (○) observed mycelial mass, (▲) observed SCOD, (−) model predictions.

wastewater is generally the most common because food-processing operations involve a number of unit operations, such as washing, evaporation, extraction, and filtration. The process wastewaters resulting from these unit operations normally contain high concentrations of suspended solid and soluble organics such as carbohydrates, proteins, and lipids, which present difficult disposal problems.

Liquid, solid, and gaseous wastes from industrial sources must ultimately be collected, treated, and recycled in an environmentally sound manner. The future will demand a higher level of natural resource conservation, which will result in stricter regulations. From the standpoint of industry, this situation has required extra investment for the treatment, disposal, and use of their own wastes. Thus, having to pay more attention to waste management has somewhat contravened

Table 12.3 Values of biokinetic coefficients using data from batch culture of *Ganoderma lucidum*

Kinetic parameters (unit)	Biokinetic values (mean ± 95% confidence interval)
μ_m (1/d)	2.27 ± 0.15
k_d (1/d)	0.05 ± 0.01
K_s (g SCOD/l)	128.0 ± 12.1
K_{si} (g SCOD/l)	49.3 ± 3.3
Y (g dry weight/g SCOD)	0.39 ± 0.03

industry's primary objective, which is to produce the best possible product at the lowest possible cost. Many food industries have used available technology to remove major pollutants, and a new dimension of food waste management is currently evolving, from a current approach of disposal to one of utilization, focusing especially on incorporating biological processes. Some key cost considerations that a food company should consider when choosing waste utilization technology include:

- Cost of treatment and disposal compared with cost of utilization
- Seasonality of waste
- Market value of recovered constituents
- Existence of profitable markets or cost of market development
- Fuel value of the waste
- Availability of appropriate technologies and the research and development efforts required to modify existing technologies, or the development of new technologies.

Optimizing a biosystem aims to improve the outcome of a process that uses biological components of living organisms. The biosystem outcome can be related to a number of different criteria, such as maximized product yield in the pharmaceutical industry, high product quality in food production, and maximum remediation performance in environmental biotechnology. Many biosystem processes still have the potential to be improved considerably, with respect to these different criteria. In order to achieve substantial improvements in biosystem outcome, a systematic approach to bioprocess optimization is required.

A mechanistic model is possible only when enough is known about the mechanism to deduce the functional relationship between input and output variables. However, there are many biological areas, like mycelial cultivation, where basic knowledge of the phenomenon is insufficient to build a mechanistic model. Empirical models and statistical analysis, in this case, play an extremely important role in elucidating basic mechanisms in complex situations and thus providing better process control.

In the optimization of biological systems, the proper selection of statistical and mathematical methods is crucial to explore such a system experimentally, with minimum effort and expense. For example, the RSM is a useful technique in building empirical models to describe a response where several independent variables influence a dependent variable in the system. Model building should be an iterative process; comparing the available data with the tentative model enables a researcher to determine whether the model is adequate or not. If the model is found to be inadequate, the model should be modified or more data should be collected. If it is adequate, the parameter values along with their precision should be estimated, usually using the regression method for biological processes. If the precision is not high enough, additional data should be collected to repeat this procedure or a new region should be explored.

12.5   References

ALLAN F M (1979), *Resource Recovery and Recycling*, John Wiley & Sons, Inc., New York.

CHEUNG P C K (1997), 'Chemical evaluation of some lesser known edible mushroom mycelia produced in submerged culture from soy milk waste', *Food Chem*, 60(1): 61–65.

CRESPO J P S G, MOURA M J and CARRONDO J J T (1990), 'Some engineering parameters for propionic acid fermentation coupled with ultrafiltration', *Appl Biochem Biotech*, 24(3): 613–625.

EYAL J (1991), *Mushroom Mycelium Grown in Submerged Culture – Potential Food Applications*, Van Nostrand Reinhold, New York.

FANG Q and ZHONG J (2002), 'Effect of initial pH on production of ganoderic acid and polysaccharide by submerged fermentation of *Ganoderma lucidum*', *Process Biochem*, 37(7): 769–774.

FAOSTAT (2001), *FAO Statistical Databases*, Food and Agriculture Organization of the United Nations, Washington, DC.

FERRY G J (1993), *Methanogenesis*, Chapman & Hall, New York.

FRITZSCHE C J (1992), in *Water Environmental Technology*, 44–51.

GONG C S (2001), 'Ethanol production from renewable resources', *Fuel and Energy Abstracts*, 42(1): 10.

GRADY JR C P L, DAIGGER G T and LIM H C (1999), *Biological Wastewater Treatment*, Marcel Dekker, New York.

GREEN J H and KRAMER A (1979), *Food Processing Waste Management*, AVI Publishing, Westport, CT.

GRIFFIN D H (1994), *Fungal Physiology*, John Wiley & Sons, New York.

HAAST J D, BRITZ T J and NOVELLOW J C (1986), 'Effect of different neutralizing treatments on the efficiency of an anaerobic digester fed with deproteinated cheese whey', *J Diary Res*, 53(3), 467–476.

HUANG Y L, WU Z, ZHANG L, CHEUNG C M and YANG S T (2002), 'Production of carboxylic acids from hydrolyzed corn meal by immobilized cell fermentation in a fibrous-bed bioreactor', *Bioresource Technol*, 82(2): 51–59.

HWANG S and HANSEN C L (1998), 'Characterization and bioproduction of short-chain organic acids from mixed dairy-processing wastewater', *Trans ASAE*, 41(3): 759–802.

HWANG S, LEE Y and YANG K (2001), 'Maximization of acetic acid production in partial acidogenesis of swine wastewater', *Biotechnol Bioeng*, 75(5): 521–529.

KANG H, LEE H, HWANG S and PARK W (2002), 'Effects of high concentrations of plant oils and fatty acids for mycelial growth and pinhead formation of *Hericium erinaceum*', *Trans. ASAE*, 45(1): 257–260.

LIM J, KIM T and HWANG S (2001), 'Growth kinetic parameter estimation of *Candida rugopelliculosa* using a fish manufacturing effluent', *Biotechnol Lett*, 23(24): 2041–2045.

MARTIN A M (1998), *Bioconversion of Waste Materials to Industrial Products*, Blackie Academic and Professional, London and New York.

PAREKH S R and CHERYAN M (1991), 'Production of acetate by mutant strains of *Clostridium thermoaceticum*', *Applied Microbiology and Biotechnology*, 36(2): 384–387.

PETRE M, ZARNEA G, ADRIAN P and GHEORGHIU E (1999), 'Biodegradation and bioconversion of cellulose wastes using bacterial and fungal cells immobilized in radiopolymerized hydrogels', *Resource, Conversion and Recycling*, 27(2): 309–332.

POUILLART P R (1998), 'Role of butyric acid and its derivatives in the treatment of colorectal cancer and haemoglobinopathies', *Life Sciences*, 63(20): 1739–1760.

ROPARS M, MARCHAL R, POURQUIE J and VANDERCASTEELE J P (1992), 'Large scale enzymatic hydrolysis of agricultural lignocellulosic biomass', *Bioresource. Technol*, 42(2): 197–203.

STRONACH S M, RUDD T and LESTER J N (1986), *Anaerobic Digestion Processes in Industrial Wastewater Treatment*, Springer-Verlag, New York.

WASSER S P and WEIS A L (1999), 'Therapeutic effects of substances occurring in higher Basidiomycetes mushrooms: a modern perspective', *Crit Rev Immunol*, 19(1): 65–96.

YANG S T (2002), 'Effect of pH on hydrogen production from glucose by a mixed culture', *Bioresource Technol*, 82(1): 87–93.

13

Waste treatment

C. L. Hansen, Utah State University, USA and S. Hwang, POSTECH, Korea

13.1 Introduction: key issues in food waste treatment

Food processing waste disposal and treatment are an important part of the overall food production and processing network. Waste products from food processing facilities include bulky solids, airborne pollutants, and wastewater. All of these cause potentially severe pollution problems and are subject to increasing environmental regulation in most countries. Generally, wastewater is the most common waste because food processing operations involve a number of unit operations; such as washing, evaporation, extraction, and filtration. The process wastewaters resulting from these operations normally contain high concentrations of suspended solids and soluble organics such as carbohydrates, proteins, and lipids, which present difficult disposal problems. Regulatory agencies such as the Environmental Protection Agency (EPA) in the United States have promulgated regulations on effluent for a variety of food processing industries. Table 13.1 summarizes pollution characteristics of typical food industry wastes. In the table, BOD_5 means biochemical oxygen demand and TSS is total suspended solids.

Technology to remove major pollutants such as total dissolved and suspended solids and organic materials in the food processing industry can generally be classified into physical, biological and chemical treatments. Biological treatment can be sub-classified into aerobic and anaerobic treatments. Under aerobic conditions, microorganisms convert carbohydrates, lipids, and proteins in wastes into microbial biomass and carbon dioxide (CO_2). Under anaerobic conditions, wastes containing those components can be digested to yield methane (CH_4), which can be burned as a fuel source. Also, ethanol or organic acids can be produced by the anaerobic microbiological

Table 13.1   Pollution characteristics of selected food processing wastes (Hwang, 1995)

| Waste | Pollutional Characteristics (mg/L) | | | |
	BOD_5	TSS	Protein	Fat
Dairy	1000–4000	1000–2000	6–82	30–100
Fish	500–2500	100–1800	300–1800	100–800
Fruit	1200–4200	2500–6700	–	–
Meat	1000–6500	100–1500	350–950	15–600
Poultry	200–1500	75–1100	300–650	100–400
Vegetable	1000–6800	100–4000	–	–
Municipal	100–300	100–500	150–530	0–40

process. Figure 13.1 is a basic outline of how organic materials are broken down by bacteria.

Key considerations in determining appropriate treatment technology are process operating cost, quantity and characteristics of waste, and market value of recovered products. Because of the diverse nature of the various food processing wastes such as differences in concentrations of carbohydrates, protein and fat, there are differences in treatment methods. Unit processes including physical, biological, and chemical or other tertiary treatment are the parts common to all waste treatment systems. Descriptions of common unit processes are covered in this chapter. The waste treatment systems most often used for the major food commodities, dairy, meat, poultry, seafood, and vegetable processing are listed. For those food products that do not have waste treatment systems listed, the reader can infer how a particular unit process would be useful to treat that waste based on similarities to those listed.

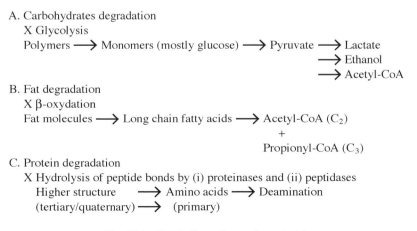

Fig. 13.1   Catabolism of organic materials.

13.2   Common food waste treatment systems

Commonly used food processing waste treatment methods include those listed below. The order of presentation has relevance to frequency of use from first to last, but this is not quantified and can vary from country to country (Hansen and Wrigley, 1997, Gosta, 1999). Later in the chapter, waste treatment methods are described in more detail.

1. Land application of untreated or partially treated waste. Can be one of the least expensive methods if inexpensive land is available with good drainage and local regulations allow it. However, cheap land is often not available and the problems with this technology include possible odour nuisance, salt buildup in the soil, standing water, and maintenance difficulties. The success of this method depends upon the use of proper application rates. Effective pretreatment of the wastewater is helpful. Wastewater high in organic matter should be screened prior to irrigation. The soil can do an excellent job of removing nutrients and pathogens from waste.

2. Sedimentation, settling and chemical precipitation. Sedimentation or settling of solids occurs in collecting ponds or lagoons. When the concentration of solids is greatest, addition of chemical coagulant will help in removing them.

3. Dissolved air flotation. DAF is extremely useful and efficient to remove fats, oils, and grease (FOG) and it can remove significant quantities of other pollutants. A DAF unit can remove up to 90% of fats, oils and grease (FOG) and 50% of biochemical oxygen demand (BOD). These high removal rates usually require pH adjustment and chemical addition. Still, DAF is often the unit process of choice for food processing waste. DAF is also often useful to achieve the desired BOD or solids concentration before final discharge.

4. Stabilization ponds. These are similar to mechanically aerated lagoons except O_2 is provided by the algal population. These ponds have a relatively large surface to volume ratio and thus sunlight reaches algae that produce O_2 by photosynthesis. Stabilization ponds require a relatively large land area and may be impractical for that reason. Part of waste treatment in ponds and lagoons occurs via physical methods i.e. setting of solids during the holding period. Odours and seepage to ground water may be problems, also insect control is necessary.

5. Aerated lagoons. The active biomass in the lagoon is low, thereby requiring longer periods of aeration for comparable performance to other aerobic processes such as the activated sludge process. Aerated lagoons are not vigorously mixed, so there is solids sedimentation. The aerated lagoon is often facilitative (part with O_2 and part without), thus, organic matter is actually removed through a combination of physical separation, aerobic, and anaerobic stabilization. Some aerated lagoons systems are designed in series flow to improve final effluent. In these cases, most of the aeration capacity is provided in the first cell. Second or third cells provide for settling and so called polishing.

6. Anaerobic lagoons. The biological degradation of organic material occurs in the absence of dissolved oxygen. Organic materials are converted to organic acids, carbon dioxide, and methane. In addition, odiferous gases including ammonia and hydrogen sulphide are produced in small quantities. Conversions of 80 to 90% of organic matter may be achieved, at a slower rate and producing less sludge than in aerobic systems.

7. Other anaerobic processes. Anaerobic digestion is often thought of primarily for solids destruction. More recently, the energy that is produced as biogas, has become a good reason to consider the process. Improved methods to co-generate electricity and heat from biogas are now available. The electricity that is generated is considered 'green' energy yielding carbon credits. Carbon credits are saleable because of the Kyoto protocols for reduction of greenhouse gas emissions. Anaerobic digestion requires control of pH value and this can be costly. Anaerobic digestion is not commonly used alone because the quality of the treatment is poorer than that required by receiving water standards. It can be combined with aerobic treatment. Rapid decomposition of organic material in an anaerobic process will result in a smaller and more effective aerobic process.

8. Activated sludge process. The process provides aerobic biological treatment employing suspended growth of bacteria. The organisms are separated from the treated effluent by means of sedimentation and returned back to the aerated chamber. This makes the solids retention time much longer than the hydraulic retention time and thus improves process results. The process produces a relatively high quality effluent, but is often too expensive for treating the large quantities of wastewater produced by food processing.

9. Membrane processes. Reverse osmosis, diafiltration, electrodialisis and ultrafiltration, are some of the methods used in this process. They are especially useful for whey treatment. This process can yield saleable products from some wastes, such as whey protein from whey.

10. Chemical methods. Often used in combination with other methods. When used singly, it has the disadvantage of relatively high cost with often nominal effectiveness in organic matter removal. Metals, such as calcium or iron may be used to aid certain unit processes. Adding polyelectrolytes improves the efficiency of other unit processes such as DAF.

11. Trickling filters. This employs a fixed support medium to maintain the active organisms within the wastewater stream. Specific plastic media have been developed for this purpose. Organic matter is absorbed onto the fixed biological film and is subsequently oxidized. Trickling filters are followed by clarification, e.g. settling basins and DAF facilities in order to intercept and remove solids from the filter.

12. Rotating biological discs. A modification of the trickling filter process whereby a fixed biological film is rotated through the wastewater. A large biological surface is provided by a series of closely spaced discs mounted on a rotating horizontal shaft. The device usually works in the presence of oxygen or it can be used for anaerobic treatment.

13.2.1   Dairy processing waste

The order of common waste treatment processes given as Section 13.2 is similar for dairy processing waste. Land availability and costs may preclude land application. DAF is one of the most useful waste treatment processes for dairy processing waste. Lagoons and stabilization ponds are used where conditions allow. Odour from any waste treatment process open to the air is becoming a more serious problem and impediment for using uncovered lagoons.

Whey is a unique part of dairy processing wastewater because of its high pollution potential and quantity. About 9 kg of whey are produced for every kg of cheese produced. Figure 13.2 summarizes various processes used in the treatment of whey that yield end products. Whenever a saleable end product can be produced, it is often going to be the treatment of choice, even if the plant only breaks even in the process. Sludge disposal is becoming an increasingly difficult problem to solve.

13.2.2   Meat, poultry, seafood processing waste

Meat, poultry, and seafood processing waste have many similarities. Typical waste loads for meatpacking wastes are given in Table 13.2. As can be seen from the table, there is a significant amount of fat and protein in these types of waste. The type of waste treatment for these wastes often includes the steps listed below (Ockerman and Hansen, 2000, Hansen *et al.*, 1984).

1. Screens and fat traps. The main task is to remove particles contained in the wastewater.
2. Flotation with some chemical addition, to remove suspended solids and emulsified fats.
3. Aeration of the waste to minimize organic and nitrogen load.

Solids fractionation/concentration processes
- Ultra filtration ⟶ retains protein and up to 40% of minerals in retentate and produces whey permeate
- Reverse osmosis ⟶ retains lactose and most of the minerals and produces relatively clean permeate that can be recycled for some useful purpose in the plant
- Evaporation ⟶ concentrated whey (from whey), concentrated whey protein solution from retentate of UF process
- Drying ⟶ dried whey, whey protein concentrate from UF retentate, lactose rich solids from whey permeate (RO)
- Desalination ⟶ desalinated or partially desalinated why byproduct
  - nanofiltration
  - Ion exchange or electro-dialysis

Fig. 13.2   Whey treatment methods yielding saleable byproducts.

Table 13.2 Meat and seafood processing waste characteristics (Ockerman and Hansen, 2000, Carawan *et al.*, 1979)

Red meat packing waste load (kg per 1000 kg live weight kill)	BOD$_5$	COD	TSS	Fat
	13.4	–	10.3	5.2
Red meat (mg/L)	1000–6500	–	100–1500	15–600
Salmon (mg/L)	250–2600	300–5500	120–1400	20–550
Blue crab (mg/L)	600–4400	1000–6300	330–620	150–220
Shrimp (mg/L)	720–2000	1200–3700	500–3000	250–700
Tuna (mg/L)	720–900	1200–2300	500	250–290
Oysters (mg/L)	250–800	500–2000	200–2000	10–30
Clams (mg/L)	500–1200	700–1500	200–400	20–25
Farm raised catfish (mg/L)	340	700	400	200

4. Anaerobic lagoons to destroy organic solids
5. Biofilters, carbon filters, and scrubbers are used to control odours and air emissions.
6. Desinfection of the final effluent, when high levels of bacteria are found. Use of chlorine compounds is one good alternative, ultraviolet light may also be used.

13.2.3 Fruit and vegetable processing waste

Table 13.3 lists fruit and vegetable processing waste characteristics Common fruit and vegetable waste treatment processes are (Jones, 1994; Carawan *et al.*, 1979):

1. Screening. An effective separation and segregation of the solids in the effluent will often be the first step in the waste treatment. The effectiveness of this depends on the physical properties of the particles, including size, density, and concentration, and on the capability of the equipment for their separation.
2. Trickling filters.
3. Sedimentation, settling and chemical precipitation.
4. Activated sludge process.
5. Aerated lagoons and stabilization basins. They are used when land is available.
6. Anaerobic lagoons.
7. Fungal treatment. Selected strains of fungi are used mostly on waste from corn and soy bean processing plants. The mycelium can be harvested by filtration and has utility as a feed product.
8. Tertiary (chemical or ultraviolet light). Is used when the bacteria content is too high and to reduce odours.

Table 13.3   Fruit and vegetable processing waste characteristics

	BOD (mg/l)	TSS (mg/l)	Reference
Mean of 11 common vegetables with flow of wastewater equal to 17.1 m³/ tonne (4100 gal/ton) of raw product	866	785	EPA, 1977
Mean of 8 common fruits with flow of wastewater equal to 14.44 m³/tonne (3500 gal/ton) of raw product	1025	230	
Citrus	<500	130	
Apples products	1000–2000	150	
Apple juice	500–1000	104	
Carrots	2000–3000	262–1540	
Cherries, brined	3000–5000	90–130	Carawan *et al.*, 1979
Olives	<500	400	
Peas	2000–3000	80–670	
Pineapples	3000–5000	840–1160	
Dehydrated vegetables	<500	300	

13.3   Physical methods of waste treatment

The following discussions give more information about specific treatment processes. These are all treatment processes used in food waste treatment.

13.3.1   Screening

Screening is usually the first unit operation in a wastewater treatment process. A screen is a device with openings, generally of uniform size, that is used to retain the coarse solids present in wastewater.

The screening element consists of parallel bars, rods or wires, grating, wire mesh, or perforated plate. The openings may be of any shape but are generally circular or rectangular slots. The materials removed by these devices are called 'screenings'. The headloss through the process is normally the parameter used to determine the efficiency of screening.

(1) Bar Racks

Bar racks are normally used to protect pumps, valves, pipelines, and other facilities from damage or clogging by rags and large objects. Industrial waste plants may need them depending on the character of the wastes. The following equation is used to calculate headloss through the bar rack.

$$h_L = \frac{1}{0.7} \left(\frac{V^2 - v^2}{2} g \right)$$

h_L: head loss [m]
0.7:   an empirical discharge coefficient to account for turbulence
V: velocity of flow through the openings of the bar rack [m/s]
v: approach velocity in upstream channel [m/s]
g: acceleration due to gravity [m/s2]

(2) Screens
Screens include the inclined disk type or cylindrical, tangential, vibrating, and drum type. An advantage of the tangential screen is that it is mostly self cleaning. The application for screening ranges from primary treatment to the removal of residual suspended solids in the effluent from biological treatment processes. The headloss can be estimated using the following equation.

$$h_L = \frac{1}{C(2g)} \left(\frac{Q}{A}\right)^2$$

h_L: head loss [m]
C: coefficient of discharge for the screen (@ 0.60)
g: acceleration due to gravity [m/s^2]
Q: discharge through screen [m^3/s]
A: effective open area of submerged screen [m^2]

13.3.2   Sedimentation (settling)
Sedimentation is the separation of suspended particles that are heavier than water by gravitational settling. In most cases the primary purpose is to produce a clarified effluent, but it is also necessary to produce concentrated sludge that can be easily handled and treated.

There are 4 types of sedimentation and it is common to have more than one type of settling occurring at a given time. In fact, it is possible to have all four occurring simultaneously during the sedimentation operation.

(1) Discrete Particle Sedimentation (Type 1)
This is the sedimentation of particles in a suspension of low solid concentration. Particles settle as individual entities, and there is no significant interaction with neighbouring particles. Sedimentation of grits and sand particles from wastewater are type 1.

(2) Flocculant Sedimentation (Type 2)
This refers to a rather dilute suspension of particles that coalesce, or flocculate, during the sedimentation operation. By coalescing the particles increase in mass and settle at a faster rate. This process is used to remove a portion of the suspended solids in untreated wastewater in primary settling facilities and in upper portions of secondary settling facilities. It is also used to remove chemical floc in settling tanks.

(3) Hindered (or Zone) Sedimentation (Type 3)
This refers to suspensions of intermediate concentration in which inter-particle forces are sufficient to hinder the settling of neighbouring particles. The particles tend to remain in fixed positions with respect to each other and the mass of particles settles as a unit. It occurs in settling devices where a solids-liquid interface develops at the top, usually used in conjunction with biological treatment facilities.

(4) Compression Sedimentation (Type 4)
Compression takes place from the weight of the particles, which are constantly being added to the structure by sedimentation from the supernatant fluid. It usually occurs in the lower layers of a deep sludge mass; such as in the bottom of deep secondary settling and sludge thickening facilities (Eckenfelder, 1961).

13.3.3 Flotation
Flotation is a unit operation used to separate solid or liquid particles, such as grease from a liquid phase. Separation may be enhanced by introducing fine gas (usually air) bubbles into the liquid phase. The bubbles attach to the particles and the buoyant force of the combined particle and gas bubbles cause the particle to rise to the surface. Particles that have a higher density than the liquid can thus be made to rise. The rising of particles with lower density than the liquid can also be facilitated (e.g., oil suspension in water).

In wastewater treatment flotation is used principally to remove suspended matter. The principal advantage of flotation over sedimentation is that very small or light particles that settle slowly can be removed more completely and rapidly. Once the particles have been floated to the surface they are removed by a skimming operation (Eckenfelder, 1961).

Dissolved Air Flotation
In this system, air is dissolved in the wastewater held under high pressure. This is followed by release of pressure to the atmospheric pressure level. In small systems the entire flow may be pressurized to 275–350 kPa with compressed air added at the pump suction. In larger units a portion of the effluent is recycled, pressurized, and semi-saturated with air. The recycled flow is mixed with the unpressurized main stream and the air finally comes out of solution.

Air Flotation
Air that is injected directly into the liquid phase through a revolving impeller or through diffusers forms air bubbles to provide the buoyant force. Aeration alone for a short period is not particularly effective in flotation of solids.

Vacuum Flotation
Wastewater is saturated with air and then a partial vacuum is applied. This causes the dissolved air to come out of solution as fine bubbles. The solid

particles attached to the bubbles rise to the surface to form a scum layer which is removed by a skimming mechanism.

Chemical Additives
Chemicals such as aluminum, ferric salt, activated silica and organic polymers are used to aid the flotation process. These chemicals function to create a surface or a structure that can easily absorb or entrap air bubbles.

13.3.4   Filtration
Filtration is usually a polishing step to remove small flocs or precipitant particles. Although filtration removes many pathogens from water, filtration should not be relied upon for complete health protection.

The most commonly used filtration processes involve passing the water through a stationary bed of granular medium that retains solids. Several modes of operation are possible in granular medium filtration including: downflow, upflow, biflow, pressure, and vacuum filtration. Among these, the most common practice is gravity filtration in a downward mode; with the weight of the water column above the filter providing the driving force.

The solids removal operation with granular medium filters involves several complicated processes. The most obvious process is the physical straining of particles too large to pass between filter grains. Removal of particles and floc in the filter bed depends on mechanisms that transport the solids through the water to the surface of the filter. Transport mechanisms include settling, inertial impaction, diffusion of colloids into areas of lower concentrations, and to a lesser extent, Brownian movement and van der Waals forces. (*Brownian motion: The random, thermal motion of solute due to their continuous bombardment by the solvent molecules*)

Removal begins in the top portion of the filter. As pore openings are filled by the filtered material increased hydraulic shear sweeps particles farther into the bed. When the storage capacity of the bed has thus become exhausted the filter must be cleaned. Hydraulic backwashing is the most used method to clean the filter. Backwash water containing the accumulated solids is disposed of and the filter can then be reused.

The smaller the size of granular media, the smaller the pore openings through which the water must pass. Small pore openings increase filtration efficiency not only because of straining but also because of other removal mechanisms. However, as size of pore openings decrease, head loss through the medium increases, resulting in a diminished flow rate. Larger media increase pore size, reduce head loss, and increase flow rate, but filtration efficiency decreases (Cheremisinoff, 1993).

(1) Slow sand filter
This was the first filter to be used for water purification during the 1800's. These filters were constructed of fine sand with an effective size of about

0.2 mm. The small size resulted in virtually all of the suspended material being removed at the filter surface. Slow sand filters have large space requirement and are capital intensive. Additionally, they do not function well with highly turbid water since the surface plugs quickly, requiring frequent cleaning.

(2) Rapid sand filter

This filter was developed to alleviate difficulties of the slow sand filter. Sizes range from 0.35 to 1.0 mm or even larger, with effective sizes of 0.45-0.55 mm. These larger sizes result in a rate of filtration an order of magnitude larger than that of the slow sand filter.

(3) Dual media filters

Dual media filters are usually constructed of silica sand and anthracite coal. The large pores in the anthracite layer remove large particles and flocs while most of the smaller material penetrates to the sand layer before it is removed. Dual media filters thus have the advantage of more effectively utilizing pore space. A disadvantage of dual media filters is that the filtered material is held loosely in the anthracite layer. Any sudden increase in hydraulic loading removes particles from the anthracite layer and transports them to the surface of the sand layer which results in rapid binding at this level.

(4) Mixed media filters

The ideal filter would consist of medium sized granules graded evenly from large at the top to small at the bottom. This can be accomplished by using three or more types of media with carefully selected size and density. Dual and mixed media filters make possible the direct filtration of water of low turbidity without settling operations.

13.4   Biological methods of waste treatment

13.4.1   Aerobic processes

The uses for aerobic biological processes are chemical oxidation of organic material, ammonium (NH_4^+), or ammonia (NH_3). These reactions are brought about with oxygen as the final electron acceptor. Organic material is mineralized to H_2O, CO_2, NH_4^+ or NH_3, and other constituents, and the NH_4^+ or NH_3 are oxidized to nitrate (NO_3^-). Each reaction also results in the synthesis of new cell mass.

Biochemical Reactions

(i) Oxidation and Synthesis

$$C_aH_bO_cN_dS_e + O_2 + nutrient \rightarrow CO_2 + NH_3 + C_5H_7NO_2 + H_2O$$
$$+ \text{ other end product}$$

(a,b,c,d and e are determined once specific end products are known)

(ii) Nitrification

$$NH_3 \xrightarrow{Nitrosomonas} NO_2 \xrightarrow{Nitrobacter} NO_3^-$$

(iii) Denitrification

$$NO_3^- \rightarrow NO_2^- \rightarrow NO \rightarrow N2O \rightarrow N2$$

(iv) Endogenous respiration

$$C_5H_7NO_2 + 5O_2 \rightarrow 5CO_2 + 2H_2O + NH_3 + energy$$

(113)   (160) (These are the molecular weights)
$1:1.42$ (These are ratios of molecular weights showing it takes 1.42 moles
            of O_2 to every mole of bacteria to undergo endogenous respiration.)

Microbiology
Achromobacter, Beggiatoa, Flavobacterium, Geotirichum, Nitrobacter, Nitrosomonas, Pseudomonas, Sphaerotilus are the common microorganisms found in aerobic systems. Protozoa such as *Vorticella, Opercularia, Epistylis* and rotifers do not stabilize waste but they consume dispersed bacteria and small biological floc particles that have not settled (i.e. protozoa and rotifers act as effluent polishers)

(i) Suspended growth system
Bdellovibrio, Lecicothrix, Mycobacterium, Nocardia, Thiothrix, Zoogloea are bacteria that are mostly found in suspended growth systems.

(ii) Attached growth system
Alcaligenes, Chlorella, Fusazium, Mucor, Penicillium, Phormicium, Sphaerotilus natans, Sporatichum, Ulothrix, and yeasts are mostly found in attached growth systems. Algae such as *Chlorella, Phormicium, Ulothrix* are not directly involved in waste stabilization. They add oxygen to the system but they can cause clogging of the system or take up O_2 when they die, which can cause bad odours.

13.4.2 Anaerobic processes
The anaerobic process mineralizes organic and inorganic materials in the absence of molecular oxygen. Because oxygen is toxic to most anaerobic microorganisms, molecules other than oxygen such as sulphur or carbon dioxide are used as the

final electron acceptors. Thus, the final products of anaerobic digestion of organic waste are mostly CH_4, CO_2, and biomass, whereas the final products of aerobic digestion are mostly CO_2, H_2O and biomass. Aerobic bacteria get more energy from the conversion of organic substrate to their end products than do anaerobic bacteria to theirs and thus they are generally faster growing.

The overall anaerobic conversion of biodegradable organic materials to the final end products, CH_4 and CO_2 can be divided into three phases which occur simultaneously. These are hydrolysis, acidification and the formation of end products. Since the major end product is mostly CH_4, this phase is often called methanogenesis. A basic outline of the pathways of anaerobic metabolism is given in Fig. 13.3. More detailed pathways through completion of the acidogenic phase are given in Fig. 13.4. Under most circumstances in treating food processing wastes, acetate is a very common end product of acidogenesis. This is fortunate because acetate is easily converted to methane in the methanogenic phase. Due to the difficulty of isolation of anaerobes and the complexity of the bioconversion processes much still remains unsolved about anaerobic processes.

Hydrolysis phase
Large insoluble molecules such as carbohydrates, proteins and fat are solubilized into their monomers. This process is mediated by extra-cellular enzymes produced by various hydrolyzing bacteria (Bryant, 1979). This step is sometimes called the liquefaction phase.

Acidogenic phase
A group of microorganisms called acidogens, which produce mainly acids convert the soluble products of the hydrolysis phase to short chain organic acids,

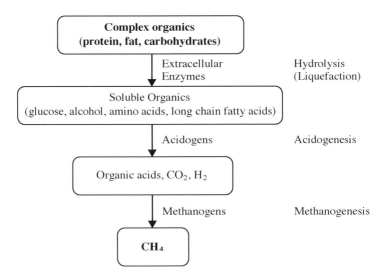

Fig. 13.3   Schema of anaerobic metabolism pathways.

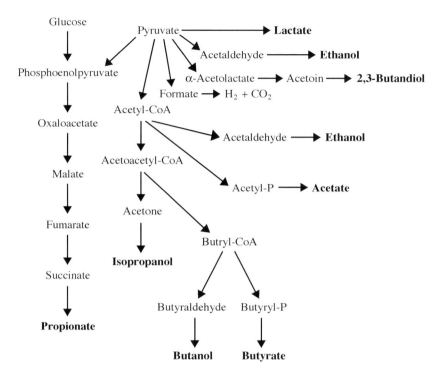

Fig. 13.4   Anaerobic metabolism pathway and end products.

CO_2, and hydrogen (McCarty and Smith, 1986). These acids normally have chain lengths from C_1 to C_4. This step is often called the acidogenic phase or acidogenesis.

End products formation
Finally, methane producing bacteria, called methanogens, convert the short chain organic acids to final products which are mainly CH_4 and CO_2. This step is called the methanogenic phase or methanogenesis (Balch *et al.*, 1979; Huser *et al.*, 1982; Zeikus, 1977).

Microbiology

(i) Hydrolyzing bacteria and acidogens
Actinomyces, Bifidobacterium spp., *Clostridium* spp., *Corynebacterium* spp., *Desulphovibrio* spp., E. coli, Lactobacillus, Peptococcus anaerobus, Staphylococcus

(ii) Methanogens
Methanobacillus, Methanobacterium, Methanococcus, Methanosarcina (Smith

and Mah, 1978; Zinder *et al.*, 1984; Robinson and Tiedje, 1984; Schmidt and Ahring, 1993).

13.5   Chemical methods of waste treatment

Disinfection refers to operations aimed at killing pathogens. Sterilization, the complete destruction of all living matter, is not usually the objective of disinfection. A good disinfectant must be toxic to microorganisms at concentrations well below the toxic thresholds to humans and higher animals. Additionally, it should have a fast rate of kill and should be persistent enough to prevent regrowth of organisms in the distribution system. In the USA, disinfection and chlorination have become synonymous terms, while ozonation has been used more widely in Europe.

13.5.1   Chlorination

Chlorine may be applied to water in gaseous form, Cl_2, or as an ionized product of solids, $Ca(OCl)_2$ or $NaOCl$. The reactions in water are as follows:

$$Cl_2 + H_2O \rightarrow H^+ + HOCl$$

$$Ca(OCl)_2 Ca^{2+} + 2OCl^-$$

$$NaOCl \rightarrow Na + OCl$$

Where: $Ca(OCl)_2$ is calcium hypochlorite, $NaOCl$ is sodium hypochlorite, $HOCl$ is hypochlorous acid, and OCl is hypochlorite ion. Both $Ca(OCl)_2$ and $NaOCl$ are more expensive on an equivalence basis than liquefied Cl_2. Chlorine gas, however, is a very strong oxidant that is toxic to humans. Since it is heavier than air, it spreads slowly at ground level. Therefore, extreme care must be provided in its manufacture, shipping, and use. This sometimes dictates the use of hypochlorites.

At low concentrations, chlorine probably kills microorganisms by penetrating the cell and reacting with the enzymes and protoplasm. At higher concentrations, oxidation of the cell wall will destroy the organism. Factors affecting the process are:

(1)   Forms of chlorine
(2)   pH
(3)   Concentration
(4)   Contact time
(5)   Types of organism
(6)   Temperature

13.5.2   Ozonation

Ozone can be produced in a high strength electrical field from pure oxygen or from the ionization of clean dry air.

$$O_2 \xrightarrow{\text{high voltage}} O + O$$
$$O_2 + O \leftrightarrow O_3$$

Ozone is a powerful oxidant which reacts with reduced inorganic compounds and with organic material including vigorous reactions with bacteria and viruses. It is reported to be more effective than chlorine in inactivating resistant strains of bacteria and viruses. Because ozone is chemically unstable it must be produced on-site and used immediately. Cost of ozonation is two to three times higher than the cost of chlorination. Ozone has a low solubility in water and thus must be mixed thoroughly with the water to ensure adequate contact. This can be a problem when air is used as the oxygen source, since large volumes of nitrogen must also be handled.

13.5.3   Ultraviolet irradiation

Irradiation with ultraviolet light is a promising method of disinfection. This method is effective in inactivating both bacteria and viruses. The most effective band for disinfection is in the shorter range of 2000 to 3000. The glass surface must be constantly cleared to provide efficient irradiation.

13.6   Land treatment of waste

13.6.1   Land application

Land application of wastes is the surface spreading or subsurface injection (10–25 cm (4–10 in) down) of liquid or solid waste material on or into soil (Ockerman and Hansen, 2000). Wastewater polluted with organics, including wastewater from a food processing facility can be land applied. This method of treatment removes a large percentage of organic and inorganic pollutants from food processing waste. Land treatment employs mechanical, biological and chemical processes which occur naturally in or on the soil in the purification of wastewater. Organic matter in food processing waste represents high BOD and is degraded by soil bacteria. Refractory matter eventually becomes part of the soil matrix. Nutrients in the waste are taken up by crops or recycled into the soil.

Commercial agricultural manure spreaders and liquid-manure injection equipment are suitable for land application of food processing waste. Subsurface injection of offensively odoriferous material will control odors, reduce insect attraction and conserve nitrogen. Surface-applied animal-processing waste can be incorporated into the soil by ploughing or discing soon after application. References which explain land application of wastes in more detail include Loehr et al., 1979a, b, Reed and Crites, 1984, Ockerman and Hansen, 2000, and McFarland, 2001.

The large amounts of wastewater produced by a food processing plant often require a very large land area for even a medium-sized plant. If land is not available or too costly, this type of treatment method is not feasible. In most cases, waste streams should be segregated and the most heavily polluting wastes can be land applied, as for example blood from meat processing or rendering that is about 160 000 mg/l BOD.

Food processing waste is often a good source of nitrogen (N), phosphorus (P) and potassium (K). Nearly all of the P and K in food processing wastes are usually available for plant use the year of application. Nitrogen in organic waste is of two forms, organic and ammonia. The nitrogen in the ammonia form is available when spread. However, if the organic waste is spread onto the surface, sizable amounts of ammonia N may be lost to the atmosphere by volatilization. Generally less than one-half of organic N is available to plants in the year it is spread. Organic N released (mineralized breakdown of organic N to available ammonia N) during the second, third and fourth years after initial application is usually about 50%, 25% and 12.5% respectively of that mineralized during the first cropping season. However, to determine precisely how much N, P or K will be available to crops from land application of food processing waste, it is necessary to have the waste analysed. This can be done at state universities (in the USA) or at a commercial laboratory. Records should be kept of each year's application. Soil receiving food processing wastes should also be tested regularly for levels of the so called macro nutrients N, P, and K. The waste and soil might also be tested for micro nutrients such as sulphur and iron and for possible toxic substance buildup such as copper, zinc, lead, or cadmium. Food waste application rates should be designed to match the nutrients removed by the crop. Table 13.4 lists approximate amounts of nutrients removed from the soil by several crops. If a particular waste is extremely high in some particular nutrient, it will limit application rate. Most often, application rate is limited by P or N. If application rate is limited due to a particular nutrient, the waste can still be used to advantage by adding commercial fertilizer to make up deficits in other nutrients.

Table 13.4 Approximate amounts of plant nutrients removed from soil by selected crops (Ockerman and Hansen, 2000)

Crop and yield goal	N	P_2O_5	K_2O
	[kg/ha (lb/acre)]		
Alfalfa (Lucerne):13.4 tonne/ha (6 ton/acre)	381 (340)	90 (80)	404 (360)
Corn: 13.9 m³/ha(160 bushel/acre)	163 (145)	67 (60)	50 (45)
Corn: 17.4 m³/ha (200 bushel/acre)	202 (180)	84 (75)	62 (50)
Corn silage: 60.5 tonne/ha (27 ton/acre)	275 (245)	95 (85)	275 (245)
Soybeans: 4.4 m³/ha (50 bushel/acre)	213 (190)	45 (40)	78 (70)
Soybeans: 5.7 m³/ha (65 bushel/acre)	275 (245)	56 (50)	101 (90)
Wheat: 0.9 m³/ha (55 bushel/acre)	78 (70)	39 (35)	22 (20)

13.6.2   Landfilling

Landfilling is the most widely used method for the long term handling of solid wastes. Important aspects in the implementation of sanitary landfills include: (1) site selection (2) landfill methods and operations (3) control of gas and leachate produced in landfills. Landfilling is becoming more and more of a problem primarily because landfill sites are very limited (i.e. land is generally expensive near the heavily populated area where large amounts of solid wastes, including food processing wastes are produced) and people's biased thinking Most people seem to understand the waste problem but nobody wants to build landfill site near their home.

13.7   Future trends

Lettinga and Hulshoff Pol, (1991) have listed criteria to select proper treatment for any given wastewater. The criteria for environmental protection technologies and methodologies are as follows:

1. They should lead to prevention of the production of additional wastewater(s), or at least a sharp reduction.
2. They should not require any dilution of the pollutants with clean water.
3. They should provide a high efficiency with respect to environmental pollution control.
4. They should lead to maximum recovery and reuse of polluting substances (e.g. to integrated systems) particularly from food processing wastewaters.
5. They should be low cost, both with respect to construction, required infrastructure (including energy requirement) and operation and maintenance.
6. They should be applicable at small as well as large scale.
7. They should lead to a high self-sufficiency in all respects.
8. They should be well acceptable for the local population.
9. The method should provide sufficient treatment efficiency for removal of various categories of pollutants, i.e. biodegradable organic matter, suspended solids, ammonia, organic-N compounds, phosphates and pathogens.
10. The system should be stable for interruptions in power supply, peak loads, feed interruptions, and/or for avoiding toxic pollutants.
11. The process should be flexible with respect to future extensions and possibilities to improve the efficiency.
12. The system should be simple to operate, maintain and control, so that good performance doesn't depend on the continuous presence of highly skilled operators and engineers.
13. Land requirements should be low, especially when little land is available and/or the price of the land is high.
14. The number of process steps should be as low as possible.

15. The lifetime of the system should be long.
16. The system should not suffer from serious sludge disposal problems.
17. The application of the system should not be accompanied with malodor or other nuisance problem.
18. The system should offer good possibilities for recovery of useful byproducts, such as for irrigation or fertilization.
19. There should be sufficient experience with the system to manage it easily.

Based on these criteria, anaerobic treatment technology is regarded as a highly effective method for the treatment of most wastewaters (Lettinga and Hulshoff Pol, 1991; Speece, 1996). It is also considered as the basis of a very promising environmental protection and resource recovery concept for the following reasons:

1. Anaerobic treatment is significantly less expensive than aerobic treatment for treatment of medium and high strength wastewater (COD > 1500 mg/l).
2. Energy is produced instead of used. This is especially of interest when treating highly concentrated waste.
3. Less space is required for an anaerobic treatment plant compared to an aerobic treatment plant.
4. Anaerobic treatment technology has relatively low equipment cost.
5. Anaerobic treatment is more suitable for the treatment of wastewaters of seasonal industries, because anaerobic sludge can be preserved without serious deterioration in activity or settleability, provided the temperature is maintained below 15°C.
6. Available anaerobic treatment systems can be applied at small as well as large scale.
7. The anaerobic treatment technology does not require the import of expensive equipment.
8. Anaerobic process stability can be easily achieved.
9. Waste biomass disposal costs are low, partly because wasted sludge volume is relatively low.
10. Nutrient supplementation costs are usually low.
11. Management requirement is low.
12. Off-gas air pollution can be eliminated.
13. Foaming of surfactant containing wastewaters can be avoided.
14. Aerobic non-biodegradable organics can often be degraded anaerobically.
15. Chlorinated organic toxicity levels can be reduced.
16. Seasonal treatment can be provided.

The development of new bioreactors is one reason that anaerobic treatment technologies will likely gain in popularity. Organic loading rates (OLR) well over $20 \, kg \, COD/m^3/day$ ($g \, COD/l \cdot d$) are possible for some types of food processing wastes in a modern high rate anaerobic digester (Table 13.5). High loading rates result in relatively low hydraulic retention times, which reduces capital costs because vessels holding the waste are smaller. More information

Table 13.5 Comparisons of various anaerobic treatment processes for cheese whey

Reactor[a]	Waste	Temp. (°C)	HRT (d)	Influent strength (g COD/l)	OLR (g COD/l.d)	Treatment efficiency (%)	Reference
UFFLR	sour whey	35	5	79	14	95	Wildenauer and Winter (1985)
DSFFR	whey	35	5	13	2.6	88	De Haast et al. (1985)
FBR	whey	35	0.4	7	7.7	90	Boening and Larsen (1982)
FBR	whey	35	0.1–0.4	0.8–1.0	6–40	63–87	Denac and Dunn (1988)
AAFEB	powder	28–31	0.4–1.1	10	8.9–27	77–93	Switzenbaum and Danskin (1982)
	whey	35	0.6–0.7	5–15	8.2–22	61–92	Switzenbaum and Danskin (1982)
AnRBC	whey	35	5	64	10.2	76	Lo and Liao (1986)
	whey	35	6–11	61–70	6.3–10	76	Lo and Liao (1986)
SDFA	whey	4.3		69.8	16.1	99	Barford et al. (1986)
UASB	deproteinized whey	35	1.5	11	7.1	94	Schroder and De Haast (1989)
UASB	whey	33	5	5–28.7	0.9–6	97–99	Yan et al. (1989)
DUHR	whey	35	7	68	10	97	Malaspina et al. (1995)
UASB	whey	35	2.3–11.6	5–77	1–28.5	95–99	Kalyuzhnyi et al. (1997)
UASB	whey	22–30	5.4–6.8	47–55	7–9.5	90–94	Kalyuzhnyi et al. (1997)
UASB	whey	20–29	3.3–12.8	16–50	1–6.7	90–95	Kalyuzhnyi et al. (1997)
UASB	whey	34	5	25–30		90–98	Yan et al. (1993)
CSTR and DUHR	whey				10	98	Malaspina et al. (1996)
CSTR and AT	whey	33–36	2	2–6	5	90–95	Ince (1998)
MPAR	whey	34	2.3–2.4	20–37	9–14.7	92–98	Guiot et al. (1995)

[a] Acronyms for the various types of reactors are as follows:

UFFLR = upflow fixed-film loop reactor; DSFFR = downflow stationary fixed-bed reactor; FBR = fluidized-bed reactor; AAFEB = anaerobic attached-film expanded-bed reactor; AnRBC = anaerobic rotating biological contact reactor; SDFA = semicontinuous digester with flocculent addition; UASB = upflow anaerobic sludge-blanket reactor; DUHR = downflow-upflow hybrid reactor. CSTR = continuously stirred tank reactor; AT = anaerobic filter; MPAR = multiplate anaerobic reactor.

about anaerobic systems for food processing waste in addition to references listed in Table 13.5 can be found in: Speece, 1996; Lettinga and Hulshoff Pol, 1991; McCarty, 1964; Malina and Pohland, 1992, and Tchobanoglous and Burton, 1991.

13.8   References

BALCH, W.E., G.E. FOX, L. J. MAGRM, C.R. WOESE and R.S. WOLFE (1979), Methanogens: Reevaluation of a unique biological group, *Microbiological Reviews* 43(2): 260–296.

BARFORD, J.P., R.G. CALI, I.J. CALLANDER and E.J. FLOYD (1986), Anaerobic digestion of high-strength cheese whey utilizing semicontinuous digesters and chemical flocculent addition, *Biotechnology and bioengineering* 28(11): 1601–1607

BOENING, P.H. and V.F. LARSEN (1982), Anaerobic fluidized bed whey treatment, *Biotechnology and bioengineering* 14: 2539–2556.

BRYANT, M.P. (1979), Microbial Methane Production-Theretical Aspects, *J. of Animal Science* 48(1): 193–201.

CARAWAN, R.E., J.V. CHAMBERS and R.R. ZALL (1979), *Water and Wastewater Management in Food Processing- Seafood Water and Wastewater Management*, Raleigh, NC, North Carolina Agricultural Extension Service.

CHEREMISINOFF, NICHOLAS P. (1993), *Filtration Equipment for Wastewater Treatment*. Englewood Cliffs, Prentice Hall.

DE HAAST, J., T.J. BRITZ, J.C. NOVELLO and E.W. VERWEY (1985), Anaerobic digestion of deproteinated cheese whey, *J. Dairy Res.* 52: 457–467.

DENAC, M. and I.J. DUNN (1988), Packed- and Fluidized-bed biofilm reactor performance for anaerobic wastewater treatment, *Biotechnology and bioengineering* 32(2): 159–173.

ENVIRONMENTAL PROTECTION AGENCY (1977), *Pollution abatement in the fruit and vegetable industry*, EPA-625/3-77-0007, Washington, DC, USEPA Technology Transfer.

ECKENFELDER, W.W. JR. (1961), *Biological waste treatment*, Oxford, Pergamon Press.

GOSTA, BYLUND (1999), *Dairy processing Hand Book*. Tetra Pak Processing System. Lum, Sweden. 331–354.

GUIOT, S.R., B. SAFI, J.C. FRIGON, P. MERCIER, C. MULLIGAN, R. TREMBLAY and R. SAMSON (1995), Performance of a full-scale novel multiplate anaerobic reactor treating cheese whey effluent, *Biotechnology and bioengineering* 45: 398–405

HANSEN, C.L., R.K. WHITE and M. OKOS (1984), Optimization of Water Management and Associated Energy in a Meat Packing Plant, *Transactions of the Am. Soc. of Ag. Engr.* 27(1): 305–319.

HANSEN, C.L. and R.J. WRIGLEY (1997), Dairy Processing Waste Management in Australia, *Proceedings of International Congress on Engineering and Food 7*, Boca Raton, FL, USA, Lewis Publishers, CRC Press.

HUSER, B.A., K. WUHRMANN and A.J.B. ZEHNDER (1982), *Methanothris soehngenii* gen. nov., a new acetotrophic non-hydrogen-oxidizing methane bacterium, *Archives of Microbiology* 132(1): 1–9.

HWANG, S. (1995), Bioprocess Models to Control Production of Short-Chain Organic Acids from Cheese-Processing Wastewater, *Dissertation, Utah State University*, Logan, UT, 164 pp.

HWANG, S. (1993), Anaerobic Treatment of Whey Permeate Using Upflow Sludge Blanket Bioreactors. *Thesis, Utah State University*, Logan, UT, 85 pp.

INCE, O., B.K. INCE and O. YENIGUN (2001), Determination of potential methane production capacity of a granular sludge from a pilot-scale upflow anaerobic sludge blanket reactor using a specific methanogenic activity test, *Journal of chemical technology and biotechnology* 76(6): 573–578.

JONES, H.J. (1994), *Waste Disposal Control in the Fruit and Vegetable Industry,* London, Noyes Data Corporation.

KALYUZHNYI, S.V., E.P. MARTINEZ and J.R. MARTINEZ (1997), Anaerobic treatment of high-strength cheese-whey wastewater in laboratory and pilot UASB reactors, *Bioresource technology* 60(1): 59–65

LAWRENCE, A.W. and MCCARTY, P.L. (1969), Kinetics of methane fermentation in anaerobic treatment. *J. WPCF.* 41: R1–R16.

LETTINGA, G. and L.W. HULSHOFF POL. (1991), *Application of modern high anaerobic treatment processes for wastewater treatment, New developments in industrial wastewater treatment,* edited by Turkman A. and O. Uslu. 33 64, Netherlands, Kluwer Academic Publishers.

LO, K.V. and P.H. LIAO (1986), Digestion of cheese whey with anaerobic rotating biological contact reactor. *Biomass* 10: 243–252.

LOEHR, R.C., JEWELL, W.J., NOVAK, J.D., CLARKSON, W.W. and G.S., FREIDMAN (1979a), *Land application of wastes, vol. 1, ,* New York, Van Nostrand Reinhold.

LOEHR, R.C., JEWELL, W.J., NOVAK, J.D., CLARKSON, W.W. and G.S., FREIDMAN (1979b), *Land application of wastes vol. 2,* New York, Van Nostrand Reinhold.

MCCARTY, P.L. and D.P. SMITH (1986), Anaerobic wastewater treatment. *Environmental Science and Technology* 20(12): 1200–1206.

MCCARTY, P.L. (1964) *Anaerobic Waste Treatment Fundamentals,* Public Works.

MCFARLAND, M. (2001), *Biosolids Engineering,* Columbus, Ohio, McGraw-Hill.

MALASPINA, F., L. STANTE, C.M. CELLAMARE and A. TILCHE (1995), 'Cheese whey and cheese factory wastewater treatment with a combined biological anaerobic-aerobic plant', *In Proc. III Int. Symp. on waste management problems in Agro-industries*, Mexico, pp. 63–76.

MALASPINA, F., C.M. CELLAMARE, L. STANTE and A. TILCHE (1996), Anaerobic treatment of cheese whey with a downflow-upflow hybrid reactor. *Bioresource technology* 55: 131–139.

MALINA, J.F., JR and F.G. POHLAND (1992), *Design of anaerobic processes for the treatment of industrial and municipal wastes,* Lancaster, PA, Technomic.

MONOD, J. (1949), The growth of bacterial cultures. ann. Rev. Microb. 3: 371–394.

OCKERMAN, H.W. and C.L. HANSEN (2000), *Animal By-Product Processing & Utilization,* Lancaster, PA, Technomic.

REED, S. C. and R.W. CRITES (1984), *Handbook of land treatment systems for industrial and municipal wastes,* Park Ridge, New Jersey, Noyes Publications.

ROBINSON, J.A. and J.M. TIEDJE (1984), competition between sulfate-reducing and methanogenic bacteria for H2 under resting and growing conditions, *Archives of Micriobiology* 137(1) 26–32.

SCHMIDT, J.E. and B.K. AHRING (1993), Effects of hydrogen and formate on the degradation of propionate and butyrate in thermophilic granules from an upflow anaerobic sludge blanket reactor, *Applied and Environmental Microbioloby* 59 (8): 2546–2551.

SCHRODER, E.W. and J. DE HASST (1989) Anaerobic digestion of deproteinated cheese whey in an upflow sludge blanket reactor, *J. Dairy Res.*56: 129–139.

SMITH, M.R. and R.A. MAH (1978), Growth and methanogenesis by *Methanosarcina* strain 227 on acetate and methanol. *Applied and Environmental Microbiology.* 36(6): 870–879.

SPEECE, R.E. (1996), *Anaerobic biotechnology for industrial wastewaters,* Nashville, Tennessee, Archae Press.

SWITZENBAUM, M.S. and S.C. DANSKIN (1982), Anaerobic expanded bed treatment of whey. *Agric. Waste* 4: 411–426.

TCHOBANOGLOUS, G. and F.L. BURTON (1991), *Wastewater Engineering Treatment, disposal, and reuse, 3rd edn*, New York, McGraw-Hill.

WILDENAUER, F.X. and J. WINTER (1985) Anaerobic digestion of high strength acidic whey in a pH-controlled up-flow fixed-film loop reactor, *Appl. microbiol. Biotechnol* 22: 367–372.

YAN, J.Q., K.V. LO. and K.L. PINDER (1989), Anaerobic digestion of whey in the UASB reactor, *ASAE paper no. 89-6570*, New Orleans, Louisiana.

YAN, J.Q., K.V. LO. and K.L. PINDER (1993), Instability caused by high strength of cheese whey in a UASB reactor, *Biotechnology and bioengineering* 41: 700–706.

ZEIKUS, J.G. (1977), The biology of methanogenic bacteria, *Bacteriological Reviews* 41(2): 514–541.

ZINDER, S.H., S.C. CARDWELL, T. ANGUISH, M. LEE and M. KOCH (1984), Methanogenesis in a thermophilic (58°C) anaerobic digester, *Water Research* 16: 303–311.

14

Assessing the safety and quality of reused packaging materials

C. Simoneau and B. Raffael, European Commission Joint Research Centre, Italy and R. Franz, Fraunhofer Institute for Process Engineering and Packaging, Germany

14.1 Introduction

Most foodstuffs are sold packaged. The packaging fulfils in particular an important role for mechanical, biological and chemical protection of perishable food products during their shelf life. Various food contact materials and constituents can be used provided they do not pose health concerns to consumers, which may occur when some substances from the food packaging migrate into the food. To ensure the safety of such materials, food-packaging regulations in Europe thus require that the packaging materials must not cause mass transfer (migration) of harmful substances to the food, by imposing restrictions on substances from the material itself that could migrate into the food. Consequently food packaging materials must comply with many chemical criteria and prescribed migration limits. The migration of substances from the materials into the foodstuffs is a possible interaction that must be minimised or even avoided, since it may affect the food or pose health concerns to the consumer. There is thus a strong need to better understand migration phenomena and develop appropriate methodologies to test migration from food contact materials. This need is even more relevant in the case of recycled food contact material, a growing field due to its importance with respect to the protection of the environment. Indeed, recycling of packaging is required by environmental Directives and a certain amount is destined towards recycling for food use.

The most common materials used for food packaging are paper and board, plastics in the form of rigid containers or films, glass and coated metals (such as tin, steel and aluminium). Novel packaging has appeared in recent years, as for example biobased, biodegradable, active, and intelligent packagings. The most commonly recycled materials for food use are currently PET as well as paper and board. This use of recycled layers in PET beverage bottles or of recycled paper fibres should however not be the cause of potential health risks to the consumer. Therefore there is a need to know on a scientific basis whether the current state of the art of both collecting and recycling in Europe for food use is adequate, i.e. does not leave or bring forth the presence of contaminants. This type of investigation in turn requires the development of fast, simple and efficient analytical screening methods. Such methodologies are necessary in order to provide an estimation based on experimental data of the nature and concentration of potential harmful chemicals either retained through or brought in via the recycling processes and thus provide a sound basis for risk assessment.

The main role of packaging is to protect the food from any possible contamination of chemical, biological or physical origin and to let the consumer identify the product content, displaying all the information of use to the consumer (such as price, ingredients, instructions for use, shelf life, and other). It also represents an essential opportunity to attract the consumer with nice shape, adequate graphics and colours, and affect positively consumer preference in the absence of a sensory impact (Anon., 2002a). The demanding consumers of today constantly request a greater variety and greater convenience (for storage and for preparation) of foods to be available on the market. Alternatively industry must constantly improve the production service and the quality level to adequately answer the public demand as well as the legislation requirements. Thus there is the need of constant improvements of existing packaging and the development of new ones.

Every type of food has special requirements for an adequate food packaging depending on its characteristics and its uses, thus the variety of products present on the market is very broad. The most common materials used for food packaging are paper and board, plastic in the form of rigid containers or films, glass and coated metals (such as tin, steel and aluminium). In recent years two new categories of food packaging appeared: active and intelligent packaging. Active packaging interacts with the contents, senses the environmental changes and acts, changing its properties, to protect, improve and prolong shelf life of the packed food. The protection might be from microbial attack, chemical contamination, oxygen, humidity, light, etc. Intelligent packaging interacts with the content to give information on the status of the content. It can consist of electronic locators, anti-theft and anti-counterfeiting devices, time-temperature integrators, sensors for the presence of spoilage or microorganisms. At the moment these types of packaging are not specifically regulated by European laws. Not being inert, they can cause the interaction of unwanted substances with the food. (Brody, 2001).

14.2   Recyclable plastic packaging: PET

Plastic comes as a by-product of oil refining, and uses only 4% of the total worlds oil production. It is a 'biogeochemical' manipulation of certain properties of oil, into polymers, that behave 'plastically'. Plastics, for their high versatility and all their properties suitable for food packaging, are used more and more extensively. The combination of the many properties of plastics makes them ideally suited for food packaging. The demand for plastic materials has increased in the last years, reaching 66.5% of all the packaging used for food contact and several leading companies that once specialised in paper and board products have decided to concentrate more effort and resources into developing their involvement in plastics, and this is a trend that others may follow. Besides polyolefines, polyethylene terephthalate (PET) is the most popular polymer for food contact as it has excellent toughness and chemical inertness both for freezing and ovening food (De A. Freire *et al.*, 1999). PET is a relatively new plastic resin; it is a lightweight, durable, and shatter-resistant form of polyester. It is known to have high heat resistance and chemical stability. It has good barrier properties against oxygen and carbon dioxide and, when melt-blown, for flavours and fats (hydrocarbons) and hence has wide ranging packaging applications in the food and drink industry, both rigid and flexible forms finding particular uses as bottles, trays, containers and films. (Sheftel, 2000; Anon., 1994; Jones *et al.*, 1995).

   PET is based on carbon-oxygen-carbon links and is made by the low pressure catalytic melt polymerisation of dimethyl terephthalate (DMT) (or terephthalic acid, TPA) and ethylene glycol. The reacting groups are $-OH$ on the glycol and $-COOCH_3$ on the DMT (or $-COOH$ on the TPA) and the formation of a long chain depends on the formation of a CH_3OH (or H_2O) molecule made from the ending of the reactants that is split out. Ethylene glycol is made by ethylene, either by controlled catalytic oxidation or via chlorohydrin. Dimethyl terephthalate is made by esterification of terephthalic acid that is prepared by catalytic partial oxidation of p-xylene, itself a by-product of the petroleum industry. The reaction takes place in three steps. In a first step a transesterification takes place. Then internal rearrangements lead to the formation of hydroxyethyl methyl terephthalate and methanol that evaporates due to the temperature at which the reaction is conducted, pushing the transesterification towards high conversion. A further transesterification forms methanol and bis(2-hydroxyethyl) terephthalate that plays as both ester and alcohol transesterificating into the final product. The molecules grow to a molecular weight of up to 20 000. In order to make the grade of PET required for soft drink bottles the molecular weight must be increased over that which is suitable for film and this is done by heating film grade resins in a separate reactor to lengthen the chains (Boccaleri *et al.*, 1996; Jenkins *et al.*, 1991; Robertson, 1993a, b).

14.2.1 Recycled PET

Recycling is an opportunity to limit environmental pollution and to make an important contribution to attaining a more sustainable economy. In addition it helps to preserve the earth's resources. Among the benefits of recycling are cited the reduction of air pollution and conservation of mineral oil and gas, the extension of disposal capacity, saving money in disposal costs, conserving earth resources, creating jobs, and providing a reliable, cost-effective feed stock to industry. Among its drawbacks are cited that plastics from 'non-packaging' sources are rarely recycled, only PET and HDPE are recycled in significant quantity, plastics cannot be recycled indefinitely and are generally not recycled into food containers. In addition, because plastics are lightweight, they are expensive to collect and sort, and automatic sorting equipment can be expensive and some resins difficult to clean, thus virgin resins can be cheaper to buy.

The recycling of one way packaging and of PET bottles with or without deposit is an increasing task for the industry to avoid burning the material due to environmental concerns. After collection the waste is sorted and, where necessary, decontaminated and scrutinised before it is bailed, compacted and shipped for reprocessing into products such as: damp proof coursing for the construction industry, bin liners, chemical sacks, carrier bags, silage stretch, food packaging again. PET is the most promising plastic for recycling, as it shows a low diffusivity and low uptake that pose difficulties to possible contaminants to migrate into food. Moreover, as it is used almost without any additives for all its packaging devices and since there is almost only food grade material on the market for all applications, the material coming from the post consumer recollection can be considered homogeneous and almost fit for purpose.

The process of mechanical recycling of PET for food containers is fairly recent: the first approval was received in 1994 with a 'no-objection' letter from the U.S. Food and Drug Administration (FDA), which represents the sanctioning authority in the US. From that moment the research to improve and ensure the quality of the recyclates began to grow. This was realised increasingly by development and application of socalled super-clean processes (Franz *et al.*, 1998; Franz and Welle, 1999; Franz, 2002; Franz and Welle, 2002).

There are several modes of plastics recycling and each one poses different concerns regarding contaminants. For plastic recycling 3 approaches are available:

- Direct refill
- Physical reprocessing, such as grinding and melting, and reformation. This process does not alter the polymer structure.
- Chemical treatment

Also the reuse of pre-consumer industrial scrap must be mentioned, as it is accepted by the Environmental Protection Agency (EPA) with the definition of 'primary recycling'. This system is considered not to pose any hazard.

The direct reuse of the bottles, after being washed and sanitised has a big disadvantage in the case of plastic materials. As plastic is more likely to absorb contaminants than other types of packaging materials, there is the concern that after some reuse the presence of contaminants could be too high to respond to the safety rules or the bottle could be damaged, losing their mechanical properties. Recent studies however demonstrated that repeated washings do not cause a major risk of contaminants absorbance (Nielsen *et al.*, 1997: Jetten *et al.*, 1999). Physical reprocessing is based on the grinding of the used bottles to get flakes or pellets that after being washed to remove possible contaminants are melted to reform the package, without any alteration of the polymer. It is important that the size of the flakes or pellets is small enough to permit a proper washing, as small parts offer a high surface to the washing process. All the components of the package that are not made by PET, such as closures, labels, safety seals and coatings, are eliminated using water-based systems that separate the different materials based on their different density, being 1 g/cc PET density (Anon., 2002b). Recycled PET is always blended with virgin material with fractions of typically no higher than 25%. The reason for that is that the recycled material can cause deterioration of the polymer mechanical or optical properties.

Recent studies have shown that PET can absorb low molecular weight chemicals, mainly volatiles, but hardly absorbs the larger non-volatile compounds (Demertzis *et al.*, 1997; Begley *et al.*, 2002). Thus aroma compounds coming from beverages, typically packed into PET bottles, can be adsorbed by the package. But also other substances that can accidentally or on purpose come in contact with the bottle when a post consumer use takes place can contaminate the PET (Serad *et al.*, 2001; Bayer, 2002; Sadler *et al.*, 1996; Jetten *et al.*, 1999). It is important thus to note that proper purification procedures are essential in the contaminants removing step. Indeed while non-volatile compounds can hardly be removed when absorbed by a PET package, a washing and drying procedure in combination with modern deep-cleansing technologies can easily remove volatile substances, while not altering the PET properties even after several reuses (Komolprasert *et al.*, 1995; Jetten *et al.*, 1999; Jetten and de Kruijf, 2002). Thus, taking into account also that the packages intended for non-food applications are generally in a much smaller percentage than the packages used for food contact, recycled PET for food packaging can include also other purposes packaging without significantly altering the purity of the containers (Begley *et al.*, 2002). Nevertheless proper sorting procedures are needed.

Chemical reprocessing consists in the depolymerisation of polymers into their basic chemical units being monomers and other raw chemicals. The resulting monomers must be purified, through washing, distillation, crystallisation and chemical treatment, and then blended with virgin materials.These chemicals can be reused for repolymerisation and regeneration of the used packaging. It is quite obvious that this kind of recycling is more expensive than producing virgin plastics.

As PET is easily separated from domestic waste it is one of the more increasingly recycled thermoplastics. The primary market for recycled PET is the fibre industry, which uses PET for carpet fibres, sweaters and other clothing. New developments in the production, processing and end use of PET mean that polyethylene terephthalate is being hailed as the new bulk thermoplastic. One of the main technological demands in PET recycling is the elimination of some impurities that are always present and that can affect the mechanical properties and the strength of the materials. PVC, esters or acidic glues, such as EVA or acrylics, used to fix the label to the container, NaOH or alkaline detergents used to remove the labels, can constitute such impurities (Pawlak *et al.*, 2000). PET is a very stable polymer even at high temperatures. However, for temperatures higher that 260°C, and thus very close to the melting point that is 270°C some degradation and chain scission can occur with release of volatile compounds and increase in the level of oligomers (De A. Freire *et al.*, 1999). The presence of PVC in particular can catalyse the chain scission, due to hydrogen chloride that is released from the macromolecules during its degradation. The hydrolysis is an autocatalytic process because the end hydroxyl groups transform into carboxyl groups that catalytically accelerate further hydrolysis (Paci and La Mantia, 1999).

One possible solution to avoid contamination problems is the recycling of in-house scrap materials as practised by the packaging industry for many years. Such materials present no potential hazard because they have never been used as packaging and their possible contamination is at a minimum. But such materials have a limited application field as they show poorer properties than before being processed. Moreover, in any plastics recollections the presence of low percentages of other polymers can always occur. And if PVC is present among these other polymers, even in low amount (50 ppm are enough), it catalyses the hydrolysis, reducing the strength of the final material (Pawlak *et al.*, 2000).

In general, the use of recycled packaging materials, other than metals and glass, after the consumer has used them is potentially linked with the problem of possible contamination from a variety of sources. Strict source control procedures on the feedstreams in combination with modern deep-cleansing technologies, however, have been shown to reduce this risk to acceptable low levels without endangering the consumers' health (Franz *et al.*, 2003a, b). However, when there are no controls on the treatment procedures or the uses to which these materials have been put, there is no control on the type of contaminants that may be present. Therefore, without special control and purification procedures it appears to be inevitable that some recycled materials would not be acceptable for use in many food packaging applications. One option to manage even those cases would be to apply a virgin PET layer as a functional barrier to prevent migration of recycling related contaminants from the core layer (Food and Drug Administration, 1992)

14.3 Recyclable paper and board packaging

Paper and board packaging can be considered the first real flexible packaging used for food. It represents the largest sector of the food and drink packaging market, being used in the 48% of all packaging (Anon., 1999; Song *et al.*, 2000). Paper and board is made from wood pulp. Pulp wood is composed of about 40–50% cellulose, 16% other carbohydrates, 25–30% lignin and the remaining is made of proteins, resins and fat content. To produce white paper, the pulp must be almost entirely free from lignin. As a result, the cellulosic part of wood, made of fibres which are finer than a piece of hair but whose length generally equals one hundred times its thickness, becomes the predominant element of a sheet of paper. (Anon., 2002d; Damant and Castle, 1999). The production process is long when taking into account that the wood must be stored for three years before being processed, to allow a proper drying. When it is dry enough, the bark is discarded and the wood is chipped and cooked at high heat and pressure, obtaining the so-called mechanical pulp, that has characteristics close to the wood. A subsequent chemical treatment then eliminates most of the lignin by digestion with a limestone and sulphurous acid for eight hours. From more than three tons of wood only one ton of pulp is obtained. This must be washed, bleached and coloured, and in this step the remaining lignin is totally removed. During a final step the water is removed using bronze wires, showers, and the pulp is rolled into finished paper. (Decker anf Graaf, 2002; Söderhjelm and Sipiläinen-Malm, 1996). The result is a network of hydrophilic cellulosic fibres and fibre fragments that are bonded together with the necessary additives to get the desired physical and mechanical properties. To get a sheet of paper 10–20 layers of fibres are necessary, while up to 50 layers are needed for a sheet of paperboard. Depending on the surface treatment to get the final sheets, the distribution of pores, that have a capillary dimension between 0.3 and 7 μm, can differ, giving several paper grades. The capillary distribution, that follows the fibre path, is highly important when paper is intended for printing, as it can affect the distribution of inks on the surface. (Söderhjelm and Sipiläinen-Malm, 1996). Even if such material has a low cost it has many advantages. If properly treated it has a good mechanical and physical strength and is very versatile: when coated with waxes or polymers, for example, it can be used even for liquids. Being a natural product it is environmentally friendly, as it is biodegradable and easily recyclable (Anon., 2002e).

According to EU Directive 85/572/EEC (EEC, 1985), foods intended to come into contact with paper and board packaging are divided into three different categories: wet and fatty foods, dry and non-fatty foods and foods that must be peeled, shelled or washed before being used. The category of frozen foods is considered dry if the food was frozen prior its packaging, wet and fatty if the freezing step took place when the food was already packed. The production of paper and board for food contact use in Europe involves 44% of virgin pulp and 40% of recycled (but depending on the type of materials this percentage can rise to 90%). Other pulp and non-fibrous materials cover the remaining 16%.

Packaging paper and board covers 40% of the whole paper and board production (Escabasse and Ottenig, 2002).

14.3.1   Paper and board recycling

Paper and board are extensively recycled, and about 30% of the used fibres for packaging use are presently recycled (Damant and Castle, 1999). Recycled fibres usually come from newspapers, magazines, cardboard boxes and other types of printed matter. These must therefore be rid of their ink if the desired product is a paper that not only boasts the aesthetic attributes, the whiteness and the homogeneity of brand-new paper, but also conserves its mechanical properties (resistance). The recovered paper must be sorted out in order to remove all impurities that cannot be used for making recycled-content pulp (for instance, the small plastic windows of envelopes, sticky tape, paper clips, bindings, etc.). The sorted paper is then sent with a large amount of water, into a pulper (a type of large mixer) from which a coarse pulp is drawn. To remove all impurities contained in the pulp, the mix is directed through screen plates with decreasing mesh, which separates out rejects of different shapes and sizes. The finer impurities that were not filtered by the screen plates, such as inks, toner, glue and other additives are removed in the flotation cells, where the impurities attach themselves to bubbles of inflated air, migrating to the water surface.

The pulp is then poured into a thickener (a disc or drum type filter) which performs a more thorough cleaning. The dry matter concentration in the pulp increases from 1–2% to 10–12%. By that time, the product at hand is a recycled, unbleached pulp (Anon., 2002d). Generally the production process of paper and board food packaging does not introduce any harmful substances to the pulp. Indeed studies on the virgin paper used for food packaging have shown that the levels of chemicals that can potentially migrate into food are very low (Castle *et al.*, 1997b; Binderup *et al.*, 2002).

The possible sources of contamination of a recycled paper package can mainly be found in additives and in components linked to the printing and labelling process undertaken for the previous use of the package. Thus, substances like dialkylamino benzophenone, suspected to be carcinogenic (Kitchin and Brown, 1994; Damant and Castle, 1999) and used in UV-cured printing inks for cartonboard, diisopropylnaphthalene (DIPN), used as solvent for carbonless copy paper and thermal paper manufacture (Boccacci Mariani *et al.*, 1999), partially hydrogenated terphenyls (HTTP), phthalates, azo colorants, fluorescent whitening agents, primary aromatic amines, polycyclic hydrocarbons (PAH) and benzophenone, all involved in the labelling process, can be introduced via the recycling process to the packaging (De Voogt *et al.*, 1984; Castle *et al.*, 1997b; Sipiläinen-Malm *et al.*, 1997; Aurela *et al.*, 1999; Binderup *et al.*, 2002). Once present in the packaging they could migrate into the food content under certain conditions, posing health concerns for the consumers.

14.4   Food contact materials: the regulatory context

Nice and attractive packaging is extremely important for the success of a packed food. But the quality of food depends somewhat on its packaging (Katan, 1996b; Harmati, 1995), as packaging of food should be mostly intended to protect the food content from environment contamination and to prevent from possible releases of the food into the environment (Piringer, 1994).

The current European Union (EU) legislation on food packaging aims to prevent migration of toxicants or contaminants to the food in unsafe levels and to avoid contamination damaging the integrity, composition or organoleptic properties of foods. These concepts are implemented via a list of authorised substances (positive list), subjected to specific migration limits (SML) or equivalent limits in the package (maximum quantity in material QM), whenever necessary from the toxicology of the substances. An overall migration limit (OM) defines 'unacceptable alteration by uptake of packaging constituents' of the food and has been set in the EU at 60 mg/kg food or $10 \, mg/dm^2$. The EU legislation in the field of packaging takes the form of Directives, guiding Member States towards harmonisation.

In order to test migration and compliance with food packaging regulations, methodologies have been developed for plastics and paper within the frame of the European Committee for Standardisation (CEN) both for OM and SM. These are based on time-temperature exposures in contact with liquid simulating food (food simulants) or, in special cases, with a highly adsorptive porous polymer (modified polyphenylene oxide, known also as Tenax®).

According to Article 2 of Framework Directive 89/109/EEC (EEC 1989), mass transfer of substances to food in quantities that could endanger human health, could bring about an unacceptable change in the composition of food and could bring about a deterioration in the organoleptic characteristics of the foodstuffs is therefore prohibited. Thus it is compulsory that, if a material comes into contact with food, it should be safe for the consumer's health.

In 1991 the total amount of plastic used for packaging in Europe was approximately 10 millions tonnes and food packaging accounted for over half of the total (Castle, 1994) and given the increasing importance of packaging on the food industry, in February 1997, the Commission approved a Decision (Commission Decision 97/138/EC) for the establishment of a European database system to collect information on packaging production, recycling and waste and on the improvements made in the Member States to comply with Directive 94/62/EC (EC, 1994a) on packaging and packaging waste. In 2001, an estimate of the total amount of packaging consumption in Europe was published in the Proposal for a Directive of the European Parliament and of the Council amending Directive 94/62/EC (EC, 1994b), based on reports on packaging and packaging waste by national competent authorities (according to Commission Decision 97/138/EC (EC, 1997)). According to such an estimate, the total amount of packaging consumption in 1997 in Europe was 58 millions tonnes, out of which 39.1% was due to paper and cardboard and 16.3% was due to plastics. In 1998 the European production slightly increased to 60 million tonnes. Both these values, without taking into account the difference in population between the Member States, lead to a value of 155 kg of packaging placed on the market per capita.

Thus the quality of packaging has a fundamental importance for the quality of the food packed in it as it preserves the food, but also because it can potentially contaminate the food with undesired substances. These substances can derive directly from the packaging materials, such as residual monomers from a polymer, or can be related to the labels necessary to inform the consumers such as inks, adhesives and solvents, (Katan, 1996a; Strandburg et al., 1990). For this reason, the Council Directive 89/109/EEC (EEC, 1989) states that materials and articles intended to come into contact with food must not endanger consumer safety nor alter the properties and the organoleptic characteristics of the packed food. Such directive specifies that all materials or articles that are intended to come into contact with food or that directly or indirectly can have contact with food must be stable and inert enough not to transfer any dangerous substance into food. Directive 90/128/EEC (EEC, 1990) establishes a list of authorised substances that can be used in the packaging preparation process and specific migration limits for the substances that can potentially migrate from the package into the food. To ensure the safety of food packed in industrially prepared packages, Directive 82/711/EEC (EEC, 1982) gives the basic rules to test migration from plastic materials, establishing specific test conditions, based on time and temperature exposure, and Directive 85/572/EEC (EEC, 1985) establishes a selection of reproducible food simulating agents. The same Directive fixes a set of simulating liquids to foodstuffs and reduction factors to correlate the specific migration limit SML values into the simulants to migration into real foodstuff. Though Directive 85/572/EEC (EEC, 1985) states that for dry foods no migration measurement is required and for milk measuring migration in water only is required, in those cases migration of unwanted substances from packages into food can occur (even with PET that has a very low migrating potential) and can alter the organoleptic properties (Schwope and Reid, 1988).

In the European Union, the safety of plastic packages is assessed on conventional worst-case scenarios assuming the maximum possible dietary intake of a component that can migrate into foods and ignoring both food consumption patterns and packaging usage factors. In contrast in the USA intake calculations use Food Consumption Factors. They make some allowance for the different type of foods in the diet as well as the relative proportions of different varieties of packaging materials that come into contact with these foods. The USA system does not have direct applicability in the EU because of differences in the regulatory process, but the principle could be applied in some circumstances (Vergnaud, 1998; Munro, et al., 2002). To comply with all these directives and requirements, it is then necessary to have practical (fast, sensitive, reliable) analytical methods to assess the chemical migration from food packaging into food, for the different categories of food and packaging. Moreover, new scientific data to strengthen the legislators' information and to support legislation improvements are always needed (Katan *et al.*, 1996b). In order to demonstrate the compliance with the overall migration limit given in Directive 90/128/EEC (EEC, 1990), overall migration tests and SML tests for all substances used in the packages are usually performed by industries on the packaging with food simulants (O'Brien *et al.*, 2000). On the other hand, specific migration testing can be done using food simulants or in contact with foods. A comprehensive overview on migration testing can be found in the literature (Franz, 2000).

14.5   Key safety issues for recycling packaging

14.5.1   Migration from plastics materials

The prevention of contamination of food by the packaging intended to protect it is the object of constant research and regulations. Any substance that migrates from the packaging into the food is of concern if it could be harmful to the consumer. Even if the migrating substance is not potentially harmful it could have an adverse effect on the flavour, the odour and the acceptability of the food and the compositional complexity of plastics (that can have several additives to modify the properties) enhances this possibility (Huber *et al.*, 2002; Ewender *et al.*, 1995).

Migration is the mass transfer from an external source (the package or the environment) into food by sub-microscopic processes, while 'negative' migration is a mass transfer from food into an external acceptor by sub-microscopic processes. The main mechanisms that underlie migration are diffusion processes in the package and the food, i.e. the macroscopic movement of molecules that results going from high to low pressure or between gases of different compositions, and partitioning effects between packages and food, the thermodynamic contribution to migration..

The migration of additives or contaminants from PET into food can be separated into 3 different stages:

- Diffusion within the polymer;
- Solvatation at the polymer-food interface;
- Dispersion into food.

In the first stage migration follows the laws for diffusion. If the rate of variation of the concentration is constant, the first Fick law is applicable:

$$F = -D_C(dC/dx)$$

where F = flux of migrant (mol/cm^{2}s), C = concentration (mol/cm^3), x = thickness of polymer (cm), D_C = diffusion coefficient (cm^2/s).

It means that the migration process is controlled by:

- the change of concentration across the polymer,
- the diffusion coefficient, that is determined by the nature of the polymer,

- the original concentration of the migrant in the packaging before the contact with food,
- the solubility of the migrant in the contacting phase or the partition coefficient between the polymer and the contacting phase,
- the temperature of the system
- the contact time.

If the diffusion coefficient is not dependent by concentration and the rate of variation of the concentration is variable then the second Fick law for diffusion is applicable:

$$dC/dt = D_C(d^2C/dx^2)$$

where C = concentration (mol/cm^3), x = thickness of polymer (cm), D_C = diffusion coefficient (cm^2/s).

At the second stage migrants move via solvatation into food and this is facilitated if the migrant is similar or even more soluble in the food than in the polymer. This is the reason why the concern of contamination is higher for fatty foods, as the majority of additives and possible contaminants are lipophilic (fat-soluble).

At the third stage, just beyond the interface between the polymer and the food, the solvated migrant molecules diffuse away from the interface and move into the food. The migrant solubility into food and the diffusion coefficient are the primary factors that govern the dispersion of the migrants. Indeed, mixing and high temperature at this stage can enhance the dispersion (Oi-Wah Lau and Siu-Kay Wong, 2000; Katan, 1996b; Brandsch et al., 2002; Begley et al., 2002).

Migration through diffusion in the polymer can be influenced by several parameters:

- Presence of functional groups in the polymer: the presence of polar and hydrophilic substituents reduces the permeability of gases and enhances the permeability of water or vapour;
- Polymer density: it can be considered as an estimation of the free volume between molecules. A higher density hinders migration;
- Glass transition: each polymer has a temperature called glass transition temperature below which it is brittle and glass-like, while above which it is softer and more rubbery. In the first case the diffusion rate is lower;
- Crosslinking: the higher the number of crosslinks is, the lower the diffusion rate is (Jenkins and Harrington, 1991; Giacin, 1995).

Moreover if the polymer is in contact with a liquid or a solid food, components from such food can migrate into the polymer. The migration of contaminants from the polymer into the food is enhanced by this phenomenon, so that it is possible to think about a double migration (Serad et al, 2001; Sadler et al., 1996; Jasse et al., 1994; Linssen and Roozen, 1994; Nielsen and Jägerstad, 1995).

Typical volatile migrants from plastic materials are: residual solvents and low molecular weight oligomers from the polymer, monomers from the polymer and from the polymeric additives, degradation products from the polymer (such as acetaldehyde for PET) and from the additives (such as benzene or toluene, coming from the polymerisation step) catalysts and impurities of monomers and additives (Feigenbaum *et al.*, 2002; Hotchkiss, 1995; Harmati, 1995). Migration of mainly low molecular weight compounds, as monomers, antioxidants, reaction products and decomposition products can affect quality of food in taste and odour. Plastics have a high potential to transmit taint or odour to a food. These taints may be residual monomer e.g. styrene. This is in many cases the likely compound responsible when consumers detect a 'plastic taste' in a food (Wilm, 2000; Figge, 1996; Watson, 1993).

14.5.2   Migration from paper and board

Migration from paper and board is less studied than the migration from polymers that in general obey Fick's law on diffusion and consequently are more predictable, and very few studies have been published on the mechanisms. The main reason is that the inhomogeneity of fibre-based materials makes modelling difficult. However it is well known that migration occurs even towards dry food. Due to the presence of high amounts of pores in the matrix, the migration of humidity coming from food into the paper can occur very rapidly. Penetration of water to the outer layers of the paper can occur in less then a microsecond. Penetration of fats and oils is slower, but they can penetrate into the paper as well. Once on the bulk of the paper such liquids can extract the eventual contaminants bringing them to the surface for diffusion. From the surface these substances can then migrate into the food. The typical type of food that is generally packed into paper and board packages is dry, thus the type of contact differs from the one described for plastic materials, being non-homogeneous. In this case, to predict the migration from the packaging into the food, the knowledge of the adsorption isotherms and the partition coefficients between paper and board and air of each of the potential migrants are necessary (Aurela and Ketoja, 2002; Söderhjelm and Sipiläinen-Malm, 1996; Katan *et al.*, 1996a). Research in the field demonstrated that chemical migration from paper and board packaging into food can take place, thus there is the strong need to develop migration tests to determine such migration and to improve the quality control testing and law compliance by industry (Nerin and Asensio, 2002).

While for plastic materials the most appropriate and adequate food simulants to be used in migration studies have been established, for paper and board packaging materials only liquid simulants are currently foreseen according to the legislation, as the migration into dry food has not been sufficiently extensively examined and evaluated to include legislative positions. Yet migration from paper and board into dry food is an established occurrence and the liquid simulants cannot simulate the real behaviour of dry food, the typical food category that is generally in contact with fibre-based materials. It is quite obvious that this type of simulant is not suitable for paper and board migration experiments as they can destroy the matrix that, without any coating, is totally permeable to liquids (Castle *et al.*, 1997a, b; Directive 82/711/EEC (EEC, 1982); Directive 85/572/EEC (EEC, 1985); Directive 90/128/EEC (EEC, 1990)). For this reason in recent years, studies on the migration of chemicals from paper and board packaging have been performed, using Tenax (modified polyphenylene oxide), known to be a strong absorber of volatiles and broadly used for trapping and extraction and recently also as dry food simulant (Peres *et al.*, 2002; Bouche *et al.*, 2002; Duckham *et al.*, 2001; Seitz and Ram, 2000; Wilkes *et al.*, 2000; Elmore *et al.*, 2001; Martin and Ames, 2001; Nerin *et al.*, 1998; Aurela *et al.*, 1999, 2001; Nerin and Asensio, 2002; Summerfield and Cooper, 2001).

14.6   Testing the safety of recycled packaging

When appropriate methodologies are developed they must help to quantify the presence and interactions of toxicants or contaminants in the packaging material itself, in the food matrix following its migration, or in simpler liquid medium used as a food simulant. The results can then be used to provide the means to test the compliance of materials for the industry, enforce the current national legislations by government laboratories, support harmonisation of legislation at the EU level by supplying the proper analytical tools to support such legislation, and in the final outcome ensure consumer safety.

14.6.1   Recycled PET

In the recycled PET field there is the strong need to have relevant analytical data on the nature and the concentration of the contaminants that can be found in recycled PET to enable a statistical overview of the European situation on the safety of reusing PET for food purposes. The knowledge of the contaminants and the information on practical and effective test methods would help the future legislation to be more conspicuous and precise, leading to a benefit for the European industry development in the field (Franz, 2002; Baner *et al.*, 1994; Nielsen and Jägerstad, 1995). Moreover a deep knowledge of the contaminants that can be released by recycled PET for food use and their behaviour is required to feed and improve the mathematical models used to predict migration. Such models are nowadays considered a very useful tool to avoid heavy difficult and expensive analytical procedures and waste of polluting chemicals. They are currently broadly studied and discussed at the European level as a quality assurance instrument and in the USA they are used to help in the regulatory process (Reyner *et al.*, 2002a; Brandsch *et al.*, 2002; O'Brien and Cooper, 2002; Rosca *et al.*, 2001; Piringer, 1994).

An EU project FAIR CT98-4312 investigated screening techniques for contaminants from recycled materials for food contact, with one part focusing on PET and the other on paper and board. The study on migration from PET was divided into two parts. In a first phase adequate screening methods were developed to allow the investigation of large numbers of samples. Thus different methods were compared, investigating separation of contaminants based on polarity and volatility. In a second phase, the most promising method was further tested, and applied to both artificially contaminated samples as well as samples originating from a collection conducted over Europe and covering various potential parameters of recycling which may have an influence on the presence and nature of contaminants (Franz *et al.*, 2003a).

14.6.2   Paper and board

The present situation for recycled paper and board knowledge is the opposite of the recycled PET already described. The potential contaminants that can be present in the recycled fibres packaging and can be released into the food content endangering consumer safety or altering the organoleptic properties of food, are already well known as they have been extensively studied in the past (Aurela *et al.*, 1999; Aurela *et al.*, 2001; Aurela and Ketoja, 2002; Binderup *et al.*, 2002; Boccacci Mariani *et al.*, 1999; Castle *et al.*, 1997a; Castle *et al.*, 1997b; Conti and Botre, 1997; Damant and Castle, 1999; Gruner and Piringer, 1999; Johns *et al.*, 2000; Lindell, 1991; Nerin and Asensio, 2002; Sipiläinen-Malm *et al.*, 1997; Song *et al.*, 2000; Söderhjelm and Sipiläinen-Malm, 1996; Sturaro, *et al.*, 1994; Sturaro *et al.*, 1995; Summerfield and Cooper, 2001). What has not been broadly studied yet is the interactions of these potential contaminants between the fibres and the food and their behaviour if a coating layer is added to the paper and board package to prevent migration into the food (Franz, 2002). Moreover, present legislation lacks the necessary tools for simulating migration into dry foodstuffs and only few studies have been performed on the adequate dry food simulants to be used in migration studies from paper and board packages. Thus, more information on the extraction power of a known dry adsorbent and suggested simulant for dry foods, such as Tenax, and its comparison with a real dry food matrix to verify its conformity with the requirements for a food simulant are needed (EEC, 1985; Aurela *et al.*, 1999; Aurela *et al.*, 2001; Nerin and Asensio, 2002). Thus, the investigation on paper and board in the European project FAIR CT98-4312 was also divided into two phases. A first phase focused on the determination of adsorption isotherms of selected contaminants from recycled samples at various conditions of time-temperature exposures. The main goal was to evaluate the ability of volatile contaminants to migrate to foodstuffs, to better understand the paper and board migration phenomena. The second phase was directed towards the comparison of the extractive power of a currently proposed potential simulant for dry foods, Tenax, with the extractive power of foods foreseen to be highly extractive. In comparison, several foods were also investigated.

14.6.3   Analytical techniques for safety screening

The majority of the substances foreseen for migration from recycled PET and paper and board food packages should be mostly volatiles or medium volatiles and low molecular weight compounds (Feigenbaum *et al.*, 2002; Bayer, 2002; Begley *et al.*, 2002; Binderup *et al.*, 2002; Bouche *et al.*, 2002; Camacho and Karlsson, 2001; Castle *et al.*, 1997b; Elmore *et al.*, 2001; Figge, 1996; Franz, 2002; Gruner and Piringer, 1999, Escobal *et al.*, 1999). Thus among the various analytical techniques available nowadays, GC-MS should be the most appropriate, even if trials with HPLC should be foreseen as confirmation of the GC-MS technique (Castle *et al.*, 1996; Harmati, 1995; Camacho and Karlsson, 2001; Frank *et al.*, 2001; Kolb and Ettre, 1997; Komolprasert *et al.*, 1994; Lindell, 1991; Linssen *et al.*, 1995).

For recycled PET screening purposes, headspace extraction can be considered one suitable candidate as it does not require the use of any solvent and is fast and adequate for volatile substances. Moreover, as migration from plastic materials into food in enhanced by high temperature, a high temperature headspace extraction should represent the most suitable extraction technique. For these reasons it is broadly used in many types of analyses of volatiles and also for food packaging materials screenings and studies (Franz, 2002; Bouche *et al.*, 2002; Elmore *et al.*, 2001; Peres *et al.*, 2002; Wenzl and Lankmayr, 2000; Wilkes *et al.*, 2000; Sadler *et al.*, 1996; Linssen *et al.*, 1995; Komolprasert and Lawson, 1995; Katan *et al.*, 1996a; Baner *et al.*, 1994; Lindell, 1991). However, the headspace extractive power must be checked comparing it with the conventional liquid extraction, using swelling and extractive solvent that can adequately stimulate the migration of potential contaminants from the recycled PET samples. Thus a comparison of different solvents, known to be good swellers and good extractants should be made (Feigenbaum *et al.*, 2002; Hernandez-Munoz *et al.*, 2002; Cooper *et al.*, 1998; Sadler *et al.*, 1996; O'Brien *et al.*, 2000, Castle 1996; Baner *et al.*, 1994).

For the first phase of the recycled paper and board studies, a selection of surrogate to impregnate the paper samples had to be established. These substances must constitute a valid example of the real possible contaminants of recycled fibres packaging. According to previous studies in the field a group of 11 chemicals was chosen (Paquette, 1998; Franz, 2002; Söderhjelm and Sipiläinen-Malm, 1996; Song *et al.*, 2000; Sturaro *et al.*, 1994; Summerfield and Cooper, 2001). For the second phase the extractive power of Tenax was compared to a real fat dry food in the simulation of the worst case scenario. Tenax is known to be a good adsorbent, used broadly in chromatographic techniques for adsorption of volatiles, and is considered a suitable candidate to represent the worst case scenario as dry food simulant. As the potential organic migrating compounds have generally a higher solubility in fatty media and thus the migration into fatty foods or simulants is generally higher than into aqueous foods or simulants, a fatty dry food (soup), was also investigated (Baner *et al.*, 1994). Once Tenax and (dry) soup have adsorbed the migrating contaminants, they can be liquid extracted and the extract analysed with GC-MS (Martin and

Ames, 2001; Duckham *et al.*, 2001; Nerin *et al.*, 1998; Peres *et al.*, 2002; Seitz and Ram, 2000; Summerfield and Cooper, 2001; Wilkes *et al.*, 2000).

14.7   Future trends

Most legislative and analytical advances in the field of food contact materials compliance have been made for plastics polymers, yet there is still a strong need to better understand migration phenomena and develop appropriate methodologies to test migration from recycled materials (a growing field for its relevant importance for the environment protection) as well as paper and board in direct or indirect contact with foods. There is a particular need to develop a screening method to provide the analytical means to support legislatively imposed migration limits both for enforcement (government laboratories) and compliance (industry) purposes. In the field of recycled materials, the fundamental question to answer is what are the contaminants from recycled materials and when used either in direct contact or between virgin layers for food contact, can they potentially migrate to the food during the lifetime of the container. If the presence of such contaminants is detected, what then are the most adequate methodologies to ensure an efficient screening to ensure consumer safety? (Franz, 2002; Franz, 1995; Nielsen and Jägerstad, 1995). As the migration control measurements are highly time-consuming and expensive, the European research in the field of food packaging in recent years has been putting a strong effort in the development of effective migration mathematical models to predict migration from food packaging without the need of extensive laboratory work. However mathematical models must be supported by scientific evidence, thus there is still a strong need of more data on migration (Reyner *et al.*, 2002b; Reyner *et al.*, 1999; Brandsh *et al.*, 2002; O'Brien and Cooper, 2002; Oi-Wah Lau and Siu-Kay Wong, 1997; O'Brien *et al.*, 1999; Chatwin, 1996). In the past 5 years in Europe many projects studied the interactions of food and packaging with many different approaches and many different conclusions. (Gilbert and Rossi, 2000). However, as still many questions have not yet an answer, more research in the field must be performed, to ensure consumer safety, to help future legislation development and to enhance the potential of European industries. (Franz, 2002; Sadler *et al.*, 1996; Jetten, *et al.*, 1999).

14.8   Sources of further information and advice

http://cpf.jrc.it/webpack/ is the web site prepared by the Sector Contact Materials of the Unit Physical and Chemical Exposure of the Institute of Health and Consumer Protection of Joint Research Centre (JRC) for the European Commission. It provides a number of resources to facilitate the access to European Commission documents and provide a public service resource on relevant legislative and analytical information.

http://www.ivv.fhg.de/fair/ is a a project on knowledge and methodologies with respect to the safe reuse of recycled PET and cellulosic fibres as well as for the applicability of functional barrier layers for protection against contamination from recycled materials. Practical recommendations and guidance to the Commission for appropriate legislation on recycled materials are an output of the project.

http://www.eco-pac.com/ is a thematic network project on recyclable and biodegradable food and beverage packaging methods used within the food packaging industry. The project seeks to establish new technologies and practices in the field of recyclable packaging and raise know-how throughout the European industry. This aims to combine the ongoing R&D activities with other initiatives that will embrace education, legislation and technology transfer.

http://www.ipe.es/gestionproyectos/proyect_biosafepaper.htm Is a project that aims at the development, validation and intercalibration of a short-term assay battery for safety assessment of food contact paper and board. The work will be carried out at a pre-normative level. By disseminating the results a new, agreed risk evaluation procedure will be created.

14.9   References

ANON (1994), 'Migration from plastic packaging materials'. *Food Science Australia Fact Sheet*, http://www.dfst.csiro.au/migpac.htm
ANON (1999), 'Key note Packaging (food & drink)', *Worldwide Business Information and Market Reports*, http://www.the-list.co.uk/acatalog/kn21029.html
ANON (2002a), 'Packaging as a Source of Information', *Food Technology Centre*, Prince Edward Island, Canad,. http://www.gov.pe.ca/ftc/cottage/b7.php3
ANON (2002b), 'From the forest to the mill', *Forest product association*, http://www.cppa.org/english/wood/tours.htm
ANON (2002c), 'Paper'. *Irish Food Packaging Information Service*, Department of Food Science, Food Technology & Nutrition, University College of Cork, Ireland. http://space.tin.it/scienza/anscioc/paper.htm
ANON (2002d), 'PET bottles', *Association of post consumer plastic recyclers*. http://plasticsrecycling.org/pet.htm
AURELA B, KULMALA H and SÖDERHJELM L (1999), 'Phthalates in paper and board packaging and their migration into Tenax and sugar', *Food additives and contaminants*, 19 (12), 571–577.
AURELA B, OHRA-AHO T and SÖDERHJELM L (2001), 'Migration of alkylbenzenes from packaging into food and Tenax', *Packaging technology and science*, 14, 71–77.
AURELA B and KETOJA J A (2002), 'Diffusion of volatile compounds in fibre networks: experiments and modelling by random walk simulation', *Food additives and contaminants*, 19 (suppl.), 56–62.
BANER A L, FRANZ R and PIRINGER O (1994), 'Alternative fatty food simulants for polymer migration testing' in Mathlouthi M, *Food packaging and preservation*, Glasgow, Chapman & Hall, 23–47.
BAYER F L (2002), 'Polyethylene terephthalate recycling for food-contact applications:

testing, safety and technologies: a global perspective', *Food additives and contaminants*, 19 (suppl.), 111–134.

BEGLEY T H, MCNEAL T P, BILES J E and PAQUETTE K E (2002), 'Evaluating the potential for recycling all PET bottles into new food packaging', *Food additives and contaminants*, 19 (suppl.), 135–143.

BINDERUP M-L, PEDERSEN G A, VINGGAARD A M, RASMUSSEN E S, ROSENQUIST H and CEDERBERG T (2002), 'Toxicity testing and chemical analyses of recycled fibre-based paper for food contact'. *Food additives and contaminants*, 19 (suppl.), 13–28.

BOCCACCI MARIANI M, CHIACCHIERINI E and GESUMUNDO C (1999), 'Potential migration of diisopropyl naphthalenes from recycled paperboard packaging into dry food', *Food additives and contaminants*, 16, (5), 207–213.

BOCCALERI E, CALOGERO A, CAPRA G, CAVIGLIASSO P, PANIZZA R, PRETE A, SFORZINI L and CASALE A (1996), 'Facciamo il poliestere', *American chemical society* http://www.polial.polito.it/cdc/macrog/petsyn.html.

BOUCHE M P, LAMBERT W E, VAN BOCXLAER J F, PIETTE M H and DE LEENHEER A P (2002), 'Quantitative determination of n-propane, iso-butane, and n-butane by headspace GC-MS in intoxications by inhalation of lighter fluid', *Journal of Analytical Toxicology*, 26 (1), 35–42.

BRANDSCH J, MERCEA P, RÜTER M, TOSA V and PIRINGER O (2002), 'Migration modelling as a tool for quality assurance of food packaging', *Food additives and contaminants*, 19 (suppl.), 29–41.

BRODY A L (2001), 'What's the hottest food packaging technology today?', *Food Technology*, 55, (1), 82–84.

CAMACHO W and KARLSSON S (2001), 'Quality-determination of recycled plastic packaging waste by identification of contaminants by GC-MS after microwave assisted extraction (MAE)', *Polymer degradation and stability*, 71, 123–134.

CASTLE L (1994), 'Recycled and reused plastic for food packaging', *Packaging technology and science*, 7, 291–297

CASTLE L (1996), 'Methodology', in Katan L L, *Migration from food contact materials*, London, Chapman & Hall, 207–250.

CASTLE L, DAMANT A, HONEYBONE C A, JOHNS S M, JICKELLS S M, SHARMAN M and GILBERT J (1997a), 'Migration studies from paper and board food packaging materials. Part 2. Survey for residues of dialkylaminobenzophenone UV-cure ink photoinitiators', *Food additives and contaminants*, 14 (1), 45–52.

CASTLE L, OFFEN C P, BAXTER M J and GILBERT J (1997b), 'Migration studies from paper and board food packaging materials. 1. Compositional analysis', *Food additives and contaminants*, 14 (1), 35–44.

CHATWIN P C (1996), 'Mathematical modelling', in Katan L L, *Migration from food contact materials*, London, Chapman & Hall, 26–50.

CONTI M E and BOTRE' F (1997), 'The content of heavy metals in food packaging paper: an atomic absorption spectroscopy investigation', *Food control*, 8 (3), 131–136.

COOPER I, GOODSON A and O'BRIEN A (1998), 'Specific migration testing with alternative fatty food simulants', *Food additives and contaminants*, 15 (1), 72–78.

DAMANT A and CASTLE L (1999), 'Literature review of contaminants in recycled fibres of paper and board food contact materials', *EU Project FAIR-CT98-4318 "Recyclability" Interim Report from 01-01-99 to 31-07-99, Annex III.*

DE A FREIRE M T, DAMANT A, CASTLE L and REYES F G R (1999), 'Thermal stability of polyethylene terephthalate (PET): oligomer distribution and formation of volatiles', *Packaging technology and science*, 12, 29–36.

DE VOOGT P, KLAMER J C and BRINKMAN U A (1984), 'Identification and quantification of polychlorinated biphenyls in paperboard using fused silica capillary gas-chromatography', *Bulletin of environmental contamination and toxicology*, 32, 45–52.

DECKER R and GRAFF A (2002), 'Paper Vs. Plastic Bags?' *Dr. Candice Bradley Ecological Anthropology 36, Lawrence University.* http://www.angelfire.com/wi/PaperVsPlastic

DEMERTZIS P G, JOHANSSON F, LIEVENS C and FRANZ R (1997), 'Development of a quick inertness test procedure for multi-use PET containers – sorption behaviour of bottle wall strips', *Packaging technology and science*, 10, 45–58.

DUCKHAM S C, DODSON A T, BAKKER J and AMES J M (2001), 'Volatile flavour components of baked potato flesh. A comparison of eleven potato cultivars', *Nahrung*, 45 (5), 317-323.

EEC (1982), European Commission Directive No. 82/711/EEC laying down the basic rules necessary for testing migration of the constituents of plastic materials and articles intended to come into contact with foodstuffs.

EEC (1985), European Commission Directive No. 85/572/EEC laying down the list of simulants to be used for testing migration of constituents of plastic materials and articles intended to come into contact with foodstuffs.

EEC (1989), European Commission Framework Directive 89/109/EEC on the approximation of the laws of the Member States relating to materials and articles intended to come into contact with foodstuffs. Article 2.

EEC (1990), European Commission Directive No. 90/128/EEC relating to plastic materials and articles intended to come into contact with foodstuffs.

EC (1994a), European Commission Directive No. 94/62/EC on packaging and packaging waste.

EC (1994b), Proposal for a Directive of the European Parliament and of the Council amending Directive 94/62/EC.

EC (1997), European Commission Decision 97/138/EC establishing the formats relating to the database system pursuant to European Parliament and Council Directive 94/62/EC on packaging and packaging waste.

ELMORE J S, PAPANTONIOU E and MOTTRAM D S (2001), 'A comparison of headspace entrainment on Tenax with solid phase microextraction for the analysis of the aroma volatiles of cooked beef', *Advanced Experimental Medical Biology*, 488, 125–132.

ESCABASSE J-Y and OTTENIO D (2002), 'Food-contact paper and board based on recycled fibres: regulatory aspects – new rules and guidelines', *Food additives and contaminants*, 19 (suppl.), 79–92.

ESCOBAL A, IRIONDO C and KATIME I (1999), 'Organic solvents adsorbed in polymeric films used in food packaging: Determination by head-space gas chromatography', *Polymer testing*, 18, 249–255.

EWENDER J, LINDNER-STEINERT A, RÜTER M and PIRINGER O (1995), 'Sensory problems caused by food and packaging interactions: overview and treatment of recent case studies', in Ackermann P, Jägerstad M and Ohlsson T, *Foods and packaging materials – Chemical interactions* Cambridge, The Royal Society of Chemistry, 33-44.

FEIGENBAUM A, SCHOLLER D, BOUQUANT J, BRIGOT G, FERRIER D, FRANZ R, LILLEMARK L, RIQUET A M, PETERSEN J H, VAN LIEROP B and YAGOUBI Y (2002), 'Safety and quality of food contact materials. Part 1: Evaluation of analytical strategies to introduce migration testing into good manufacturing practice', *Food additives and contaminants*, 19 (2), 184–201.

FIGGE K (1996), 'Plastics', in Katan L L, *Migration from food contact materials* London,, Chapman and Hall, 77–108.

FOOD AND DRUG ADMINISTRATION (1992), 'Points to consider for the use of recycled plastics in food packaging: chemistry considerations'.

FRANK M, ULMER H, RUIZ J, VISANI P and WEIMAR U (2001), 'Complementary analytical measurements based upon gas chromatography-mass spectrometry, sensory system and human sensory panel: a case study dealing with packaging materials', *Analytica chimica acta*, 431, 11–29.

FRANZ R (1995), 'Permeation of flavour compounds across conventional as well as biodegradable polymer films', in Ackermann P, Jägerstad M and Ohlsson T, *Foods and packaging materials - Chemical interactions* Cambridge,, The Royal Society of Chemistry, 45–50.

FRANZ R, HUBER M and WELLE F (1998), 'Recycling of post-consumer poly(ethylene terephthalate) for direct food contact application a feasibility study using a simplified challenge test', *Deutsche Lebensmittel-Rundschau*, 94, 303–308.

FRANZ R and WELLE F (1999), 'Analytical screening and evaluation of post-consumer PET recyclates materials from the market with respect to reuse for food packaging applications', *Deutsche Lebensmittel-Rundschau*, 95, 94–100.

FRANZ, R (2000) 'Migration of plastic constituents'. Chapter 10, pp. 287–357. In: Piringer, O.-G. and Baner, A.L. (eds): *Plastic Packaging Materials for Food-Barrier Function, Mass Transport, Quality Assurance and Legislation*. Wiley-VCH Verlag GmbH Weinheim.

FRANZ R (2002), 'Programme on the recyclability of food-packaging materials with respect to food safety considerations: polyethylene terephthalate (PET), paper and board, and plastics covered by functional barriers', *Food additives and contaminants*, 19 (suppl.), 93–110.

FRANZ R and WELLE F (2002), 'Recycled poly(ethylene terephthalate) for direct food contact applications: challenge test of an inline recycling process', *Food additives and contaminants*,19 (5), 502–511.

FRANZ R., MAUER A and WELLE F (2003a), 'European survey on post-consumer poly(ethylene terephthalate) materials to determine contamination levels and maximum consumer exposure from food packages made from recycled PET'. *Food Additives and Contaminants*, submitted for publication.

FRANZ R, BAYER F and WELLE F (2003b), 'guidance and criteria for safe recycling of post consumer polyethylene terephthalate (PET) into new food packaging applications'. Report from EU project FAIR-CT98-4318 'Recyclability' (Duration: 1999–2002).

GIACIN J R (1995), 'Factors affecting permeation, sorption, and migration processes in package-product systems', in Ackermann P, Jägerstad M and Ohlsson T, *Foods and packaging materials – Chemical interactions* Cambridge, The Royal Society of Chemistry, 12–22.

GILBERT J and ROSSI L (2000), 'European priorities for research to support legislation in the area of food contact materials and articles', *Food additives and contaminants*, 17 (1), 83–127.

GRUNER A and PIRINGER O (1999), 'Component migration from adhesives used in paper and paperboard packaging for foodstuffs' *Packaging technology and science*, 12 (19), 28–38.

HARMATI Z (1995), 'Safety of food by packaging', in Ackermann P, Jägerstad M and

Ohlsson T, *Foods and packaging materials – Chemical interactions* Cambridge, The Royal Society of Chemistry, 23–30.

HERNANDEZ-MUÑOZ P, CATALÁ R and GAVARA R (2002), 'Simple method for the selection of the appropriate food simulant for the evaluation of a specific food/packaging interaction', *Food additives and contaminants*, 19 (suppl.), 192–200.

HOTCHKISS J H (1995), 'Overview on chemical interactions between food and packaging materials', in Ackermann P, Jägerstad M and Ohlsson T, *Foods and packaging materials – Chemical interactions* Cambridge, The Royal Society of Chemistry, 3–11.

HUBER M, RUIZ F and CHASTELLAIN F (2002), 'Off-flavour release from packaging materials and its prevention: a foods company's approach' *Food additives and contaminants*, 19 (suppl.), 221–228.

JASSE B, SEUVRE A M and MATHLOUTHI M (1994), 'Permeability and structure in polymeric packaging materials', in Mathlouthi M, *Food packaging and preservation*, Glasgow, Chapman & Hall, 1–22.

JENKINS W A and HARRINGTON J P (1991), 'Packaging Foods with Plastics', Lancaster, USA, Technomic Publishing Co. Inc.

JETTEN J, DE KRUIJF N and CASTLE L (1999), 'Quality and safety aspects of reusable plastic food packaging materials: a European study to underpin future legislation', *Food additives and contaminants*, 16 (1), 25–36.

JETTEN J and DE KRUIJF N (2002), 'Quality and safety aspects of reusable plastic food packaging materials: influence of reuse on intrinsic properties', *Food additives and contaminants*, 19 (1), 76–88.

JOHNS S M, JICKELLS S M, READ W A and CASTLE L (2000), 'Studies on functional barriers to migration. 3. Migration of benzophenone and model ink components from cartonboard to food during frozen storage and microwave heating' *Packaging technology and science*, 13, 99–104.

JONES K M, KERR G P and TINDALE N (1995), 'Refillable bottle of polyethylene terephthalate copolymer and its manufacture', Imperial Chemical Industries Plc., London, England. http://www.rapra.net/absdocs/petpackaginglert.htm.

KATAN L L (1996a), 'Effects of migration', in Katan L L *Migration from food contact materials*, London, Chapman & Hall, 11–25.

KATAN L L (1996b), 'Introduction' , in Katan L L *Migration from food contact materials*, London, Chapman & Hall, 2–10.

KATAN L L, SCHWOPE A D and ISHIWATA H (1996a), 'Real life and other special situations' in Katan L L *Migration from food contact materials*, London, Chapman & Hall, 251–276.

KATAN L L, ROSSI L, HECKMAN J H and ISHIWATA H (1996b), 'Regulations', in Katan L L *Migration from food contact materials*, London, Chapman & Hall, 277–291.

KITCHIN K T and Brown J L (1994), 'Dose-response relationship for rat liver DNA damage caused by 49 redent carcinogens', *Toxicology*, 88, 31–49.

KOLB B and ETTRE L S (1997), 'Static Headspace-Gas Chromatography Theory and Practice', New York ,Wiley-VCH.

KOMOLPRASERT V, HARGRAVES W A and ARMSTRONG D J (1994), 'Determination of benzene residues in recycled polyethylene terephthalate (PETE) by dynamic headspace-gas chromatography' *Food additives and contaminants*, 11 (5), 605–614.

KOMOLPRASERT V and LAWSON A R (1995), 'Residual contaminants in recycled poly(ethylene terephthalate): effects of washing and drying', *ACS Symposium Series No. 609: 435–444*, Washington DC, American Chemical Society.

LINDELL H (1991), '*A study of odour and taste originating from food packaging board analysed by chromatographic techniques and sensory evaluation*', Imatra, FI, Åbo Academic Press, 11–15.

LINSSEN J and ROOZEN J P (1994), 'Food flavour and packaging interactions' in Mathlouthi M, *Food packaging and preservation*, Glasgow, Chapman & Hall, 48–61.

LINSSEN J, REITSMA H and COZIJNSEN J (1995), 'Static headspace gas chromatography of acetaldehyde in aqueous foods and polythene terephthalate', *Lebensmittel Untersuhung Forshung*, 201, 253–255.

MARTIN F L and AMES J M (2001), 'Formation of Strecker aldehydes and pyrazines in a fried potato model system', *Journal of agricultural and food chemistry*, 49 (8), 3885–3892.

MUNRO I C, HIYWKA J J and KENNEPOHL E M (2002), 'Risk assessment of packaging materials', *Food additives and contaminants*, 19 (suppl.), 3–12.

NERIN C and ASENSIO E (2002), 'Selection of solid food simulants for migration test from paper and board', *Worldpak 2002 Conference New York*, University Proceedings, Washington D.C., Boca Raton, Michigan State, CRC.

NERIN C, BATLLE R and CACHO J (1998), 'Design of a test for migration studies in the vapour phase', *Food additives and contaminants*, 15 (1), 89–92.

NIELSEN T and JÄGERSTAD M (1995), 'Supercritical fluid extraction (SFE) coupled to capillary gas chromatography for the analysis of aroma compounds sorbed by food packaging materials', in Ackermann P, Jägerstad M and Ohlsson T, *Foods and packaging materials – Chemical interactions*, Cambridge, The Royal Society of Chemistry, 59–64.

NIELSEN T, DAMANT A P and CASTLE L (1997), 'Validation studies of a quick test for predicting the sorption and washing properties of refillable plastic bottles', *Food additives and contaminants*, 14 (6–7), 685–693.

O'BRIEN A, LEACH A and COOPER I (2000), 'Polypropylene: establishment of a rapid extraction test for overall migration limit compliance testing', *Packaging technology and science*, 13, 13–18.

O'BRIEN A and COOPER I (2002), 'Practical experience in the use of mathematical models to predict migration of additives from food-contact polymers', *Food additives and contaminants*, 19 (suppl.), 63–72.

O'BRIEN A, GOODSON A and COOPER I (1999), 'Polymer additive migration to foods – a direct comparison of experimental data and values calculated from migration models for high density polyethylene (HDPE)', *Food additives and contaminants*, 16 (9), 367–380.

OI-WAH LAU and SIU-KAY WONG (1997), 'Mathematical model for the migration of plasticisers from food contact materials into solid food' *Analytica chimica acta*, (347), 249–256.

OI-WAH LAU and SIU-KAY WONG (2000), 'Contamination in food packaging materials', *Journal of Chromatography A*, 882, 255–270.

PACI M and LA MANTIA F P (1999), 'Influence of small amounts of polyvinylchloride on the recycling of polyethyleneterephthalate', *Polymer degradation and stability*, 63, 11–14.

PAQUETTE K (1998), 'FDA regulations for paper products for food contact', *Food, cosmetics and drugs packaging*, February, 35–39.

PAWLAK A, PLUTA M, MORAWIEC J, GALESKI A and PRACELLA M (2000), 'Characterization of scrap poly(ethylene terephthalate)', *European polymer journal*, 36, 1875–1884.

PERES C, DENOYER C, TOURNAYRE P and BERDAGUE J L (2002), 'Fast characterization of cheeses by dynamic headspace-mass spectrometry', *Analytical chemistry*, 74 (6), 1386–1392.

PIRINGER O (1994), 'Evaluation of plastics for food packaging', *Food additives and contaminants*, 11 (2), 221–230.

REYNER A, DOLE P and FEIGENBAUM A (1999), 'Prediction of worst case migration: presentation of a rigorous methodology' *Food additives and contaminants*, 16 (4), 137–152.

REYNER A, DOLE P and FEIGENBAUM A (2002a), 'Integrated approach of migration prediction using numerical modelling associated to experimental determination of key parameters', *Food additives and contaminants*, 19 (suppl.), 42–55.

REYNER A, DOLE P and FEIGENBAUM A (2002b), 'Migration of additives from polymers into food simulants: numerical solution of a mathematical model taking into account food and polymer interactions', *Food additives and contaminants*, 19 (1), 89–102.

ROBERTSON G L (1993a), 'Structure and related properties of plastic polymers', in Hughes H A, *Food Packaging – Principles and practice*, New York, Marcel Dekker Inc., 9–62.

ROBERTSON G L (1993b), 'Permeability of thermoplastic polymers', in Hughes H A, *Food Packaging – Principles and practice*, New York, Marcel Dekker Inc., 73–110.

ROSCA I D, VERGNAUD J M and BEN ABDELOUAHAB J (2001), 'Determination of the diffusivity of a chemical through a polymer', *Polymer testing*, 20, 59–64.

SADLER G, PIERCE D, LAWSON A, SUVANNUNT D and SENTHILL V (1996), 'Evaluating organic compound migration in poly(ethylene terephthalate): a simple test with implications for polymer recycling', *Food additives and contaminants*, 13 (8), 979–989.

SCHWOPE A D and REID R (1988), 'Migration to dry foods', *Food additives and contaminants*, 5 (suppl. n. 1), 445–454.

SEITZ L M and RAM M S (2000), 'Volatile methoxybenzene compounds in grains with off-odours', *Journal of agricultural and food chemistry*, 48 (9), 4279–4289.

SERAD G E, FREEMAN B D, STEWART M E and HILL A J (2001), 'Gas and vapour sorption and diffusion in poly(ethylene terephthalate)', *Polymer*, 42, 6929–6943.

SHEFTEL V O (2000), *'Indirect food additives and polymers: migration and toxicology'*, Boca Raton, Florida, Lewis Publisher, 1132–1134.

SIPILÄINEN-MALM T, LATVA-KALA K, TIKKANEN L, SUIHKO M L and SKYTTA E (1997), 'Purity of recycled fibre-based materials', *Food additives and contaminants*, 14 (6–7), 695–703.

SONG Y S, PARK H J and KOMOLPRASERT V (2000), 'Analytical procedure for quantifying five compounds suspected as possible contaminants in recycled paper/paperboard for food packaging', *Journal of agricultural and food chemistry*, 48, 5856–5859.

SÖDERHJELM L and SIPILÄINEN-MALM T (1996), 'Paper and board', in Katan L L, *Migration from food contact materials*, London, Chapman and Hall, 159–180.

STRANDBURG G, DE LASSUS P T and HOWELL B A (1990), 'Diffusion and sorption of linear esters in selected polymers film', in Koros W J, *Barrier, polymers and structure*, American Chemical Society.

STURARO A, PARVOLI G, RELLA R, BARDATI S and DORETTI L (1994), 'Food contamination by diisopropylnaphthalenes from cardboard packages', *International journal of food science and technology*, 29, 593–603.

STURARO A, PARVOLI G, RELLA R and DORETTI L (1995), 'Hydrogenates terphenyl contaminants in recycled paper', *Chemosphere*, 30 (4), 687–694.

SUMMERFIELD W and COOPER I (2001), 'Investigation on migration from paper and board into food – development of methods for rapid testing', *Food additives and contaminants*, 18 (1), 77–88.

VERGNAUD J M (1998), 'Problems encountered for food safety with polymer packages: chemical exchange, recycling', *Advances in colloid and interface science*, 78, 267–297.

WATSON D H (1993), 'Safety of chemicals in food: chemical contaminants', in Watson D H, *Food Science and Technology*, New York, *Ellis Horwood series*.

WENZL T and LANKMAYR E P (2000), 'Reduction of adsorption phenomena of volatile aldehydes and aromatic compounds for static headspace analysis of cellulose based packaging materials', *Journal of chromatography A*, 897, 269–277.

WILKES J G, CONTE E D, KIM Y, HOLCOMB M, SUTHERLAND J B and MILLER D W (2000), 'Sample preparation for the analysis of flavours and off-flavours in foods', *Journal of chromatography A*, 880 (1–2), 3–33.

WILM K H (2000), 'Packaging', in *Our Food, Database for food & related sciences*. http://www.ourfood.com/Packaging.html

15

Environmental training for the food industry

B. Weidema, Technical University of Denmark

15.1   Introduction

This chapter presents a view on the current, international research in workplace learning. Little research has been done with particular reference to environmental training, and even less with reference to such training in the food industry. However, the chapter seeks to relate the general state-of-the-art in workplace learning to the specific field of environmental training, drawing upon the author's experience from 20 years as an environmental consultant to the food industry. The chapter is introduced (Section 15.2) by quoting the requirements on employee participation and training in the European regulation on environmental management and audit systems, and outlining the value of employee participation and training in the development, implementation and operation of an environmental management system.

Section 15.3 analyses the different training needs for the different departments in a food company, while Section 15.4 introduces the concept of the learning organisation, which has been applied with success in the food industry, although also noting some barriers to its application. The advantages and difficulties in involving suppliers and customers in the training are briefly touched upon in Section 15.5. The external training situation, including the use of experimental situations, is treated in Section 15.6, while learning at the workplace and the design requirements for a workplace curriculum is covered by Section 15.7. Section 15.7.2 outlines the specific advantages of Internet-based training. The maintenance of a learning organisation is the topic of Section 15.8, focusing especially on the problems related to personnel turnover, and the crucial role of management. Finally, Section 15.9 outlines some current and future trends in workplace learning, especially regarding training evaluation, and

Section 15.10 provides a guide to the major research groups, books and journals in the area.

15.2   The importance of environmental training

The objective of environmental training for the food industry is to *build competence* to understand how the industry and its products impact on the environment and how these impacts can be continuously reduced through application of an environmental management system. In the food industry, the environmental management system will typically be integrated with a quality management system and a Hazard Analysis Critical Control Point (HACCP) food safety management system, and the training may therefore integrate aspects from all of these systems. Decisions that affect the environment are made at many different levels in a company, and environmental training should therefore not be reserved for the personnel with explicit responsibility for environmental issues. A successful environmental management system relies on the positive forces of responsibility and creativity of all employees.

Employee participation is a requirement in an environmental management system registered according to the European EMAS II regulation (Annex I-B(4)):

employees shall be involved in the process aimed at continually improving the organisation's environmental performance.

with the following specific references to training (Annex I-A(4)(2)):

The organisation shall identify training needs. It shall require that all personnel whose work may create a significant impact upon the environment, have received appropriate training.

It shall establish and maintain procedures to make its employees or members at each relevant function and level aware of:

a)   the importance of conformance with the environmental policy and procedures and with the requirements of the environmental management system;
b)   the significant environmental impacts, actual or potential, of their work activities and the environmental benefits of improved personal performance;
c)   their roles and responsibilities in achieving conformance with the environmental policy and procedures and with the requirements of the environmental management system, including emergency preparedness and response requirements;
d)   the potential consequences of departure from specified operating procedures.

Beyond these formal requirements, the *real challenge* of environmental training is to ensure that environmental management becomes and remains a productive force and a continuous source of innovation, rather than another burden on top of other daily procedures. The value of employee participation lies in:

- Alertness to important causes of inefficient use of inputs or emissions, that may else go unnoticed. As inputs are also directly linked to costs and emissions signal a possible wasted input, sometimes even costly to handle, such alertness is of direct economic benefit, besides its merit for the environmental performance. We have numerous examples that employees in their mapping of material and energy flows within their unit discover electrical appliances running without apparent purpose, 'hidden' water losses, and even a 'forgotten' heater in a cooling tunnel.
- Preparedness to accept changes, when new procedures have to be implemented as part of the environmental management system or to improve environmental performance.
- Spreading responsibility for the environment to all employees that take operational decisions is the best guarantee that problems are minimised and eventually entirely prevented.
- Alertness to opportunities for reaping benefits through communicating the improvements in environmental performance already achieved.
- A more stable workforce that takes more pride in their work and acts as ambassadors for their company in the community.
- Embedding the investment in environmental training inside the company rather than letting the investment leak to external consultants.

It should be clear from the above that learning the procedures of the environmental management system is the least part of environmental training. The major part of environmental training lies in developing a commitment to continuous investigation of the structures and activities constituting the food production chain and its interaction with the environment, and a competence to respond to the result of this investigation. Such training must be interactive; investigating and challenging the participants' attitudes to the environment and their understanding of their roles in relation to their work, the product they produce, and their opportunities in affecting its environmental impact.

Such training is best performed in groups, composed of participants from different parts of the company, preferably representing all activities from the raw materials entering into the company till the product reaches its customer. It is important that both middle and top management also take part in the training. More specific tasks may be solved as homework or in more homogeneous groups in each department.

15.3   Environmental training needs in differing departments

The understanding of what the environment is, and what environmental performance is, and how it is influenced, may differ widely between the different departments and job functions in a food company. Environmental training should take into account such differences in perception, in order to appear meaningful to the personnel. The pre-conceived perceptions may be challenged and altered as a part of the training, but the personnel must feel that their perception is taken seriously and investigated, in order to maintain and stimulate their active involvement.

The *purchasing department* will typically see environmental issues in terms of differences between suppliers and how environmental performance may become part of the purchase requirements. It may be less obvious that a specific choice of raw material supply may also have environmental consequences in the production and later stages of the value chain. Training should give the purchasers the ability to raise environmental issues with the suppliers, but also an understanding of the possible environmental consequences in the production and later stages.

The *production and logistics departments* will typically focus on their own emissions and raw material and energy use. Energy is typically the most important factor, especially when dealing with dried, heated or cooled products. Work environment may be a cause of concern that some employees may regard as part of the environmental work, and solutions to environmental issues may at times conflict with concerns for the working environment. Such positions need to be taken seriously, and integrated into the training. Product losses may also be an area that is in focus in the production and which – like energy use – has important direct economic consequences. Often it is found that the causes of product losses are not fully known, and often blamed on suppliers or other factors outside the influence of the production and logistics departments. This may in fact make it even more important to include such issues in the training.

Sales and marketing departments typically see environmental issues as a complicating factor in sales, if at all relevant. Only in a few cases, environment issues are seen as an opportunity in the food industry. Thus, a starting point for the training should be the options for turning environment into a benefit for the sales. From this starting point, the sales and marketing may become important advisers for the overall development of the environmental policy of the company. An environmentally better product does not improve the environment unless it is sold, thus replacing less environmentally friendly alternatives. Therefore, increase in competitiveness and market shares must be part of any environmental business strategy. Sales and marketing departments need motivation and awareness of the tools available for handling environmental issues raised by customers, and for communicating the environmental work already done by the company.

Administrative functions, such as *personnel, legal and financial administration* may likewise have each their view on environmental issues. Often, they see their role as less directly linked to the environmental performance of the company. However, a closer examination may reveal that the signals coming from the administration plays a major role in shaping the views of the entire organisation. The training for these groups should therefore focus on understanding the possible role they may play in the shaping of an environmentally conscious company. It could include issues such as recruitment and training plans (for the personnel administration), environmental management accounting and total cost accounting (for the financial administration), and inclusion of environmental issues in the formal relations to suppliers, sub-contractors, licensees and local authorities (for the legal administration).

Top management has a large influence on the long-term environmental impacts of a food company. Location of a new plant, or expanding or closing of an old one, a decision on a merger, entering a new market, deploying a new process, or developing a new product, are all examples of decisions which can be taken with or without a view to the environment. The building of an environmental management system can be completely compromised if decisions at the strategic level are not judged on their environmental merits and communicated to the rest of the organisation with this in mind. Training for top management should include the use of long term planning and scenario techniques to include environmental issues in strategic decisions.

The individual needs of each personnel group should not blur the need for joint training events *across departments*, which serve to create a common understanding, sharing of views and integration of solutions. Especially, the involvement of top management in joint training events is essential for signalling commitment to the implementation and maintenance of the environmental management system.

15.4   The concept of the learning organisation in environmental training

The concept of the learning organisation was coined by Senge (1990) who outlined what he named *five disciplines* of a learning organisation: personal mastery, mental models, shared vision, team learning and systems thinking. Senge states that these five disciplines in combination comprise a critical mass to build a learning organisation. The principles have been followed successfully in many small and large companies, also in the food industry.

The discipline of *personal mastery* is the individual foundation of learning. It is the will to learn, the desire to learn, or as Senge puts it: 'approaching one's life as creative work.' A company that wishes to support learning, needs to support the individual employees in their search for what is important to themselves as individuals. The motivation for learning needs to be founded deeply in each employee. This also implies openness towards new ideas and serious feedback to the employees that show initiative. Personal mastery cannot be forced. It cannot be taught. It is basically achieved by creating an atmosphere of openness, where 'it is safe for people to create visions, where enquiry and commitment to the truth are the norm, and where challenging the status quo is expected' (Senge 1990). While personal mastery focuses on the individual, *empowerment* is a term that covers the same basic meaning, but is used more often in relation to groups.

When establishing environmental data monitoring procedures as part of an environmental management system, it is important that the data collection procedures and the data formats do not alienate employees, but are flexible enough to encompass their own perception of importance.

The understanding that individuals have different learning styles (see e.g. Felder 1996) implies that training must be designed to accommodate such differences. Self-directed learning is one way to ensure that the individual learning styles are respected (Fisher 1995). In a study involving 67 employees in the fish industry, Straka (1997) concluded that self-directed learning is in itself correlated with a feeling of competence, success and efficiency – although it is unclear whether this feeling is a prerequisite for the training or a result of the training. The most productive interpretation is probably to see this as a positively reinforcing loop.

The discipline of *mental models* is the skill to surface, investigate, challenge and modify the underlying 'pictures' that each individual has of a particular issue. We may all think that we know what 'the environment' is, but in fact some will think of personal health, others of nature, others again on pollution, each giving different weight and perspective to the issue. To create a learning organisation, it is essential to surface such different concepts that may else be a barrier to training and a hidden cause of conflict. A traditional business reaction to environmental issues has been 'we don't have any problems here'. This is a mental model that is certain to hamper implementation of environmental management and create conflicts in the training situation. Less extreme versions of this mental model are those placing the responsibility elsewhere, as the slaughterhouse that received the information that the main environmental impacts in the product chain were in agriculture, with a sigh of relief: 'Then we don't have to do anything.' However, most often, mental models are more or less subconsciously shaping the way we view the world and as 'hidden agendas' they shape our attitude to new initiatives. Awareness of mental models is the key to ensure that new procedures and the environmental management system itself are shaped to fit the current understanding among all involved. Through examination and sharing, mental models are developed and improved. Improving mental models *is* learning in essence.

The discipline of *shared vision* builds on the two previous disciplines, emerging from personal visions and sharing of mental models. Commitment to a shared vision is the key to liberate the creative forces in the employees to the benefit of the whole company. Like personal mastery, commitment cannot be taught. It must be *grown* from openness and feedback. Building a shared vision in the area of environmental work may be difficult if the company is not already a learning organisation. Small steps may be needed, starting with issues where commitment can be more easily achieved, such as e.g. the work environment. And even there, commitment to a shared vision cannot be forced. It requires freedom of choice. The discipline of *team learning* is the skills of productive dialogue and discussion: to suspend mental models, seeing every colleague as a contributor, facilitating dialogue, and balancing dialogue and discussion.

The discipline of *systems thinking* is the core discipline of Senge's concept and gave the title to his book: 'The fifth discipline' (Senge 1990). Systems are essentially composed of positive and negative feedback loops, with or without time delays, which reinforce or balance each other. Business and its environment are systems that we can only truly understand and manage when we discover the 'hidden' rules of systems dynamics, and how these govern our actions. What we learn from systems thinking is that things are often not as simple as they may appear, and that sustainable solutions – both for the business and for the environment – often requires that we embark on a long tough haul rather than jumping to quick, but temporary solutions.

Without an understanding of systems dynamics, solving a problem may involve shifting the problem to somewhere else, as when we 'solved' environmental problems by building longer effluent pipes and higher chimneys in the 1960s or flue gas filters and wastewater treatment plants in the 1970s. The realisation that those were only temporary solutions came slowly, evolving into current cleaner production practices. Systems thinking reveals such feedback loops where our own solutions may turn into later problems, e.g. through technological lock-in. For example, an early focus on biological cleaning of wastewater may make it more difficult to apply more fundamental, far-reaching and cheaper waste-preventive and water-saving solutions later on, since the biological treatment plants are dimensioned to a certain minimum inflow of organic matter in the wastewater. Cleaner production solutions are more often based on an understanding of the fundamental problems in a system.

An easy solution may turn out to be less of a solution in the long term. The pressure to introduce lines of ecological ('organic') food has led many a food company to accept reduced efficiency, resulting from the many stops and cleanings that are needed to run small ecological quantities in between the conventional batches. The environmental effects of the reduced efficiency (increased energy use, more product waste), which are caused by the ecological food, are seldom weighed against the environmental benefits that the ecological foods may involve. A systems view might have prevented such inefficient solutions, for example by running the ecological production on separate smaller production units that better fit the smaller quantities.

Systems thinking may also teach us to avoid 'death by data' – the situation where we focus on collecting environmental data from all over the production and the product life cycles, without the necessary understanding of which data are essential and which are just adding to the confusion. Systems thinking may teach us to focus where the bottleneck is. Realising that the bottleneck in their bakery operation was the packaging machine, Interbakery obtained overall energy efficiency gains of 25%, simply by increasing output through additional packing capacity, while keeping the bakery energy consumption constant. The improvement in energy efficiency was obtained as a consequence of the efficiency improvement, not as a consequence of focusing on the most obvious options, namely to reduce the energy consumption directly, e.g. through better insulation of the ovens. Such improvements in product flow and capacity utilisation are often possible and often have larger environmental consequences than the more obvious specific improvements that only affect a smaller part of the product chain.

Understanding the value chain – also known as the product life cycle – is also a fruitful application of systems thinking. Realising that freezing and consumer reheating of their bakery products was a major cause for their products poor life cycle performance, led Interbakery to invest in controlled-atmosphere packaging – a strategic decision to start phasing out frozen products from those areas where the consumer preferences could be shifted to the non-frozen alternative.

Using value chain thinking – and environmental product life cycle assessments – without taking a true systems perspective may lead to wrong focus, i.e. focusing on what appears as immediate big issues rather than at what *determines* the overall long-term impact of the chain. For many food products, the agricultural part of the life cycle may be identified as a main source of environmental impacts. Adding more rapeseed oil to a butter spread may therefore immediately appear an environmentally sensible development. However, if the milk production is determined by quotas, the reduced consumption of butter will simply mean that more butter is ending up as luxury consumption in Russia; the net result being that environmental impact is increased by the rape seed production, while not being compensated by a reduction in dairy farming.

15.5   Barriers to effective environmental training

Senge (1990) points to a number of behavioural patterns, so-called 'learning disabilities', which constitute barriers to implementing a learning organisation, and certainly apply to environmental training too:

- Inability to see beyond your own job description and see the meaning in learning new skills ('What does my job have to do with the environment?' 'We never had to care about that before')
- Reacting aggressively or evasively to challenges, rather than calmly analysing your own contribution to the problem and what you can do about it ('Why don't you make them do something about it?' 'You'll just have to communicate our position better')
- Focusing on specific short-term events, rather than slow, long-term processes ('We don't need an environmental management system, we already took our precautions when that problem came up last year')
- Focusing on experience, when the problems are in fact too large or long-term to experience and rather calls for logical analysis and reflection ('Let us now try this and then wait and see if that doesn't solve the problem')
- Believing or presuming that you know the answer instead of questioning your own prejudices ('We all learned that at school, you don't need to go over it again').

Opposition to systems thinking may be found among those who can benefit from a simple solution. As an example, the managers at Rose Meat strongly opposed building a costly washing unit for recyclable crates, which was becoming a requirement from the most important retail chains. Although a life cycle assessment could show the environmental advantages of recyclable crates over cardboard packaging, they avoided the issue by hiring a consultant that could provide an inconclusive life cycle study by using outdated and therefore more uncertain environmental data. Five years later, Rose Meat anyway had to build a much more costly separate cleaning unit for recyclable crates, which by then had become an industry standard.

Lack of time for reflection is one of the most important barriers to workplace learning. Lack of time may be especially problematic for the production departments where it can be difficult to take time off from the line. It is essential to consider how all employees can be given free time and credit for participating in training events and for implementing the new skills in their job situation. Motivation to participate in training depends on the initial attitude towards the environmental issues and environmental work. When motivation is not obviously present, the trainer should consider linking the subject to other subjects that are higher on the agenda for the employees in question, i.e. finding *synergies* between environmental work and the other success criteria for the personnel group in question. It may be issues such as work environment, quality performance, or reward systems – not to mention economic performance. It may also be important to stress the element of personal competence development.

Environmental issues may for some employees appear a very complicated topic, and initial training may therefore need to emphasise simple tools for identification and prioritisation of environmental aspects. It is important to convey a sense of success by drawing attention to solutions that can be easily implemented, so-called low-hanging fruits, even when these may be of minor importance in the overall company context. Cleaning without the use of water is an example of such a successful concept that is easy to grasp and introduce and can then be referred to when motivation is needed in dealing with the larger, more complicated, and less visible aspects of environmental concern.

Delegating *responsibility* is an essential key to creating motivation. Both Ellström (1996) and Becket and Hager (2000) point to the relationship between learning and the possibility to make judgements in the work process: The more complicated the task, and the more possibilities the employees have to control its solution, the higher the quality of learning. In a study including also dairy workers, Larsson *et al.* (1986) concluded that the employees expect training to be related to changes in the workplace organisation. In the food industry, the traditional workplace organisation often constitutes a barrier to workplace learning. In a study of slaughterhouse workers, Jørgensen (1999) points to the limited options for communication and the inflexible nature of the work as a barrier to experimentation and innovation.

15.6   Environmental learning across the supply chain

Placing suppliers and customers together in training situations may at first seem difficult, due to their opposed roles in the context of trade negotiations. However, experience show that the parties soon see their mutual advantage from the interaction. By looking beyond the company gates, the competitiveness of the entire value chain can be enhanced, which in the long run will benefit all the involved parties.

Product chains seem to be natural starting points for co-operation, since products already constitute the common ground for the economic exchange between the companies. Their mutual interest – the product – will give the training situation a natural focus. A problem that the customer used to blame on the supplier may suddenly be investigated in an atmosphere of mutual curiosity - often resulting in surprising leaps in mutual understanding and problem solving.

When the suppliers and customers are placed together, and preferably even from several steps of the value chain, they can no longer avoid tackling their mutual problems: to increase efficiency in the whole chain. No one can lean back and say that it is not their problem – because as soon as they do, it implies that the problem is owned by one of the other parties present. Training events may be held on 'neutral ground', but as soon as confidentiality issues can be avoided, it improves the learning if the training can be held in-house at one or more of the companies involved. To see directly how the product is respectively produced and used increases the mutual ability to clear up misconceptions and suggest improvements in the interaction between the parties in the supply chain. The objective should be to feel at home not only in your own company but in the value chain.

In this context, it is important to note that it may not always be the immediate suppliers that are affected by changes made at or required by the customers. The environmental performance of the supplier is only relevant when the supplier is both willing *and able* to change the production in response to a demand from the supplier. Due to long-term production constraints at the local suppliers of fertiliser and fodder protein, the main environmental impact of additional outputs of most West European food products will actually be caused by suppliers of fertilisers in East Europe and suppliers of soy protein in South America. Finding ways to include these suppliers in the chain dialogue is the real challenge, which cannot be substituted by involvement of the immediate or current suppliers.

15.7   External, workplace and internet-based environmental training

15.7.1   External training

The choice of external training versus on-job (workplace) training may be governed by practical considerations such as training opportunities, timing, transport and costs. However, from a pedagogical perspective, the most important difference lies in the level of abstraction that typically accompanies an external training situation as opposed to workplace learning. The external training situation is preferable when the purpose of the training is to provoke openness towards new ideas, to initiate new ways of social behaviour, or to place the everyday workplace in a new perspective. The external training situation loosens the feeling of control – both that of being under control and that of being in control. It allows the participants to voice opinions or observations that they may not venture in their normal job surroundings, and to accept other opinions for investigation that they would have immediately rejected if they had been presented 'at home'.

These characteristics of the external training situation make it ideal for building shared vision. Bold new ideas, such as a completely new way of distributing the product or servicing the customers, can be put forward in the open atmosphere of a course, without running the risk of being immediately reminded of all the practical obstacles. Criticism can be considered calmly and responded to positively. New alliances can be tested and the way the idea is presented to the decision makers can be improved.

The learning lab is a more formalised form of experimenting with new ideas. A small model world is constructed – typically first on paper – in which the different parts of the idea are detailed and the interactions can be explored. *Role-play* may be a way to make the model come alive and test the psychological mechanisms that may be involved in its implementation. Computer models allow more advanced forms of experimenting with a model world. Several *software* packages have been designed for this particular purpose, for example 'ithink' from High Performance Systems Inc. (www.hps-inc.com), 'Vensim' from Ventana Systems Inc. (www.vensim.com) and 'Studio' from Powersim (www.powersim.com).

Some ideas need to be tested in a more production-like environment. Most technical schools and some larger companies have their own pilot production environments that allow full scale experimenting without having to place the entire production at risk. But it may also be possible to isolate the testing to a specific product line where it can show its merits. Allowing new ideas to be tested in practice, either in training situations, in computer simulations or in pilot plants, is the ultimate touchstone for a learning organisation. Employees are encouraged by seeing their ideas being taken seriously, and even a failure is a success – both because of what may be learned from it, but also because it emphasises the spirit of experimentation. In this context, it is obviously important that no one is blamed for a failure – on the contrary: the unlucky employees should be praised for their courage to experiment so daringly. Having made a mistake is punishment enough in itself.

15.7.2   Workplace training

Compared to external training, workplace training provides better opportunities for integrating direct experience and real life experimentation into the training situation. Furthermore, it becomes possible to embed the training results immediately in the work situation.

Beside these pedagogical advantages, workplace training can more easily be adjusted to the pace of the individual employee and to the pace of the work situation, e.g. utilising periods of downtime.

Lave and Wenger (1991) suggested that effective workplace learning should be seen as a planned process of increased involvement. The idea is that the unskilled should continuously be provided with tasks that are at an adequate level of challenge, to provide a skill that can be learned with a not too overwhelming effort, while the learner is continuously part of the larger context in which the value and necessity of the skill is clearly visible. Thereby, the unskilled will move, step by step, from the periphery of the workplace to full participation. What Lave and Wenger point out is that this process does not happen automatically, but needs to be planned.

However, by focusing on the specific context of the workplace, the training runs the risk of being limited to a transfer of 'know how', rather than a less context specific 'know what' that allows the employee to translate the know how into other contexts and apply the acquired understanding there, and 'know why' that allows the employee to motivate or question the rationale behind the know how. For example, the 'know how' of how cleaning without the use of water is implemented in one specific situation should be translated into the more abstract 'know what' that mixing materials (here waste and water) reduces the solution space, a knowledge that can be transferred to e.g. separation of solid waste into different recyclable fractions. And the 'know why' is reached when the employee can see the limits where the principle of separation no longer makes sense, e.g. when the effort does not match the (environmental) value of keeping the waste fraction separate. Such 'robust transfer' (Billet and Rose 1997) of know how into know what and know why, requires theoretical reflection to be a planned part of the workplace training. This theoretical reflection is best ensured by creating situations where the learner is required to explain the reasons for particular procedures, e.g. to a supervisor or trainer, or as a trainer for others. Marienau (1998) and Boud (2000) argue that self-assessment may also be an adequate way to create such situations of reflection.

Summarising these understandings, Billet (1999) outlines a model for the design of workplace curricula, composed of four themes:

- the provision of a learning pathway from the periphery to full participation,
- insight in the total production, so that one's own work can be seen as a part of the overall process,
- direct support or guidance from experts that 'force' the learner into potential situations of learning,
- indirect support and guidance from colleagues and support by the physical workplace situation.

The fundamental implication is that the training must be systematically planned to be part of the work processes.

15.7.3   Internet-based training

The main difference between internet-based training and off-line training is its additional options for interaction over long distance and for automated interaction. In relation to course planning, internet-based training gives the possibility of access to highly skilled teachers, who would not else be accessible for personal interaction. It can also potentially reduce costs of training by substituting live teachers with automated interaction. In relation to training content, it is often seen that internet training – like much early computer-based training – is just a transfer of a traditional face-to-face course with paper-based handouts into the electronic media. The real interactive options of the media are seldom exploited to any significant degree.

Nevertheless, more creative use of the interactive options does appear. Creating learning groups across physical barriers is one important option. While some physical meetings may be required to create the right atmosphere in a learning group, the continued training may well be performed via internet interaction. Internet-based training may include a forum for group exercises and discussions among the employees. It may also include a chat-room, where teachers may be available at specified hours.

Creating learning groups with participants from a physically segregated supply chain, is an option to benefit from supply chain training without prohibitive costs. Unique to internet-based training is the option to create training sessions that are based on *real-time* interaction between employees at very different locations and job situations. For example, employees may contribute in real time on real cases of product development, e.g. to design the most environmentally benign version of a specific product and carry it through all design phases including consumer tests and negotiations with stakeholders and critical company board members. The training sessions may be designed as a competition between different teams, thus providing a case of 'Learning by play'. Such training sessions provide opportunities for proving the value of the training in a real life context without the pressure of a real life work situation.

Another example is an environmental management module, offered as on-job training for groups of employees with different job functions in the same company or product chain. The module may feature group exercises in which the employees play the same roles as in their normal job situation, and learn to understand their own role in the larger company and/or product chain context, as well as the role, cultures, tools and language of the other participants. Such a module should be centred on the understanding that all actors share the same objective, in the form of the product output, but have different viewpoints, cultures and languages with which to address it (Weidema 2001). The group exercises engage the employees in assisting each other to improve their mutual performance, identifying bottlenecks, removing friction, and adjusting procedures to optimise the value chain of the product they have in common.

15.8   Maintaining environmental awareness

When the environmental management system has become routine, there can be a tendency to professionalise the environmental work, i.e. to let a few selected persons take over the responsibility. Although this may be an appropriate allocation of resources, it carries the danger of stagnation of the environmental management system, since it may discourage employees from addressing further environmental issues. In this phase, it is important to maintain a learning environment that allows everyone to contribute to the further development.

When hiring new personnel, special attention is needed to introduce the newcomers to the spirit that has been obtained during implementation of the environmental management system. Specific introduction sessions may be required, where the focus should not be the 'bringing in line' of the newcomers, but rather to alert them to the openness with which these subjects are tackled. At the same time special attention should be given to take immediate advantage of the possible expertise or experience that the newcomers bring along, also noting that new eyes sometimes see things that go unnoticed by those who have been in the same position for years.

Special care is needed when hiring senior personnel, that they are introduced to the history and spirit of the management system, since they may else quickly disrupt the investment that has been made in both procedures and employee training. It is crucial that new senior personnel show respect for the environmental competence that has been built up and the individual employees that represent this intellectual capital. The alternative will typically be disappointment and apathy on behalf of the employees.

In fact, senior personnel and top management play the most crucial role in maintaining the learning organisation. It is from here that the spirit is originating and it is from here that the spirit is maintained. In the words of Nanda (1988): 'perhaps changes in attitude among top managers are key to the skill development of supervisors'. Only when managers show that they care about the environment and care about the learning organisation, is it possible to convince the employees that this is truly the company spirit. Thus, continuous attention and training is needed at the management level to keep the spirit of the learning organisation alive. Active reward systems can be a good way to provide a continuous signal that environment matters and that continuous alertness and improvement is appreciated.

15.9   Future trends

The type of participatory training that has been the focus of this chapter is still not commonplace, but is gaining ground in workplace learning. In professional training, e.g. at technical schools, the lecture form is unfortunately still prevalent. Thus, the current trend is the spreading of the concepts outlined in this chapter, while future trends are mainly seen in the increased use of computer-based training, notably with the aid of system dynamics software and internet-based training, as mentioned in Section 15.6 and 15.7.2. Other current or future trends are:

- Teaching training skills to domain experts and other employees.
- Less scheduled classroom teaching; more customised, individualised just-in-time training.
- Improving the measurement of the value of training.

Employees take over the responsibility for training, as a natural next step to participatory training. This does not mean that external expertise becomes superfluous, but that the local adaptation becomes more important. Local employees quickly gain an advantage over the external trainer in better understanding the local context, so that the role of the external expert rather becomes that of providing inspiration and material support to the local trainers. This also implies that development of training skills in the local employees, design of the training situations, and alerting local trainers to new trends in environmental management, become important responsibilities for the external trainer.

As employees gain confidence through participation, they require more individualised training. Individualised development plans means less of classroom training and more customised training. Impatience will also grow, increasing the need for just-in-time training, at the expense of scheduled courses.

As importance of training increases, management will want to know that it is effective. Kirkpatrick (1979) has provided one of the most widely referenced models for *training evaluation*, listing four levels: Reaction, learning, transfer and results. Today, most training sessions are only evaluated on the first level: on the participants' evaluation of the training, e.g. on a scoring form. The more formal evaluation of what the participants actually learned, through tests before and after the training session, is less common in workplace learning, but gaining ground. The two last levels of evaluation are more difficult and thus not widely used. However, the trend is clearly that also these forms of evaluation are on the increase.

Kirkpatrick's third level of evaluation is *transfer of learning* as measured through pre- and post-training assessment of behavioural changes, e.g. whether employees after training react differently to situations where they become aware of wasteful practices. Transfer of learning measurements may include participants' own assessment, which may even contain an important element of learning in itself (Boud 2000), as well as the assessment by subordinates, peers and superiors. The post-training measurement should not be performed earlier than 3 months after the training in order that the participants have had an opportunity to practice what they have learned.

Kirkpatrick's fourth level of training evaluation is *results evaluation*, measured e.g. in terms of improved productivity, reduced energy use, reduced waste or scrap, reduced water use, reduced biological content in waste water, reduced discard, improved safety record, reduced days of illness, improved compliance with regulations, number of employee suggestions, number of new ideas implemented, reduced turnover in the workforce, improved qualifications of job applicants, reduced downtime, reduced need for supervision, improved customer satisfaction (less complaints), and eventually increased market share. Results evaluation is not possible without the use of control groups, due to the many disturbing influences that may also affect the mentioned measurement criteria. Some additional effects of training may be less measurable, such as the ability to predict and avoid future problems.

15.10   Sources of further information and advice

Good reviews of current research in workplace learning can be found in Boud (1998) and Boud and Garrick (1999). A good introduction in Swedish is also found in Ellström *et al.* (1996). The concept of participatory learning developed by Lave and Wenger (1991) is still one of the fundamental sources of inspiration for the design of workplace training. Gerber (1998) provides a list of eleven ways to learning, which can serve as more specific inspiration for designing workplace learning. Current research is published in e.g. *Adult Education Quarterly, Human Resource Development Quarterly, International Journal of Lifelong Education, Journal of Environmental Education, Studies in Continuing Education, Studies in the Education of Adults* and *Outlines: Critical Social Studies.*

Peter Senge's *The fifth discipline* (Senge 1990) is still an indispensable introduction to the concept of the learning organisation, and one of the few works that still stands out more than a decade after its publication. Other important works from this school are Argyris (1990) and Argyris and Schön (1978, 1996). Further viewpoints on implementation and exercises 'from the field' can be found in Senge *et al.* (1994) and ideas on how to maintain a learning organisation can be found in Senge *et al.* (1999). Current research is published in e.g. *The learning organization: An international Journal* and *Reflections: The SoL Journal on Knowledge, Learning and Change.*

15.11   References

ARGYRIS C. (1990). *Overcoming organizational defenses. Facilitating organizational learning.* Boston: Allyn & Bacon.

ARGYRIS C, SCHÖN D. (1978). *Organizational learning: A theory of action perspective.* Reading: Addison Wesley.

ARGYRIS C, SCHÖN D. (1996). *Organizational learning II: Theory, method and practice.* Reading: Addison Wesley.

BECKETT D, HAGER P. (2000). Making judgements as the basis for workplace learning: Towards an epistemology of practice. *International Journal of Lifelong Education* 19(4): 300–311.

BILLET S. (1999). Guided learning in work. In Boud & Garrick (1999).

BILLET S, ROSE J. (1997). Securing conceptual development in workplaces. *Australian Journal of Adult and Community Education* 37(1): 12–26.

BOUD D. (ed.) (1998). *Current issues and new agendas in workplace learning.* Leabrook: National Centre for Vocational Educational Research.

BOUD D. (2000). Sustainable assessment: Rethinking assessment for the learning society. *Studies in Continuing Education* 22(2): 151–167.

BOUD D, GARRICK J. (eds.) (1999). *Understanding learning at work.* London: Routledge.

ELLSTRÖM P-E. (1996). Rutin och refektion. In Ellström et al. (1996).

ELLSTRÖM P-E, GUSTAVSSON B, LARSSON S. (eds.) (1996). *Livslångt lärande.* Lund: Studentlitteratur.

FELDER R M. (1996). Matters of style. *ASEE Prism* 6(4): 18–23.

FISHER T D. (1995). Self-directedness in adult vocational education students. *Journal of Vocational and Technical Education* 11(2) (http://scholar.lib.vt.edu/ejournals/JVTE/v11n2/)

GERBER R. (1998). How do workers learn in their work? *The learning organization* 4(4): 168–175.

JØRGENSEN C H. (1999). *Uddannelsesplanlægning i virksomheder for kortuddannede.* Roskilde: Roskilde Universitetscenter.

KIRKPATRICK D L. (1979). Techniques for evaluating training programs. *Training and Development Journal* 33(6): 78–92.

LAVE J, WENGER E. (1991). *Situated learning: Legitimate peripheral participation.* Cambridge: Cambridge University Press.

LARSSON S, ALEXANDARSSON C, HELMSTAD G, THÅNG P O. (1986). *Arbetsupplevelse och utbildningssyn hos icke facklärda.* Göteborg: Acta Univesitatis Gothoburgensis.

MARIENAU C. (1998). Self-assessment at work. *Adult Education Quarterly* 49(3): 135–146.

NANDA R. (1988). Organizational performance and supervisor skills. *Management Solutions* 33(6): 22–28.

SENGE P. (1990). *The fifth discipline. The art and practice of the learning organization.* New York: Doubleday/Currency.

SENGE P, KLEINER A, ROBERTS C, ROSS R B, ROTH G, SMITH B J. (1994). *The fifth discipline fieldbook: Strategies and tools for building a learning organization.* New York: Doubleday/Currency.

SENGE P, KLEINER A, ROBERTS C, ROSS R B, ROTH G, SMITH B J. (1999). *The dance of change: The challenges of sustaining momentum in learning organizations.* New York: Doubleday/Currency.

STRAKA G A (1997). Self-directed learning in the world of work. *European Journal of Vocational Training* 12: 83–87.

WEIDEMA B P. (2001). LCM a synthesis of modern management theories. Invited lecture for LCM2001 – The 1st International conference on life cycle management, Copenhagen, 2000.08.26-29. (http://www.lca.dk/publ/lcm9.pdf)

16

Comparing integrated crop management and organic production

H. van Zeijts, G.J van den Born and M.W. van Schijndel, National Institute for Public Health and the Environment (RIVM), The Netherlands

16.1 Introduction

Worldwide demand for food has grown rapidly. In 1900 the number of consumers was 1.6 billion; in 1950 this had grown to 2.5 billion and, currently, 6 billion people are demanding food. Forecasts show that agriculture and the food processing industry will have 9 billion customers in the year 2050 (UNEP, 2002). Moreover, consumption patterns tend towards more processed food and higher consumption of animal products. This growing food demand has led to a strong incentive to develop new technologies. Important innovations in agriculture are the introduction of chemical fertilizers, mechanization of farms and the development of chemical pesticides. The continuous process of breeding new varieties, which has been going on for centuries and has led to growing yields, has accelerated in recent decades via intensification of R&D activities. The technologies introduced in agriculture have, in most world regions, been a success story, when it comes to the capacity of agriculture in feeding the world. Food security has been improved for most people. However, intensification of agriculture goes along with environmental impacts. Groundwater and surface waters have become polluted with nitrates, phosphates and pesticides; pesticide residues have entered the food chain, soils have been degraded by erosion and water use by agriculture has increased rapidly. Cultural landscapes and biodiversity are also put more and more under pressure. Fortunately, there is a growing concern about these threats to the environment. The solution to the environmental problems in agriculture can be seen in two areas, which differ in their use of technologies: integrated crop management and organic farming.

Integrated crop management (ICM) uses the full range of man-made technologies. In ICM, farmers integrate different production strategies in order to benefit from the interactions between them. Technologies are not used in isolation or with a solely economic perspective. Farmers combine technologies in order to amplify the effect of the separate technologies so as to serve both economic and ecological goals. The features and environmental impacts of ICM as compared to conventional farming will be discussed in Section 16.2 which looks at current research in this area.

Organic farming makes very restricted use of new technologies. Biodynamic farming, the first branch of the organic movement established in 1908, was a reaction to the introduction of chemical fertilizers. Organic farmers consider chemical inputs as a threat to soil health, leading to deterioration of soil production capacity; this input also disturbs the circular course of farming. The rejection of genetically modified crops is a recent example of the view of organic farmers towards new technologies: technologies represent more often intensification of than a solution to environmental problems. Organic farming is discussed in Section 16.4.

The goal of this chapter is to provide information on the environmental aspects of integrated crop management and organic farming. The chapter merely describes the systems; it does not conclude which type of farming is 'best'. For pragmatic reasons we have concentrated on crop production in Europe. The indicators for environmental aspects included in this chapter are nutrients (in particular nitrate leaching), pesticides, energy use, soil (soil organic matter, biological activity, structure and erosion), and biodiversity. An impression of yields will be given to provide insight into the differing capacities of these forms of agriculture. Conclusions on the comparison of environmental aspects of integrated crop management and organic farming are provided in Section 16.6. Section 16.7 provides a short commentary on likely future trends.

16.2   Integrated Crop Management

Integrated Crop Management (ICM) systems differ from conventional farming systems in the way environmental issues are approached. Farmers who apply ICM have modified the conventional production system in order to minimize environmental effects, while remaining economically viable. ICM has its roots in integrated pest management (IPM), which has been strongly promoted by the Food and Agricultural Organization (FAO) of the United Nations. IPM was a response to the development of resistant types within target species in conventional pest management. IPM has been broadened to include ICM: a whole rotational approach involving consideration of soil management, crop nutrition, wild life and landscape management, all of which have an effect – directly or indirectly – on crop growth. In ICM farmers integrate different agricultural production strategies in order to benefit from the interactions between them. It is not a prescriptive approach but involves a set of principles and procedures that should be applied, taking into account the specific circumstances of the farm and its surroundings.

There is no commonly applied ICM standard with a legislative basis and formal certification procedures. A conceptual framework on the principles has been developed by the European division of the International Organization for Biological Control (IOBC; Boller *et al.*, 1998; Boller *et al.*, 1999). According to IOBC, integrated production aims at:

- The use of natural resources and regulatory mechanisms in order to minimize farm inputs like fertilizers, pesticides and fuel
- High quality food production, safe for human health and taking into account animal welfare
- Sustaining farm income
- Elimination or reduction of present environmental pollution
- Sustaining multiple functions of agriculture, like landscape diversification, conservation of wildlife and maintenance of local cultural traditions.

The preservation and improvement of soil fertility and of a diversified environment are essential components. Biological, technical and chemical methods are balanced carefully, taking into account the protection of the environment, profitability and social requirements. Box 16.1 outlines the 11 principles on which the IOBC conceptual framework is based.

Box 16.1   The principles of integrated production according to IOBC

1. Integrated production is applied only holistically.
2. External costs and undesirable impacts are minimized.
3. The entire farm is the unit of integrated production implementation.
4. The farmers' knowledge of integrated production must be regularly updated.
5. Stable agro-ecosystems are to be maintained as key components of integrated production.
6. Nutrient cycles are to be balanced and losses minimized.
7. Intrinsic soil fertility is to be preserved and improved.
8. IPM is the basis for decision making in crop protection.
9. Biological diversity must be supported.
10. Product quality must be evaluated by ecological parameters of the production system as well as by the usual external and internal quality parameters.
11. Animal production: a) specific requirements for the welfare of each species of farm animals; b) animal density should be maintained at levels consistent with principles 7 and 9.

In 2001, the European Initiative for Sustainable development in Agriculture (EISA) developed a Common Codex for Integrated Crop Management. EISA was founded by various national research bodies promoting ICM. Their objectives are fairly similar to the objectives of the European division of the IOBC mentioned above, although there are slight differences in the importance of the individual elements (Bradley *et al.*, 2002). The FAO has decided to start working on the promotion of good integrated farming standards. A strategy report proposes use of the Common Codex of EISA as a starting point (FAO, 2001).

These general objectives and principles have been translated into a wide range of working definitions by various research, industrial, retailer and governmental organizations. Some of these organizations function at national level, others at European or global level. An example is the retailer-driven protocol of EUREP-GAP: Euro-Retailer Produce Working Group – Good Agricultural Practice. EUREP–GAP contains several ICM elements.

Within the European Union, large differences in the level of application of ICM are found between countries. Bradley *et al.* (2002) provide data based on the European Crop Protection Association on areas under integrated crop management related to total Utilizable Agricultural Area (UAA). In Denmark, Austria, the UK and Sweden, Integrated Crop Management accounts for 23, 17, 10 and 5% of UAA, respectively. In other EU countries ICM accounts for less than 2% of the UAA. Note that these figures apply to crop production only, whereas UAA includes grassland. Moreover, these figures only give an impression of application of ICM, as many farmers apply (elements of) ICM without participating in certification schemes.

General agricultural production strategies followed in ICM are summarized by Jordan *et al.* (1997) and Morris and Winter (1999). They include:

- Multifunctional crop rotation: an effective method for crop protection by reducing the carry-over of pests and preserving soil fertility and soil structure.
- Minimum soil cultivation, conserving soil biological activity (which favours natural control of pests), soil nitrogen and soil organic matter, decreasing soil erosion and minimizing emissions of nutrients to ground and surface water.
- Integrated nutrient management: balancing nutrient supply with nutrient removal, instead of applying an excess of nutrients. This reduces growth of weeds and prevents too much vegetative growth of the crop, which makes crops less susceptible to pests and diseases.
- Integrated crop protection: a strategy which aims at lowering the input of pesticides in addition to the effects of the other strategies. This lowered input results from the use of disease-resistant cultivars, modification of sowing times, use of mechanical weed control, avoidance of prophylactic spraying through crop monitoring, and the use of thresholds to determine the most appropriate time of application.
- Ecological infrastructure management aims at maintenance and continuity of a network of linear elements within the farm mosaic (hedges and field margins). It may provide a habitat and food source for beneficial fauna,

Table 16.1 Protocol elements and occurrence in ICM schemes in the European Union (Bradley *et al.*, 2002)

Protocol elements	Proportion of schemes with protocol element
Fertilization strategies	95%
Plant protection prescriptions	93%
Soil husbandry and soil tillage restrictions	53%
Crop rotation and crop varieties	53%
Harvest and post-harvest restrictions	35%
Irrigation	33%
Record keeping	28%
Farm equipment	25%
Crop establishment	23%
Ecological infrastructure	18%
Quality	18%
Hygiene	15%
Soil cover and cover crops	15%
Education and training	10%
Traceability	10%

prevent ingress of problem weeds from field boundaries and minimize diffuse pollution into surface water.

In EU countries there is a wide range of ICM-type production systems to be found in practice, with different schemes and requirements varying per region. Protocols with requirements on integrated production of about 40 research and commercial production systems in the EU have been examined more closely by Bradley *et al.* (2002). Table 16.1 shows to what extent requirements are found in the schemes studied. Nearly all schemes studied include regulations and standards on the use of fertilizer and on prescriptions on plant protection, which all are more restricted than what is legally permitted.

In ICM, crop yields and produce quality are usually comparable with those in conventional farming. Inputs of mineral fertilizer and pesticides will be minimized without significantly reducing crop yields. Many studies indeed find no differences in crop yields (Wijnands, 1997; Janssens *et al.*, 1998; Köster, 1998; Smid *et al.*, 2001; MacKerron *et al.*, 1999; Ogilvy, 2000; Frick *et al.*, 2001). Sometimes crop yields can be up to 20% lower in ICM (MacKerron *et al.*, 1999; Blake, 2001; Frick *et al.*, 2001). Nevertheless, gross margins are usually comparable or higher for Integrated Crop Management systems (MacKerron *et al.*, 1999). This can be explained by the lower variable costs of ICM related to pesticides and fertilizer application. Integrated Crop Management requires extra labour costs through increased management time and mechanical weed control. Costs for education and on-field measurements will be greater; however, ICM usually has lower variable production costs (Bradley *et al.*, 2002; Blake, 2001; Janssens *et al.*, 1998).

Table 16.2 Qualitative environmental score of integrated crop management as compared with conventional farming

	++ much better	+ better	0 the same	– worse	– – much worse
Nutrients			light grey	dark grey	
Pesticides		light grey	dark grey		
Energy			dark grey		
Soil		light grey	dark grey		
Biodiversity		dark grey	light grey		

Dark grey colour: majority of the observations; light grey: few observations.

16.3   The environmental impact of integrated crop management

For this section, grateful use has been made of an overview study by Bradley *et al.* (2002). A qualitative overview of the environmental aspects of ICM as compared to conventional farming can be found in Table 16.2. The scoring for the five indicators is based on the described findings and done by the authors.

16.3.1   Nutrients

ICM uses relatively small amounts of chemical fertilizer, which is often combined with organic fertilizer (manure) and/or nitrogen fixing cover crops. In most ICM systems practised in the EU nitrogen application is 10 to 50% lower than in conventional farming (Bradley *et al.*, 2002; Holland *et al.*, 1994; Wijnands and Vereijken, 1992; Janssens *et al.*, 1998). In a four-year Dutch study on 38 arable ICM farms, the N and P_2O_5 surplus calculated from an input/output balance were around 20% and 40% lower, respectively, compared to conventional farms (Janssen *et al.*, 1998). Minimization of fertilizer application is reached by crop rotational management of fertility, utilization of residual soil available reserves (e.g. from crop residues left in and on the soil) – which are estimated by chemical analysis of soil nutrient content – and by supplementing appropriate fertilizer amounts when the (weather) conditions are most favourable for nutrient uptake by the crops.

As a consequence of these measures, lower mineral nitrogen levels in soil can be reached, which could lead to lower nitrate leaching rates. Besides the amount of nitrogen applied, an indication of the potential for nitrate leaching is also given by measuring the mineral nitrogen content in soil in autumn. Studies which provide data on mineral N content in soil show in some – but not all – cases a 30 to 60% lower value for Integrated Crop Management systems compared to conventional ones (Bradley *et al.*, 2002; Alphen, 2002; Ogilvy, 2000). Occasionally lower values are found, but under certain circumstances, e.g. only on sites with deep groundwater tables. In several studies nitrate leaching rates have also been derived from measured nitrate concentrations in leaching water. Nitrate concentrations in leaching water are 10 to 60% lower in Integrated Crop Management systems than in conventional farming (Wijnands and Vereijken, 1992; Holland *et al.*, 1994; Bradley *et al.*, 2002; Hack-ten Broeke *et al.*, 1993).

16.3.2   Pesticides

In Integrated Crop Management use of pesticides is minimized. In most ICM systems total active ingredient input is less than half of the amount used in conventional farming. In some cases total input has been reduced by more than 90% (Bradley *et al.*, 2002; Wijnands and Vereijken, 1992; Wijnands, 1997; Janssens *et al.*, 1998). Measures taken are crop rotational control of pests, mechanical control of weeds, selection of crop varieties for pest resistance and prescriptions on type, amount and method of application of pesticides. Crop monitoring and use of thresholds to determine the most appropriate time of application are also methods used to minimize pesticide use. Consequences of these measures are lower amounts of relatively hazardous pesticide residues and the presence of relatively harmless, non-leachable compounds of pesticide residues in soil. This leads to lower pesticide leaching to ground water and surface waters and to lower risk of pesticide residues harming soil fauna. Reductions in pesticide leaching by 60 to 100% have been reached due to protocols that only allow the use of non-leachable compounds. Actual reduction of environmental risk to fauna has also been reported (Holland *et al.*, 1994; Bradley *et al.*, 2002; Wijnands, 1997).

16.3.3   Energy

In ICM the application of artificial fertilizer and of soil tillage is minimized, leading to a lower energy use. However, the extended use of mechanical weed control counterbalances this. Few studies explicitly report data on energy use of ICM compared to conventional production. Ogilvy (2000) mentions an energy saving per ha of 8.5% (+1 to −17%) in the integrated production system. Though yields can be lower in ICM in some crops, in many cases they are comparable, as previously mentioned. Reganold *et al.* (2001) found energy input in apple production to be slightly lower on crop yield basis. In all, energy use (MJ/per kg product) tends to be comparable or slightly lower in ICM compared to conventional farming.

16.3.4   Soil

Soil condition determines to a great extent the yields to be reached in crop production. Important indicators are soil organic matter, biological activity (soil fauna and soil micro-organisms), soil structure and soil erosion. Few quantitative data comparing ICM and conventional crop production on soil organic matter are available. In one Italian study soil organic matter content was found to be 50% higher in ICM (Bradley *et al.*, 2002). The use of organic fertilizer or cover crops in combination with minimal soil tillage is important for conserving or improving soil organic matter content and biological activity. Mäder *et al.* (2002) show the importance of organic manure. Measured soil organic carbon and soil biological activity were higher in an integrated system based on manure in combination with artificial fertilizer than in an integrated system based on artificial fertilizer only.

As far as biological activity is concerned, several ICM systems show an increase in soil fauna species richness and/or density (beetles, spiders, earthworms) compared to conventional systems. However, sometimes no differences are found (e.g. Ogilvy, 2000). Soil biological activity measured as degradation of cellulose material was approximately 60% higher in the Integrated Crop Management system in a UK study. In another study, microbial biomass was about 30% higher than in conventional farming (Bradley *et al.*, 2002). Restricted tillage and increased soil cover could lead to less erosion of the soil and to improvement of soil structure. Few quantitative data are available; for example a decrease in soil erosion of 70–80% was reached in three studies summarized by Bradley *et al.* (2002). Very few data are available on soil structure, but one could imagine positive effects on soil structure in the case of non-inversion of soils.

16.3.5   Biodiversity

Some integrated crop management systems include measures to conserve or enrich existing biodiversity, habitat diversity and landscape by means of an ecological infrastructure. High levels of biodiversity can be reached in field margins and natural vegetation between fields. So, if this is part of the farming practice of an ICM farming system, it explains – at least in part – an increase in species richness and/or diversity. Boundary feature management is also effective with regard to landscape. Especially hedge rows form a mosaic of more varied fields, thereby reducing the monotony of the farmed landscape. Improvements in ecological infrastructure have been established in some ICM systems, as summarized by Bradley *et al.* (2002). In fields, crop diversity is usually higher in ICM than in conventional production. It seems that there is not much difference in floral species diversity. Considering invertebrate (soil) fauna, species richness and soil biological activity can improve in ICM systems, as mentioned in the Section on soils.

Very few data are available on vertebrate faunal diversity. One study (Morris *et al.*, 2001) mentions 3 to 6 times higher bird populations in ICM compared to conventional production. Morris *et al.* (2001) conclude that ICM (and organic farming) benefits farmland ecology in the case of specific management of non-cropped areas such as hedgerows and field margins. Particular benefits associated with ICM include those for non-migratory birds associated with cultivation of stubbles immediately after harvest that stimulates decomposition and weed volunteer germination. Diversity data on habitat are scarce. Morris *et al.* (2001) mention a production system which can be considered as a whole-farm integrated production system. Here increases of 40 to 600% in breeding bird territories were found for six species. Hausheer *et al.* (1998, in: Stolze, 2000) found a higher proportion of ecologically diversified areas in relation to total farm land on integrated farms compared to conventional ones. However, they do state that higher habitat diversity in ICM might be due to historical reasons. Farmers in less favoured areas are more often converted to integrated and organic farming than those in more productive areas. The less favoured areas are grassland, mountain or low-yield regions where habitats such as woodlands, hedgerows or wetlands etc. might traditionally be established. All in all, the available data indicate that an ICM approach can result in an increase in biodiversity compared to conventional systems, provided that management practices are aimed at maintaining more natural areas on the edges of the cropped areas.

16.4   Organic crop production

Organic agriculture is best known as a method of agriculture where no synthetic fertilizers and pesticides are used. The essence of this form of agriculture is to manage farms in such a way that problems with soil fertility and pests are prevented (FAO, 1998). The focus is on maintaining and improving the overall health of the soil-microbe-plant-animal system of the organic farm, which will affect present and future yields. Organic farmers use inputs (including knowledge) in a way that encourages the biological processes of making nutrients available and for defence against pests. The special values and principles of organic farming stem from the recognition that human society is an integrated part of nature and that our knowlegde of the far and future consequences of our actions is incomplete (Alrøe and Kristensen, 2001). Alrøe and Kristensen discuss three principles. The first is the cyclical principle, referring to the cyclical processes and the reproduction of crucial elements, such as soil fertility, crops, livestock, nature and human institutions. The second is the precautionary principle, involving a self-reflective awareness of the limits of knowledge, control and strategies for handling ignorance and uncertainty. The third principle is called the nearness principle and states that good social relationships between producers and consumers can be promoted through direct interactions and communication. According to Alrøe and Kristensen the further development of organic farming will depend on the discussions on and inquiries into the values and principles of organic farming, and how they are to be implemented in practice.

An elaborated definition is given by IFOAM, the International Federation for Organic Agricultural Movements. IFOAM sets 'Basic Standards', a list of equally important principles and ideas, showing not only the above-mentioned management focus but also referring to nutritional quality, genetic diversity of the farming system and its surrounding habitats, and environmental, ecological and social responsibility (see Box 16.2) (IFOAM, 2002). To unite the organic world, IFOAM has established an Organic Guarantee System, a common system of standards, verification and market identity. The definition of the European Commission on organic agriculture is focused on protecting the environment and seeks to promote sustainable agricultural development (Stolze et al., 2000). Compared to IFOAM, the regulations in the EU (EC, 1991) do not cover as many production areas; furthermore, they are based on a few fundamental regulations only, focusing on prohibition of chemical fertilizers and pesticides. The regulation is to a certain extent flexible with respect to adaptation, supplementation and precision of technical details for each EU member state, so as to better fit in with national conditions (Stolze et al., 2000).

To distinguish organic food products from integrated or conventional produced food products the products are labelled. There are many different labels or seals for organic food products with regional or continental validity. An example of a worldwide seal or label is the IFOAM seal (IFOAM, 2002). This seal is a market-oriented trademark and ensures wholesalers, retailers and consumers that a product and its producer are organically certified within the IFOAM Organic Guarantee System.

Organic farming is practised in many countries in the world, and its area is growing. According to IFOAM's world survey more than 23 million hectares worldwide are certified as being organic (Yussefi and Willer, 2003). The major part of this area is located in Australia (10.5 million ha), Argentina (3.2 million ha) and Italy (1.2 million ha). The precentages, however, are highest in Europe, e.g. Austria has 11.3%, Switzerland 9.7%, and Italy 7.9%; around 6.5% in Finland, Denmark and Sweden, 5% in Czech Republic and around 4% in the UK and Germany. For the whole of Europe this is 2%, in North America the organic farmed area amounts to 0.25%, and Latin America it is 0.5%. In Asian and African countries the area under organic labels is low, although much of the area is managed in an extensive way due to lack of possibility of farmers to purchase chemical inputs. Probably less than half of the global organic land area is dedicated to arable land, since in Australia and Argentina most of the organic land is extensively grazed.

By avoiding the use of artificial fertilizers and pesticides, organic farmers are forced to work with nature's own nutrient cycles and to become more self-reliant. In organic agriculture animal husbandry and crop cultivation are part of an integrated farming system, although not necessarily integrated into one farm. Organic animal husbandry incorporates several distinctive aspects of animal welfare, for example, access to outdoor areas and freedom of choice as a means of satisfying the natural behaviour preferences of individual animals.

Box 16.2   Definition of organic farming according to IFOAM

The International Federation for Organic Agricultural Movements (IFOAM), established in the early 1970s, represents over 600 members and associate institutions in over 100 countries. IFOAM (2002) defines the 'organic' term as referring to the particular farming system described in its Basic Standards. The 'Principle Aims of Organic Agriculture and Processing' are based on the following equally important principles and ideas:

- to produce food of high nutritional quality in sufficient quantity;
- to interact in a constructive and life enhancing way with all natural systems and cycles;
- to encourage and enhance biological cycles within the farming system, involving micro organisms, soil flora and fauna, plants and animals;
- to maintain and increase long-term fertility of soils;
- to promote the healthy use and proper care of water, water resources and all life therein;
- to help in the conservation of soil and water;
- to use, as far as possible, renewable resources in locally organized agricultural systems;
- to work, as far as possible, within a closed system with regard to organic matter and nutrient elements;
- to work, as far as possible, with materials and substances which can be reused or recycled, either on the farm or elsewhere;
- to give all livestock conditions of life which allow them to perform the basic aspects of their innate behaviour;
- to minimize all forms of pollution that may result from agricultural practices;
- to maintain the genetic diversity of the agricultural system and its surroundings, including the protection of plant and wildlife habitats;
- to allow everyone involved in organic production and processing a quality of life conforming to the UN Human Rights Charter, to cover their basic needs and obtain an adequate return and satisfaction from their work, including a safe working environment;
- to consider the wider social and ecological impact of the farming system;
- to produce non-food products from renewable resources, which are fully biodegradable;
- to encourage organic agriculture associations to function along democratic lines and the principle of division of powers;
- to progress towards an entire organic production chain, which is both socially just and ecologically responsible.

There is a continuous debate about whether technologies can be accepted or should be banned. For example there is debate about the use of conventional manure, conventional crop varieties, steaming of the top soil, greenhouse horticulture without soil, and the use of slurry, although the latter seems to have become more and more accepted. Banned are genetically modified organisms (GMOs) and the use of synthetic fertilizers and pesticides. These discussions are also ongoing in the animal production field. Only a few new techniques have been accepted without much discussion (automatic feeding, use of silage, use of milking robots, use of varieties from traditional breeding). In general, there is a broad acceptance of machine technology whereas there is concern or a ban on technology connected with synthetic inputs.

Yields in organic farms are generally around 50–70% of those of the same crops on conventional farms (Stockdale *et al.*, 2000). This is the result of incompletely controlled pests and disease attacks and deficiency of nitrogen and possibly other elements. Insufficient nutrient availability is seen as one of the serious problems organic agriculture faces. The most suitable and achievable option is to increase the number of restorative crops (leguminous crops and temporary grassland), but this means that fewer arable crops can be grown. Mäder *et al.* (2002) found crop yields to be 20% lower in the organic systems. This figure is based on 21-year comparison trials in Switzerland, one of the few well-documented long-term trials.

Labour costs in organic farming are much higher compared to conventional and integrated farming. Padel and Lampkin (1994) estimate that increases in labour input per ha range between 10 and 25%. This is mainly due to use of mechanical treatments instead of less labour intensive chemical treatments. In regions with high employment in the industrial and services sector, lack of labour can seriously hinder the further development of organic agriculture. However, higher costs in organic farming and lower yields are often counterbalanced to a certain extent by higher prices paid by the consumer for organic products.

Consumers shopping for organic foods look for many of the same qualities that are appreciated in non-organic products: taste, appearance and freshness, followed by convenience and price, and after that certain qualities unique to organic products (Dimitri and Richman, 2000). Additional original attributes in organic foods are the absence of pesticides, artificial ingredients or preservatives and synthetic fertilizers and the compliance of organic food production with environmental standards. On food quality and nutritional value of organic products, the public intuitively favours the organic product compared to conventional. The dominant scientific opinion however is that there is insufficient evidence to support the claim that organic food is better. Some reviewers state confidently that organic food is better and has significant health benefits (Heaton, 2001), and that insufficient evidence is due to the inclusion of methodologically flawed studies in the reviews (Heaton, 2002). Brandt and Molgaard (2001) are more reserved. They conclude that organic foods may in fact benefit human health more than corresponding conventional ones because of a higher content of defence-related secondary metabolites in organic vegetables and fruits. They also conclude that nutritionally important differences relating to content of minerals, vitamins, proteins and carbohydrates are not likely; neither do present levels of pesticide residues in conventional products give cause for concern. Other reviewers conclude that there are no differences at all (Williams *et al.* 2000, in: Bordeleau *et al.*, 2002). Identification of strong points and risks with regard to food safety by Wolfswinkel *et al.* (2001) show that exceeding of residue standards in organic vegetable and animal food products is significantly lower than in conventional products because of restricted use of pesticides and medicines. It was also concluded that especially the smaller organic producers have some lack of knowledge about hygienic practices when compared to larger organic producers, causing insufficient guarantees regarding food safety and food quality. From the above it can be concluded that the differences in nutritional value, food quality and food safety between conventional and organic foods may differ somewhat, sometimes in favour of organic and sometimes in favour of conventional, but that there is no strong evidence that one of the two farming systems produces much better or much safer food.

16.5   The environmental impact of organic farming

For this section, grateful use has been made of an overview study by Stolze *et al.* (2000). A qualitative overview of the environmental aspects of organic farming as compared to conventional farming can be found in Table 16.3. The scoring

Table 16.3 Qualitative environmental score of organic crop production as compared with conventional farming

	++ much better	+ better	0 the same	− worse	−− much worse
Nutrients		▨	▨	▨	
Pesticides	▨	▨			
Energy		▨	▨		
Soil	▨	▨	▨		
Biodiversity	▨	▨	▨		

Dark grey colour: majority of the observations; light grey: few observations.

for the five indicators is based on the described findings and has been carried out by the authors.

16.5.1   Nutrients

Nutrient management in organic soils is fundamentally different from conventionally managed soils, but the underlying processes supporting soil fertility are not (Stockdale *et al.*, 2002). Nutrient availability in organic farmed soils is more dependent on soil processes as compared to conventional, where nutrients are added directly to the pool of immediately available nutrients. In organic soils the fertility is strongly based on a high organic matter content and high microbial activity in the soil. Manure, compost, leguminous crops and in some cases also the use of mineral fertilizers (natural rock phosphate, wood ash, dolomitic limestone, ground limestone and dried seaweed) are the only nutrient resources. Soil organic matter in organic soils is equal to conventionally farmed soils or up to 20% higher (Stolze *et al.*, 2000). Other characteristics of organically managed soils are the lower nitrogen input, lower stocking densities, more leguminous crops in rotation and a greater use of catch crops in autumn and winter (Hansen *et al.*, 2001).

Low inputs in organic farming lead to lower surpluses on the nutrient balance than on conventional farms, although nutrient balances differ enormously between organic farms (Stolze *et al.*, 2000). Swiss studies show that phosphorus surpluses hardly exist, while only 14% of the organic farms has a nitrogen (N) surplus (Freyer, 1997, in: Stolze *et al.*, 2000). In the Netherlands, where sufficient (mainly conventionally produced) manure is available for organic farming, excedance of acceptable phosphorus levels in sandy soil is mentioned as an additional risk of applying manure as the sole source of nitrogen (Köster, 1998). Figures from Stolze *et al.* (2000) show that nitrate leaching in organic soils is 40–50% lower compared to conventional. The N-mineral content, which is an indicator for the potential nitrate leaching during the non-cropping period, is lower in organic soils (up to 60% lower), but under high nitrogen imports the opposite occurs (50% higher). Poudel *et al.* (2001) conclude from comparison studies that the lower N mineralization under organic and low-input farming systems lowers the potential risk of N leaching, while Drinkwater *et al.* (1998) state that also the type of nitrogen inputs in organic farming has a significant impact on long-term soil nitrogen retention. Others found no evidence of difference in nitrate leaching if the goal is to maintain the same crop yield levels as in conventional farming systems (Kirchmann and Bergstrom, 2001). Whether a soil leaches nutrients strongly depends on the farm management. Post-harvest nitrogen management strategies are necessary to conserve nitrogen for future crop use while minimizing the risk of nitrate leaching. Poor management of organic farms can lead to high levels of nitrogen or phosphate leaching in some organic systems and under particular conditions (Hansen *et al.*, 2001). When the surpluses of nitrogen and phosphorus are low, the risk of polluting ground water is absent. On the other hand, the risk of soil depletion increases, causing negative economic and productivity effects in the long-term, which is in particular a concern for developing countries. Stockdale and Watson (2002) conclude that there are opportunities to improve the efficiency of nutrient cycling on the organic farm, and increase short-term productivity as well as long-term sustainability.

16.5.2 Pesticides

The use of synthetic pestides is not allowed in organic farming. However, pesticides of natural origin are allowed: silicates, copper sulphate, or extracts of medicinal plants. Not much is known about the impact of natural pesticides; so far water contamination by these substances has not been reported. Stolze *et al.* (2000) conclude that organic farming does not pose any risk to ground and surface water pollution from pesticides and that organic farming is by far superior as compared to conventional farming. Others are more reserved, suggesting that the compounds are not without toxicological hazards to ecology or humans (Edwards-Jones and Howells, 2001). Leake (1999) states that some natural chemicals are more toxic and persistent than their synthetic equivalents. The question also arises on whether organic food products are free of pesticides. The UK Institute of Food Science & Technology (IFST, 2001) concludes that organic products can never be defined as pesticide free, but they almost certainly contain lower residues of agricultural chemicals. Overall it can be concluded that organic farming in most cases leads to less harm by pesticides to the environment than conventional farming.

16.5.3 Energy

The usage of direct and indirect energy in organic farming is much lower compared to conventional farming (Stolze *et al.*, 2000). This is mainly caused by not using artificial ammonia fertilizers, of which the production requires a large energy input. Figures of lower total energy usage per ha range from 30–50%, with exceptions to lower and higer values (Refsgaard *et al.*, 1998; Mäder *et al.*, 2002). Even though the yields are lower in organic farming, it was concluded that this did not counterbalance the difference in energy usage. When expressed in GJ per tonne, the energy use for potatoes and wheat showed a variation: around 20% lower to 30% higher energy use for potatoes and a 20–40% lower energy use for wheat (Stolze *et al.*, 2000). With regard to permanent crops, La Mantia and Barbera (1995, in: Stolze *et al.*, 2000) found 33% lower energy use per tonne in citrus fruit and 45% lower use in olive. A comparison study of three apple production systems in the US shows that the organic system has the highest efficiency when compared to integrated and conventional systems (Reganold *et al.*, 2001). Others found in long-term cropping systems that energy efficiency was up to twice as high in organic compared to conventional farming (Endtz *et al.*, 2002; Zarea *et al.*, 2000).

16.5.4   Soil

The supply of soil organic matter plays an important role in the maintenance of soil fertility in organic farming. Organically managed soils tend to have higher total soil organic matter contents compared to conventional (Stolze *et al.*, 2000). A high organic matter content improves nutrient availability, high biological activity, high water-holding capacity and greater soil stability and resistance to structural damage, such as compaction and erosion. A high biological activity promotes metabolism between soil and plant, and is an essential part of fertilizer management. Even in comparison to soils farmed under integrated crop management, Mäder *et al.* (2002) found that soils that are farmed organically exhibit greater biological activity whereas the soil chemical and physical parameters show fewer differences. Organically managed soils show a higher abundance of earthworms (needed for decomposition of plant constituents), higher microbial activity, higher content of microbial biomass and humic substances, higher microbial diversity and a higher capacity to cleave protein and organic phosphorus when compared to conventionally managed soils. A well structured soil supports high biological activity and vice versa, which is an asset for high crop yields and controlling physical erosion. Observations of the soil structure in both organic and conventionally managed soils did not show significant differences however (Stolze *et al.*, 2000), and if at all, only in the top soils. A distinct difference was only observed in plots that are managed organically for many decades (Malinen, 1987 in: Stolze *et al.*, 2000). Arden-Clarke (1988) and Hausheer *et al.* (1998, in: Stolze *et al.*, 2000) conclude that organic farmed soils are better protected from erosion than conventional ones. This is due to sustained supply of manure (although this can be the case in conventional farming too), long crop rotations, year-round soil cover and fewer row crops (e.g. potatoes and sugar beet). From these and other studies Stolze *et al.* (2000) conclude that, even though quantitative research results are somewhat scarce, organic farming has a high potential for reducing soil erosion risk.

16.5.5   Biodiversity

Organic farming provides more positive effects on wildlife conservation and landscape than conventional farming systems (Stolze *et al.*, 2000). For faunal and floral diversity, the assessment was positive and umambiguous, for habitat diversity and landscape, the assessment by Stolze *et al.* leads to the conclusion that organic farming has the potential to provide positive effects. The Soil Association (2000) found that biodiversity in organic agriculture in the UK generally has substantially higher levels of both abundance and diversity of species (plants, birds and invertebrates) compared to conventional agriculture. Especially for bird species, several studies support that bird populations perform better on organic farms. This is mainly caused by the diversity of crops grown, mixed farming practices, and the prevalence of non-crop habitats such as hedgerows and grass strips (Bradbury *et al.*, 2002). The Centre for Rural Economics Research (2002) concluded from various comparative studies that organic farming practices have characteristics that are beneficial to fauna and flora in terms of provisions of habitat and abundance of food. The positive impact of organic farming compared to conventional is clearer when the natural diversity of the surrounding area is high, and is less distinct when the natural diversity is low. Also it seems that there are significant differences between organic and conventional in terms of the numbers of weed species on fields (Norton, 2002). The reasons for organic farming being benign for biodiversity might be the absence of chemical treatments, occurrence of nutrient gradients within organic farmed plots, avoidance of large monocultures, fallow land and leguminous crops playing a role in crop rotation and stimulation of soil fauna. A question here is, however, whether the long crop rotation as a replacement for synthetic pesticides disrupts fauna and flora communities because of continuous changes in the crops grown (Arden-Clarke, 1988). Others conclude that organic farming does not automatically lead to biodiversity conservation, and thus requires good guidance to maximize the biodiversity benefits of organic agriculture (Stolton and Geier, 2002). Nevertheless, the available information justifies the conclusion that organic farming creates comparatively more favourable conditions on the species and ecosystem level for floral and faunal diversity than conventional farming systems and has the highest potential to contribute to the conservation of natural resources.

16.6   Comparing the environmental impact of integrated crop management and organic farming

With respect to environmental impacts of ICM and organic farming, one can conclude that both systems overall score better than conventional farming. Nutrient emissions and energy use between ICM and organic farming are only slightly different. Soil properties improve in both systems as compared to conventional farming, but in particular organic matter and biological activity are usually higher in organic farming than in ICM. Biodiversity and landscape usually improve via both farming systems, with a slight tendency in favour of organic farming.

The most striking differences are the use of pesticides and crop yields. Synthetic pesticides are not used in organic farming, whereas ICM permits the limited use of pesticides. Although some natural pesticides are allowed in organic farming, for most cases organic farming performs far better with regard to pesticide emissions and environmental impacts than both conventional and integrated farming. Risks of leaching to groundwater and surface waters and risks for animals are much lower in the case of organic agriculture. This does not necessarily mean however that organic products are healthier for humans than ICM or conventionally produced food, as other aspects than pesticide residues play a role too.

Yields in organic farming are generally 50 to 80% of those of conventional farming, whereas ICM has the same or only slightly lower yields. That is the result of relatively low levels of nitrogen application and the impacts of pests and disease attacks, because the organic produced crop is not allowed to be treated with chemical pesticides. A worldwide shift to organic production will lead to lower yields. Land take by agriculture may be larger in an organic farming world, but this depends on possible simultaneous shifts from meat to grain-based diets. Can organic farming feed the world? Opposite views exist. According to Donella Meadows (2000), organic farming can feed the world sustainably, but requires willingness to share, care and commit to the health of ourselves, neighbours and the planet. In her view there are sufficient alternatives: reducing birth rates, healthful eating and reducing spoilage of food. Others show that the world potentially can produce enough food, under a low population scenario, medium diet, low input agriculture, but that under those conditions some regions in the world (e.g. Asia, some parts of Europe) are not selfsupporting and imports are required (WRR, 1994). Scenarios with high population and/or effluent diet, show that many more regions do not become selfsupporting (e.g. Asia, Europe, Northern and Western Africa) and import of food is required. A worldwide use of low input agriculture – like organic farming – will require all suitable land available, causing a strong pressure on biodiversity and almost no potential left for the production of biomass for energy production (Wolf *et al.*, 2003).

16.7   Future trends

It is the farmer that has to do the job. The way he or she manages the farm significantly influences environmental impacts, regardless of which label has been put on his or her farm. A positive attitude towards environmentally-friendly production, ability to acquire knowledge on ICM or organic farming, and the willingness to exchange results with colleagues are important assets for today's farmer. However, transition to environmentally friendly farming also needs 'setting the prices right'. Governments can put environmental taxes on nitrogen and phosphate surpluses and harmful pesticides, or stimulate ICM and organic farming via subsidies or fiscal advantages. Trade policies can be used to stabilize prices and to ensure a 'level playing field', i.e. avoid unfair competition caused by differences in production standards between farmers in different world regions. Food processing companies and finally consumers are to pay a premium price for environmentally friendly products.

So far, ICM labels do not yield much extra income for farmers, but merely serve as a way of consolidating the relationship between farmer and food-processing industry or traders. As conventional farming moves more and more towards integrated farming forced by environmental regulations and the need for continuous reductions of costs, it is questionable whether any premium price will be paid by consumers in the future for ICM products. ICM will become a state-of-the-art. It will become the most dominant system for the production of raw materials for the food-processing industry: safe, relatively cheap and meeting environmental standards. For developing countries, also labour conditions – in particular those, related to pesticide use – may be an important reason to switch to ICM. Food processors and retailers have considerable power in stimulating ICM via payment of premium prices and by communication of information on ICM to farmers.

Because of its lower yields and higher labour input, organic farming can only flourish when a premium price is paid by consumers. It is therefore likely that it will remain a niche market in most countries, mainly concentrated in developed countries. However, organic food is no longer found solely in farmers' markets and health food stores, but also shows up in conventional supermarkets. 'Fresh products' is the top selling category, followed by non-dairy beverages, breads and grains, packaged food and dairy products (Dimitri and Greene, 2002). Developments differ considerably from market to market. In some countries in Europe the markets have stagnated while others went through a more positive development, although in several cases there has been a slowdown during the past year (Yussefi and Willer, 2003). It is expected that the demand for organic products and commodities around the world will increase, although somewhat less compared to previous years, but still high compared to most other food categories traded internationally. Expectations about annual growth rates in Europe for the period 2003–2005 are between 0 and 20%. In North America the expected rates are between 15 and 20%. The total retail sales for 2003 is estimated at 23–25 billion US$ and will probably be around 29–31 billion US$ in 2005 (Yussefi and Willer, 2003).

We have shown differences between integrated crop management and organic production. We have demonstrated that there is not one 'right' choice: ICM or organic: it is 'a matter of taste'. When emphasis is put on reaching very low pesticide use and the best soil conditions organic farming performs better. However, when one is concerned with efficient use of land, ICM performs better because of higher yields. The large variation within the population of ICM and organic farmers leaves enough scope for further environmental improvement.

16.8   Sources of further information and advice

Further information on organic farming or integrated crop management can be retrieved from the internet. Some web site suggestions are given here.

Web sites on integrated crop management:

http://www.sustainable-agriculture.org/
http://www.iobc-wprs.org/index.html
http://www.fao.org/prods/index.asp?lang=en
http://europa.eu.int/comm/environment/agriculture/studies.htm
http://eea.eionet.eu.int:8980/Public/irc/envirowindows/infin/home
http://www.ippc.orst.edu/sare/sal/

http://www.eurep.org/sites/index_e.html
http://www.ecpa.be/
http://www.unilever.com/images/3_4579.pdf

Web sites on organic farming:

http://www.ifoam.org/
http://www.organic-research.com/
http://www.fao.org/organicag/frame2-e.htm
http://www.organic-europe.net/
http://www.defra.gov.uk/farm/organic/
http://www.darcof.dk/organic/
http://www.soel.de/
http://www.boku.ac.at/oekoland/
http://www.nal.usda.gov/afsic/ofp/

16.9   References

ALPHEN, B J VAN (2002), 'A case study on precision nitrogen management in Dutch arable farming', *Nutrient Cycling in Agroecosystems*, 62, 151–161.

ALRØE, H F and KRISTENSEN, E S (2001), 'Values in organic farming and their implications', in: *EurSafe 2001, "Food Safety, Food Quality and Food Ethics", Proceedings of the third Congress of the European Society for Agricultural and Food Ethics, 3–5 October 2001, Florence, Italy*, Milan, University of Milan, 115–116.

ARDEN-CLARKE, G (1988), *The environmental effects of conventional and organic – biological farming systems: Part 1: Soil structure and erosion. Part 2: Soil ecology, fertility and nutrient cycles*, Oxford, Political Ecological Research Group.

BLAKE, J (2001), High input cropping *systems versus integrated cropping systems in Europe. Crop updates 2001: Farming systems*, Department of Agriculture – Western Australia. Available: *http://www.agric.wa.gov.au/cropupdates/2001/ Farm_systems/Blake.htm* (Accessed: 2003, March).

BOLLER, E F, AVILLA, J, GENDRIER, J P, JÖRG, E and MALVOLTA, C (1998), 'Integrated Production in Europe: 20 years after the declaration of Ovrannaz', *IOBC West Palaearctic Regional Section Bulletin*, 21 (1). Available: http://www.iobc.ch/ download_docs.html (Accessed: 2003, March).

BOLLER, E F, EL TITI, A, GENDRIER, J P, AVILLA, J, JÖRG, E and MALAVOLTA, C (1999), 'Integrated Production. Principles and Technical Guidelines', *IOBC West Palaearctic Regional Section Bulletin*, 22 (4). Available: http://www.iobc.ch/ download_docs.html (Accessed: 2003, March).

BORDELEAU, G, MYERS-SMITH, I, MIDAK, M and SZEREMETA, A (2002), *Food Quality: A comparison of organic and conventional fruits and vegetables*, Available: *http:// www.kursus.kvl.dk/shares/ea/03Projects/32gamle/_2002/FoodQualityFinal.pdf* (Accessed: 2003, March).

BRADLEY, B D, CHRISTODOULOU, M, CASPARI, C and DI LUCA, P (2002), *Integrated Crop management Systems in the EU*. Available: *http://europa.eu.int/comm/ environment/agriculture/pdf/icm_finalreport.pdf* (Accessed: 2003, March).

BRANDT, K and MOLGAARD, J P (2001), 'Organic agriculture: does it enhance or reduce the nutritional value of plant foods', *Journal of the Science of Food and Agriculture*, 81, 924–931.

CENTRE FOR RURAL ECONOMICS RESEARCH (2002), *Economic Evaluation of the Organic Farming Scheme, Annex 6: Evidence of the wider public benefits of organic agriculture – literature review*, Cambridge, Centre for Rural Economics Research, Department of Land Economy. Available: *http://www.defra.gov.uk/esg/economics/econeval/organic/annex6.pdf* (Accessed: 2003, March)

DIMITRI, C and RICHMAN, N J (2000), 'Organic Foods: Niche Marketers Venture into the Mainstream', in: *Agricultural Outlook*, Washington, U.S. Department of Agriculture, Economics Research Service, June–July, 11–14.

DIMITRI, C and GREENE, C (2002), *Recent Growth Patterns in the U.S. Organic Foods Market*, Washington, U.S. Department of Agriculture, Economics Research Service.

DRINKWATER, L E, WAGONER, P and SARRANTONIO, M (1998), 'Legume-based cropping systems have reduced carbon and nitrogen losses', *Nature*, 396, 262–265.

EC (1991), 'Council Regulation (EEC) No 2092/91 on organic production of agricultural products and indicators referring thereto on agricultural products and foodstuffs', *Official Journal of the European Communities*, L198, 1–15. Available: *http://europa.eu.int/eur-lex/en/consleg/pdf/1991/en_1991R2092 do_001.pdf* (Accessed: 2003, March).

EDWARDS-JONES, G and HOWELLS, O (2001), 'The origin and hazards of inputs to crop protection in organic farming systems: are they sustainable', *Agric. Syst.*, 67, 31–47.

ENDTZ, M H, SCHOOFS, A, HUMBLE, S M, HOEPPNER, J, HOLLIDAY, N J, MOULIN, A and BAMFORD, K C (2002), 'Glenlea Long-term Crop Rotation Study: A Comparison of Organic and Conventional Systems', in: Thompson, R E (ed.), *Proceedings of the 14th IFOAM Organic World Congress, 'Cultivating Communities', IFOAM, Victoria, Canada*, Ottawa, Canadian Organic Growers, 119.

FAO (1998), 'Evaluating the Potential Contribution of Organic Agriculture to Sustainability Goals', *IFOAM's Scientific Conference, November 16–19 1998, Mar del Plata, Argentina*.

FAO (2001), *Good Farming Practice; Codes of Practice to lead the Transition to Sustainable Agriculture*. Available: *http://www.fao.org/prods/PP6401/seminar.htm* (Accessed: 2003, March).

FRICK, C, DUBOIS, D, NEMECEK, T, GAILLARD, G and TSCHACHTLI, R (2001), Burgrain: Vergleichende Ökobilanz dreier Anbausysteme, *AGRAR Forschung*, 8 (4), 152–157.

HACK-TEN BROEKE, M J D, BOER, W A DE, DEKKERS, J M J, GROOT, W J M DE and JANSEN, E J (1993), *Stikstofemissie naar het grondwater van geïntegreerde en gangbare bedrijfssystemen in de akkerbouw op de proefboerderijen Borgerswold en Vredepeel*, Wageningen, SC-DLO.

HANSEN, B, ALRØE, H F and KRISTENSEN, E S (2001), 'Approaches to assess the environmental impact of organic farming with particular regard to Denmark', *Agriculture, Ecosystems and Environment*, 83, 11–26.

HEATON, S (2001), *Organic farming, food quality and human health: a review of the evidence*, Bristol, Soil Association of the United Kingdom.

HEATON, S (2002), 'Assessing Organic Food Quality; Is it better for you?', in: Powell *et al.*, *UK Organic Research 2002: Proceedings COR Conference, March 26–28*

2002, Aberystwyth, Wales, 55-60.

HOLLAND, J M, FRAMPTON, G K, CILGI, T and WRATTEN, S D (1994), Arable acronyms analysed a review of integrated arable farming systems research in Western Europe, *Annals of Applied Biology*, 125, 399–438.

IFOAM (2002), *NORMS, IFOAM Basic Standards for Organic Production and Processing, and IFOAM Accreditation Criteria for Bodies certifying Organic Production and Processing including Policies related to IFOAM Norms*, Tholey-Theley, IFOAM.

IFST (2001), *Organic Food*, London, UK Institute of Food Science & Technology. Available: *http://www.ifst.org/hottop24.htm* (Accessed: 2003, March)

JANSSENS, S R M, GROENEWOLD, J G and WIJNANDS, F G (1998), *Innovatiebedrijven geïntegreerde akkerbouw, Bedrijfseconomische en milieutechnische resultaten 1990-1993*, Den Haag, LEI-DLO.

JORDAN, V W L, HUTCHEON, J A and DONALDSON, G V (1997), The role of integrated arable production systems in reducing synthetic inputs, *Aspects of Applied Biology*, 50, 419–429.

KIRCHMANN, H and BERGSTROM, L (2001), 'Do organic farming practices reduce nitrate leaching?', *Commun.soil sci.plant anal.*, 32(7&8), 997–1028.

KÖSTER, H (1998), *Milieubewuste vormen van akkerbouw, een momentopname*, Bilthoven, RIVM.

LEAKE, A (1999), *Organic Farming and the European Union*, London, HMSO.

MACKERRON, D K L, DUNCAN, J M, HILLMAN, J R, MACKEY, G R, ROBINSON, D J, TRUDGILL, D L and WHEATLY, R J (1999), 'Organic Farming: Science or Belief', in: *Annual Report 1998–1999*, Invergownie Dundee, Scottish Crop Research Institute, 60–72.

MÄDER, P, FLIESSBACH, A, DUBOIS, D, GUNST, L, FRIED, P and NIGGLI, U (2002), 'Soil Fertility and Biodiversity in Organic Farming', *Science*, 296, 1694–1697.

MEADOWS, D (2000), *Can organic farming feed the world.* Available: *http://www.natsoc.org.au/html/papers/meadows.pdf* (Accessed: 2003, March)

MORRIS, C and WINTER, M (1999), 'Integrated farming systems: the third way for European agriculture?', *Land use policy*, 16 (1999), 193–205.

MORRIS, C, HOPKINS, A and WINTER, M (2001), *Comparison of the social, economic and environmental effects of organic, ICM and conventional farming.* Available: *http://www.glos.ac.uk/ccruimages/documents/P/60_1.pdf* (Accessed: 2003, March).

NORTON, L R (2002), 'Factors influencing biodiversity within organic and conventional systems of arable farming – methodologies and preliminary results', in: Powell *et al.*, *UK Organic Research 2002: Proceedings COR Conference, March 26–28 2002, Aberystwyth, Wales.*

OGILVY, S (2000), *LINK integrated farming systems (a field-scale comparison of arable rotations). Volume 1: experimental work*, HGCA, London.

PADEL, S and N H LAMPKIN (1994), Conversion to organic farming: an overview, In: Lampkin, NH and Padel, S, *The Economics of Organic Farming – an International perspective*, Wallingford, CAB International, 295–311.

POUDEL, D D, HORWATH, W R, LANINI, W T, TEMPLE, S R, BRUGGEN and A H C VAN (2002), 'Comparison of soil N availability and leaching potential, crop yields and weeds in organic, low-input and conventional farming systems in northern California'. *Agriculture, Ecosystems and Environment*, 90, 125–137.

REFSGAARD, K, HALBERG, N and KRISTENSEN, E S (1998), 'Energy utilization in crop and dairy production in organic and conventional livestock production systems', *Agric. Syst.*, 57, 599–630.

REGANOLD, J P, GLOVER, J D, ANDREWS, P K and HINMAN, H R (2001), Sustainability of three apple production systems, *Nature,* 410 (10 April 2001), 926–930.

SMID, J, DEKKERS, W A and DEKKING, A J G (2001), 'Geintegreerde akkerbouw: vooral lagere milieubelasting', *PPO-Bulletin Akkerbouw*, 1, 22–25.

STOCKDALE, E A, LAMPKIN, N H, HOVI, M, KEATINGE, R, LENNARTSON, E K M, MACDONALD, D W, PADEL, S, TATTERSALL, F H, WOLFE, M S and WATSON, C A (2000), 'Agronomic and environmental implications of organic farming systems', *Advances in Agronomy,* 70, 261–327.

STOCKDALE, E A and WATSON, C A (2002), 'Nutrient budgets on organic farms: a review of published data', in: Powell *et al.*, *UK Organic Research 2002: Proceedings COR Conference, March 26–28 2002, Aberystwyth, Wales,* 129–132.

STOCKDALE, E A, SHEPHERD, M A, FORTUNE, S and CUTTLE, S P (2002), 'Soil fertility in organic farming systems – fundamentally different?', *Soil Use and Management,* 18, 301–308.

STOLTON, S and GEIER, B (2002), 'The relationship between biodiversity and organic agriculture, Defining appropriate policies and approaches for sustainable development', in: *Pan-Europe Biological and Landscape Diversity Strategy, High-level Biological and Landscape Diversity Strategy: towards integrating biological and landscape diversity for sustainable agriculture in Europe, 5–7 June 2002,* Paris, Unesco.

STOLZE, M, PIORR, A, HARING, A and DABBERT, S (2000), *The Environmental Impacts of Organic Farming in Europe,* Stuttgart, University of Hohenheim.

THE SOIL ASSOCIATION (2000), *The Biodiversity Benefits of Organic Farming,* The Soil Association, Bristol. Available: *www.thesoilassociation.org* (Accessed, 2003, March).

UNEP (2002), *World Population Prospects: The 2002 Revision,* Available: *http://www.un.org/esa/population/publications/wpp2002/WPP2002-HIGHLIGHS.PDF* (Accessed: 2003, March).

WIJNANDS, F G and VEREIJKEN, P (1992), 'Region-wise development of prototypes of integrated arable farming and outdoor horticulture', *Netherlands Journal of Agricultural Science*, 40, 225–238.

WIJNANDS, F G (1997), Integrated crop protection and environment exposure to pesticides: methods to reduce use and impact of pesticides in arable farming, *European Journal of Agronomy*, 7, 251–260.

WOLF, J, BINDRABAN, P S, LUIJTEN, J C and VLEESHOUWERS, L M (2003), 'Exploratory study on the land area required for global food supply and the potential global production of bioenergy', *Agric. Syst.,* 76, 841–861.

WOLFSWINKEL, M VAN, LEFERINK, J, BOK, R and AALDERS, T (2001), *Voedselveiligheid van producten uit de biologische landbouw,* Expertisecentrum LNV, Ede.

WRR (1994), *Duurzame risico's: een blijvend gegeven,* Den Haag, Wetenschappelijke Raad voor het Regeringsbeleid.

YUSSEFI, M and WILLER, H (2003), *The World of Organic Agriculture 2003 – Statistics and Future Prospects,* Tholey-Theley, IFOAM. Available: *http://www.soel.de/oekolandbau/weltweit.html* (Accessed: 2003, March)

ZAREA, A, KOOCHEKI, A and NASSIRI, M (2000), 'Energy efficiency of conventional and ecological cropping systems in different rotations with wheat', in: *Proceedings of 13th IFOAM Organic World Congress, 'The World Grows Organic', Convention Center Basel, IFOAM,* Zürich, vdf Hochschulverlag AG an der ETH, 382–385.

17

Life cycle assessment (LCA) of wine production

**B. Notarnicola and G. Tassielli, University of Bari, Italy and
G. M. Nicoletti University of Foggia, Italy**

17.1 Introduction: key issues

Wine is one of the oldest and well known alcoholic beverages. The 2001 worldwide production was 27.77×10^6 t, mostly in Europe (70% with France 20%, Italy 19%, Spain 12%). In the decade 1991–2001, production has grown by 18%. In the market there are different typologies of wine: red, white, rosè and others as sparkling and sweet wines. The life cycle of the wine system is made up by two main phases: an agricultural one, producing grapes, and an industrial one, enabling the grapes transformation to must and then to wine. Like most the conventional agricultural operations, viticulture is characterised by high environmental burdens due to the use of pesticides and synthetic nutrients. The main issues in vinification are: the use of electric energy in equipment; the generation of different solid co-products during the process, which are potentially reusable in other industrial sectors; the releases in water due to the cleaning operations and spillage occurring during the process; emissions to the atmosphere of CO_2 (about 80 kg/t grapes) and VOC (0.93 kg/t grapes) during the fermentation process.

In the following section, after a description of the most important and common wine-making processes, pointing out the differences and the high level of knowledge which is required to produce a quality wine, the main issues linked with the application of LCA to wine system will be described. The case study described in section 17.3 relates to a company in the South of Italy producing different typologies of high and average quality wine. In the fourth section, the opportunities supplied by the LCA approach to improve the environmental performance of wine, both in the agricultural and in the industrial phase, will be identified. The chapter will conclude with a description of the technological and environmental future trends of winemaking.

17.2   Wine production

The application of LCA to the case of wine production is not a simple task because of different problems. Apart from those related to any agro-industrial product which involves both an agricultural and an industrial phase with all the difficulties linked to the assessment of environmental burdens, the main problems of applying LCA to wine production relates to the production process itself.

Wine production really is an 'art' in which the technology plays the same role as winemaker skills and the processes of storage and ageing. Even if the raw materials are just grapes, yeast and some chemicals, the ways of using them are enormously varied. Although the basic steps used in wine making are very similar for small producers of the highest quality wines, large producers of inexpensive wines and home wine makers, the size and sophistication of the equipment vary enormously. In a modern winery, the grape route to become wine goes through different thermal, clarification, filtration, stabilizing and ageing processes, which may lead to very different qualities of wine. Therefore, it is quite difficult to state a 'formula' or a raw material mix for wine production on which to built up an inventory table of the relative inputs and outputs, since it varied according to the grape variety and to wine quality and price. Consequently, it is necessary to provide an overview of the most common winemaking processes: red and white vinification. The main difference between the two processes is that in the red vinification the alcoholic fermentation occurs through maceration in presence of the solid components (pomace), while the white wine is obtained just by the fermentation of the filtered must without maceration.

17.2.1   Raw materials

'Wine is the product obtained exclusively from the total or partial alcoholic fermentation of fresh grapes, whether or not crushed, or of grape must' (EC, 1999). The vine belongs to the genus Vitis among which there are about 60 known varieties. Vitis vinifera, the European vine, is one of the most important and from it a large number of varieties stem (Macrae *et al.*, 1993). Vine cultivation requires specific planting systems and forms of training in order to obtain regular and high-quality crops. The most important planting systems adopted in Italy are the so-called *alberello, cordone speronato, tendone, guyout*, which basically differ in the way trees are planted according to vine variety.

The first factor in determining wine quality is the grape maturity level that changes depending on many parameters mainly related to the pulp and skin maturity. Consequently, different levels of maturity are necessary for producing different kinds of wine: an early harvest of the grapes is required in the production of dry white wines; grape development has to be continued when producing a quality red wine; slightly acid grapes are required for sparkling wines. In the context of LCA of wine, the different harvesting times obviously

Table 17.1 Typologies of wines

Table wines	without other indication, with geographical indication, with geographical indication and vine;
Quality wines produced in specified regions (psr) Liqueur wines	In Italy: controlled origin denomination (DOC), and controlled and warranted origin denomination (DOCG); spiritous wines, syrupy wines,
Sparkling wines	sparkling wines. aerated sparkling wines

imply a different quantities of pesticides, diesel and lube oil used. The vine pruning aims to reach a good fructification and it has to be very intense. A mixed mineral and organic fertilisation is required by using appropriate quantities of manure and triple fertilizer (N_2, P_2O_5, K_2O). Depending on the case, the quantity of manure can be minimised. Almost always an artificial irrigation is required to optimise the vine yields. Downy mildew, oidium and tignola represent the most important vine pests; different formulations of pesticides are necessary to deal with them. On average the yields vary between 6 and 12 t ha^{-1} $year^{-1}$ depending on the vine variety, climate soil and care. In Table 17.1 the different typologies of wine are shown.

17.2.2 Red winemaking

The classic steps in red vinification are: crushing and destemming of grapes, fermentation with maceration, maturation, refining operations. Grapes are crushed to break the skin, in order to release the pulp and the juice. Quality wine grapes are lightly crushed to burst the berries without lacerating the solid parts. Subsequently, in the destemming phase, the rasps are totally or partially removed depending on rasps quality, which is related to variety and maturity level. Then the pressed grapes are sulfited as they are transferred to the fermentor. In conventional red winemaking, fermentation occurs together with maceration. Maceration is the breakdown of the solid parts of grapes which generates the diffusion of the phenolic compounds through the must during fermentation (Navarre, 1995).

Fermentation occurs in tanks in which the must is pumped and it is initially activated by yeast inoculation. A particular operation is necessary during fermentation due to the production of CO_2. The CO_2, going up to the top of the tank to be emitted in atmosphere, brings the pomace with it. Since a cap of pomace is formed to the top, the main operation consists in pumping the must from the bottom over the cap. The immersion of the floating cap is necessary to diffuse the yeasts from the cap to the must and to facilitate the aeration of the must. During the fermentation it is necessary also to supply oxygen to the must;

in fact the need for oxygen is high to facilitate the yeasts growth and the complete demolition of sugar. Many operations of pumping over are carried out to properly realize the process. Another important factor to take into account is the control of temperature. Fermentation is an exothermic process, but the temperature of must should not increase too much, because it is dangerous for the yeasts of the malolactic fermentation (Ribererau-Gayon *et al.*, 2000). The optimal temperature for fermentation of red wines is 25–30°C. Therefore, in Italy and in the warm countries, fermentation is conducted together with refrigeration which is one of the highest electric energy consumption phases.

After the fermentation, the free run wine is recovered directly by the fermentor. The pomace, which contains about a 60–70% of wine, is pressed and the first pressed wine is added to the previous wine. Two kinds of press wine are generally separated: the first press wine (generally slightly more than 50% of the press wine), relatively similar to free run wine; the second press wine, very different from free run wine. The second pressed wine has a minor qualitative level than the main production and so it is processed and sold separately. Wine is then subjected to various treatments to separate the lees through decanting and centrifugation after another step of refrigeration. Subsequently the red wine of high quality is stored in wooden barrels (barriques) to maturate and to complete the malolactic fermentation. Before bottling, a physico-chemical stabilization occurs to refine the red wine which is, afterwards, ready to be drunk.

Carbonic maceration
Carbonic maceration is based on the biological phenomena which occur in the grapes berries when they are in anaerobic conditions. This technique is carried out without crushing the grapes, but filling the closed tank with the grapes and then saturating the environment with CO_2. In these conditions intra-cellular fermentation takes place and the berry develops an anaerobic metabolism producing 1.2 to 2.5% volume ethanol from sugars. After 2–3 days of maceration the rasps are separated, the berries are crushed and then the pomace is pressed. The free run wine and the first press wine are now blended and directed to the alcoholic fermentation.

Carbonic maceration requires less time than traditional winemaking. It is best applied in making new wine,[1] essentially red wine. After fermentation there is the separation of lees through decanting, refrigeration and centrifugation. The new red wine needs also many other treatments such as clarification, filtration and physico-chemical stabilization which make use of bentonite and fossil flours whose waste are difficult to dispose of.

17.2.3   White winemaking
Differently from red wines, which generally have a similar composition, there are many kinds of white wines with a great variety of tastes. This result is

---

[1] New wine is the eqiuvalent of the French 'noveau' or 'primeur', or the Italian 'novello'.

possible since the technique of white winemaking has a great impact on the quality of the obtained product, while in red winemaking the main factor determining the production of a good quality wine is the quality of the harvested grapes. White wine vinification is a process which consists in the fermentation of the must without the presence of the solid parts of the grapes.

White vinification is more complex and sophisticated than the process for red. Even the harvesting and the transport operations have to be carried out with great attention, in order to avoid the crushing of berries which could increase the tannin extraction from stems and seeds. Then the quality of the must and of the subsequent wine depends on the care used in the extraction operations. The main phases of white wine production are crushing, dripping, pomace pressing, sulfiting and clarification from which the clarified must is obtained and sent to fermentation. The best white wine is obtained from a must in which the winemaker has succeeded to separate almost all the solids from the juice or to limit the contact of must with them before fermentation. In fact the main principles to follow in white winemaking are the absence of maceration of the solid parts of grapes and the protection of must from the risks of oxidation.

Different techniques can be used to limit juice oxidation: sulfiting, adding ascorbic acid, cooling grape and musts, heating musts at 60°C, handling grapes in the absence of air, clarification. In any case must after pressing is still turbid, and it needs a step in which the separation of lees takes place, first of all through the decanting of the solids; the filtration of must using fossil flours is then carried out. Must is now ready for the alcoholic fermentation which occurs in barriques for the wines of high quality. Selected yeasts are inoculated in the must to favourite the fermentation of white wines too. But, also in this case, the theoretical yield of ethanol formation in wine is not reached because of the loss of ethanol and other alcohols in atmosphere during fermentation. As the red wines, white wines production requires a strong control of the temperature in different phases; so different refrigerations take place during the operations. After fermentation the other operations are clarification, refrigeration, filtration and physico-chemical stabilization which make use of various inputs such as bentonite, gelatine, fossil flours, ascorbic acid and so on depending on the purposes. Finally, white wine is bottled.

Cryo-maceration

This technique consists in cooling the crushed and destemmed must at 5–8°C for 10–24 hours. Then the normal white winemaking is carried out by pressing the pomace, dripping the first run must, blending the two musts and so on. The main aim of this low temperatures treatment is to have low emissions of polyphenols from skins and to leave the odours of the grapes into the must. Another advantage of this operation is the decrease of loss of alcohol during fermentation. All these characteristics make the use of cryo-maceration suitable for white wines of high quality.

17.3   Applying LCA to wine production

17.3.1   Allocation problems

Apart from small producers, modern Italian wineries usually produce different qualities of wine (about ten, between red, white, rosè, sweet, new wine). This production variety leads to allocation problems; for instance, a high quality wine is made just by the free run wine blended with the first pressed wine, while the second pressed wine is used for lower quality wines. On average, every winery, even the smallest with quality production, will be producing about 1 bottle of second pressed wine every 10 bottles of premium wine. For instance, if the functional unit was the annual winery production, then there will be no allocation problems between the two wine typologies; but, if the functional unit was the quality wine, the solving of allocation on mass, rather than price, would take off from the quality wine environmental profile about the 10% of the total burden, while running it on the basis of price would mean to transfer more of the environmental burden to the quality wine.

Another allocation problem involves all the wine co-products not used for wine production: rasps, marc and lees. Even in this case their market price, differently from their quantities, is very low. Most of the time, in Italy wine producers do not have any gain in selling these co-products (rasps to compost producers, marc and lees to tartaric acid and alcohol producers). The alternative will be to dispose of them paying the charge. Consequently, they prefer to give them free of charge to the compost makers and distilleries. This Paretian efficient example of the industrial ecology philosophy leads to allocation problems: how to treat this co-products? Are they waste for which there is no final disposal, or co-products leaving the system?

17.3.2   Carbon dioxide emissions during fermentation

Another problem dealing with in the vinification process is the treatment of the CO_2 emissions; are this emissions to be included in the inventory table or not? They represent one of the hot spots of the whole wine system but these emissions are linked to the natural carbon cycle. It is exactly the same CO_2 extracted from the atmosphere by the vine during photosynthesis and which is, afterwards, emitted during winemaking.

17.3.3   Dispersion models for pesticides and nutrients and cut-off processes

Other difficulties raising from the application of LCA to wine deal with the agricultural steps, in particular: boundaries definition, cut off criteria, relevant and irrelevant processes, pesticides and nutrients dispersion. Among these topics, the main problem is the choice of the way to model the dispersion of pesticides and nutrients in the environment. If one should prefer not to model them because of the lack of standardised dispersion model, the wine environmental profile resulting would be underestimated; on the contrary, their inclusion in the analysis will undoubtedly lead to identification as 'hot spots' because of their respectively high eco and human toxicity and nutrification potentials. The choice of the relevant and irrelevant processes to include or not in the system boundaries could represent a problem in wine LCA. There are some substances used as clarifying agents or for other uses, which are utilized in a very limited quantities (ascorbic acid, bentonite, pectolitic enzymes, gelatin, yeasts, fossil flours), and whose data are very difficult to collect. By the case, almost always their exclusion from the system shouldn't affect the study results.

17.4   Case study

In this section a cradle-to-gate LCA has been carried out on a winery localised in the South of Italy. The analysis is limited to four wines, two premium (red and white) and two medium quality wines (new and white). Aim of the study is to identify the energy and environmental hot spots of the four systems and the options for their improvement. The LCA has been made as stated by the ISO 14040 rules (ISO, 1996).

17.4.1   Goal and scope definition

Core of this LCA is a Southern Italian winery, which is localised near Castel del Monte, Apulia, a traditional area devoted to the production of premium wines. This company, which is certified Vision 2000 and ISO 14001, produces fourteen different wines, both medium and premium. Its annual production is more than 60,000 hL, the typical medium-size Italian winery.

Purpose
Goal of the study is to build up the environmental profile of the four wines, in order to identify the hot spots of the systems. Even though there is no comparative aim, since white and red wine or high quality and average quality wine don't represent 'perfect substitutes' but completely different wines, the reasons of the differences in each wine environmental profile will be examined.

Functional unit and description of the systems
The functional unit chosen is one bottle by 0.75 L of wine. The four systems under study are:

- S1: premium red wine 0.75 L bottle with a selling price to the consumer of 25€.
- S2: new red wine 0.75 L bottle with a selling price to the consumer of 5€.
- S3: premium white wine 0.75 L bottle with a selling price to the consumer of 25€.
- S4: medium white wine 0.75 L bottle with a selling price to the consumer of 5€.

The analysis is from 'cradle-to-gate', covering all the life cycle phases, starting from the production of the input used in the agricultural phase (fertilisers and pesticides) until the production of the bottled wine, including production of the barrique, glass bottle, cork tap, aluminium capsule and paper label.

The following ancillaries have been excluded from the system because of poor quality data and difficulties in their collection: production of clarification agents – bentonite, gelatin, enzymes, yeast, antioxidant agents – ascorbic acid, fossil flours. A sensitivity analysis, based on very similar products, has shown the irrelevant burden of this products and/or processes on the total system burden.

Data quality, assumptions, allocations and impact assessment methods
The foreground data have been directly supplied by the company analysed, which makes use exclusively of grapes produced in its estate vineyards. The data have been compared with those taken from the scientific literature (Muccinelli, 2002; Ribaudo, 1997; De Vita, 1997; EPA, 1995). The background data are taken from the LCA databases. The following distances for the transports have been considered:

- Distance from the vineyard to the winery: 10 km.
- Distance from the bottle manufacturing plant to the winery: 100 km.
- Distance from winery to the landfill for the disposal of the sludge: 30 km.
- Distance from the fertilisers and pesticides producing plant to the winery: 300 km.

The emissions of N_2O, NH_3, $NO_3{}^-$, due to the use of nitrogen fertilisers have been modelled respectively following Houghton (Houghton, 1997), ECETOC (ECETOC, 1994) and Brentrup (Brentrup *et al.*, 2000) methodologies. The emissions of pesticides (in particular, dicobutrazol, folpet, propiconazole, metalaxil, phosalone, oxyfluorfen) during their use have been assessed following the model developed by Hauschild (Hauschild, 2000). The allocation problems, in particular the one relative to wine and pressed wine, have been solved on the basis of mass. The co-products leaving the systems, rasps, lees and marc have been considered as solid waste for which there is no disposal treatment, since they become free of charge raw materials for other productions, respectively compost for rasps and tartaric acid for marc. The emissions of carbon dioxide occurring during the fermentation process have not been taken into account.

The inventory results expressed in physical units have been assessed by the CML 2000 assessment method (Guinèe *et al.*, 2002) and for the categories of Ecotoxicity and Human Toxicity by the 1992 CML method (Heijungs *et al.*, 1992); moreover, the category of Energy Consumption (EC) has been added to the method. The examined impact categories are the following: Energy Consumption (EC), Global Warming Potential (GWP), Ozone Depletion Potential (ODP), Human Toxicity Potential (HTP), Eco-toxicity Potential (ECA+ECT), Acidification Potential (AP), Nutrification Potential (NP), Photochemical Oxidant Creation Potential (POCP). The assessment method has been stopped to the characterisation, without going through the normalization and weighting steps.

17.4.2   Inventory analysis

The quantitative inputs of the cradle-to-gate systems are shown in Table 17.2. The four systems make use of the two different grape varieties: Aglianico for the two red wines, Chardonay for the two white wines, both typical of the area in which the winery operates. For both the varieties the yield is about 12 t ha^{-1}

Table 17.2   Inputs of the four systems

Input	S1 (g)	S2 (g)	S3 (g)	S4 (g)
Agricultural phase				
Diesel	7.17	7.03	8.04	8.04
Lube oil	0.21	0.21	0.24	0.24
Fertilisers				
N	8.24	8.08	9.24	9.24
P_2O_5	1.93	1.90	2.16	2.16
K_2O	4.82	4.73	5.40	5.40
Pesticides				
Copper oxicloride	0.88	0.86	0.78	0.78
Dicobutrazol	0.10	0.10	0.09	0.09
Folpet	0.34	0.33	0.30	0.30
Wet sulphur	0.53	0.52	0.47	0.47
Sulphur	2.43	2.39	2.18	2.18
Propiconazole	0.05	0.05	0.04	0.04
Metalaxil	1.05	1.03	0.94	0.94
Phosalone	0.35	0.34	0.31	0.31
Oxyfluorfen	0.15	0.14	0.13	0.13
Industrial phase				
Grapes	1070	1050	1200	1200
Electric energy (kWh)	0.027	0.032	0.043	0.040
GPL (m^3)	0.0000015	0.0000015	0.0000015	0.0000015
Diesel	3.8	3.8	3.8	3.8
Water (L)	1.13	1.10	1.27	1.27
SO_2 (mg)	120	65	108	108
Selected yeasts	0.30	0.25	0.24	0.24
Pectolitic enzymes	0.02	0.02	0.01	0.01
Ascorbic acid	0.03	0.03	0.04	0.04
N_2	0.47	0.47	0.47	0.47
Benthonite	-	0.40	0.43	0.43
Gelatin	-	0.04	0.04	0.04
Fossil flours	-	0.45	1.90	1.90
CO_2	-	20	-	-
Others				
Barrique	60		60	
Bottle	500	500	500	500
Cork	3.4	3.4	3.4	3.4
Aluminium capsula	1.07	1.07	1.07	1.07
Paper label	1.24	1.24	1.24	1.24

year^{-1}. The main difference between the two varieties is their maturity stage which corresponds about to the end of August for Chardonay and middle October for Aglianico. This difference implies that in the case of Chardonay there are about two pesticides treatments less than those carried out in the Aglianico (which are about ten) with a 20% less of use of pesticides, diesel and lube oil. Nevertheless, the minor quantity of these inputs can't be observed in Table 17.2 because it is counterbalanced by the different yields in the two vinifications. In fact, one bottle of red wine requires from 1.05 to 1.07 kg of grapes, while one of white wine about 1.2 kg.

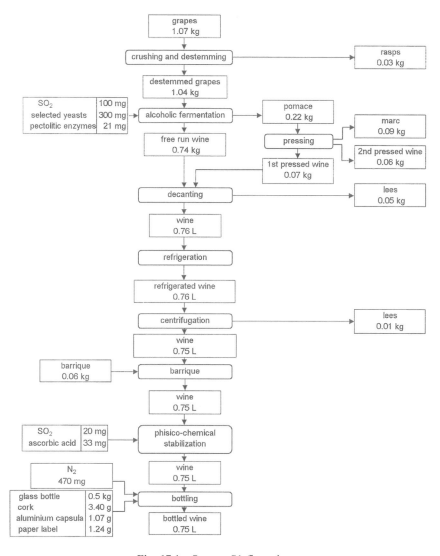

Fig. 17.1   System S1 flow chart.

All the industrial operations of the system S1 are quantitatively shown in Fig. 17.1. It can be found out that in this vinification process the operations of centrifugation, clarification, fining, sulphiting, filtration are relatively few, because of the permanence of the wine in the barrique for a period of ageing of 15 months.

The system S2, new wine, (Fig. 17.2) is characterised by the occurrence of the carbonic maceration. Since there is no permanence in barrique several refining operations are required. The system S3, premium white wine (Fig. 17.3) is characterised by a cryo-maceration of the grapes and by the permanence in

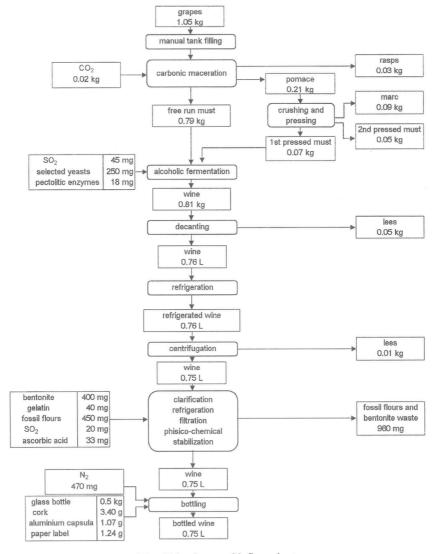

Fig. 17.2   System S2 flow chart.

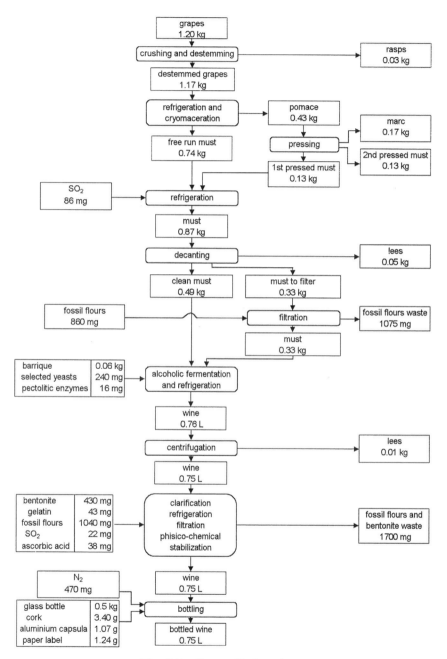

Fig. 17.3   System S3 flow chart.

Table 17.3   Electric energy consumption of the four systems.

Electric energy consumption	S1 (kWh)	S2 (kWh)	S3 (kWh)	S4 (kWh)
Industrial phase				
Weighting	0.000031	0.000030	0.000034	0.000034
Sugar analysis	0.000107	0.000105	0.000120	0.000120
Stemming	0.000808	-	0.000912	0.000912
Manual discharge	-	0	-	-
Carbonic maceration	-	0.000299	-	-
Refrigeration, cryo-maceration	-	-	0.005161	-
Vinificator discharge	-	-	0.000206	0.000206
Vinification	0.001646	-	-	-
Pomace pressing	0.000915	0.000926	0.001548	0.001548
Fermentation	-	0	-	-
Decanting	0.000122	0.000120	-	-
Refrigeration	0.005274	0.005259	0.003440	0.005161
Static decanting	-	-	0	0
Vacuum filtration	-	-	0.001720	0.001720
Fermentation with refrigeration	-	-	0.005161	0.005161
Decanting	-	-	0.000103	0.000103
Centrifuging	0.001463	0.001464	0.001479	0.001479
Clarification	-	0.000090	0.000086	0.000086
Refrigeration	-	0.006544	0.006537	0.006537
Filtration	-	0.000388	0.000378	0.000378
Physico-chemical stabilization	0	0	0	0
Bottling	0.016402	0.016375	0.016376	0.016376
Total	0.026768	0.031599	0.043263	0.039822

barrique for 15 days. The system S4, medium white wine (Fig. 17.3) follows the typical steps of the white vinification. All the steps are similar to those of S3 with the exception of the cryo-maceration which is substituted by a conventional juice pressing and of the lack of the barrique passage.

In Table 17.3 a detailed inventory of the energy consumption occurring in every single phase of the wine manufacturing for the four systems has been reported. It can be found out that the step with the highest energy consumption is the bottling, with about the 60% of the total energy consumption, followed by the refrigeration phase.

17.4.3   Impact assessment

In Figs 17.4 and 17.5 the results coming from the characterisation phase of the S1 and S2 systems are shown. The systems S3 and S4 are respectively very similar to S1 and S2. It can be easily found out that also the two systems environmental performances are very similar: the only difference is that in

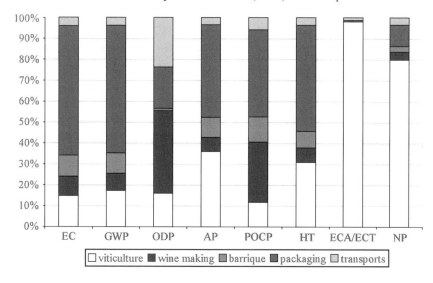

Fig. 17.4   S1 impact assessment

system S1 (as S3) there is the contribution of the barrique manufacturing which, on the contrary is not used in the systems S2 and S4. The most burdening phases in all the systems and in most of the categories are the agricultural activities and the bottle production. The agricultural activities burdens particularly on the categories of ECA/ECT (more than 97%), HT (about 30%) NP (about 80%), AP (slightly less than 40%). The first two categories depend on the use of pesticides, whose releases end up in all the media, affecting water and soil toxicity and the human toxicity of the workers spraying them.

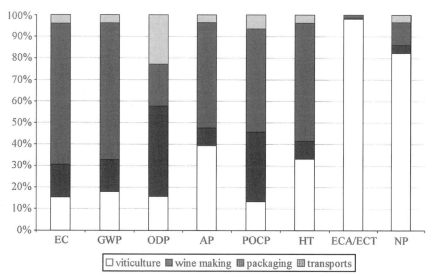

Fig. 17.5   S2 impact assessment.

The impact of the agricultural operations of vine on the NP and AP categories basically depends on the use of nitrogen and phosphatic fertilisers: the first category is due to the waterborne releases of phosphates and nitrates and to the airborne emissions of NO_x and NH_3. The categories of AP is due to the emissions of NO_x occurring during the fertilisers use.

The production of the glass bottle is the second most contributing phase in the wine life cycle. It burdens especially on the impact categories relative to the use of fossil fuels for the production of energy: EC (slightly more than 60%), GWP (about 60%) HT (about 50%), AP (about 40%).

The vinification operations in the winery burden significatively only on the category of POCP, due to the emissions of VOC occurring during the alcoholic fermentations. Among these, the most burdening one is the ethyl alcohol. Other impact categories affected by the vinification processes are those linked to electric energy production, but in a less degree compared to the phase of the glass bottle production. The other phases of the wine life cycle, as the production of the barrique, cork tap, aluminium capsule and paper label, show very irrelevant impacts on the different categories.

On the basis of these results it has seemed interesting to go deeper in the impact assessment of the winery, trying to find out the environmental hot spots among the inputs and outputs of the process itself. The results, shown in Figs 17.6 and 17.7, have been aggregated in the following inputs and outputs: electric energy, other fuels, chemicals (SO_2, N_2, CO_2), VOC (due to the alcoholic fermentation), waste water. The two figures are respectively relative to the systems S1 and S2. Systems S3 and S4 are very similar to S1. In the systems S1, S3 and S4 the consumption of electric energy and of other fuels (LPG and diesel) are the most burdening activities in the impact categories of EC, GWP,

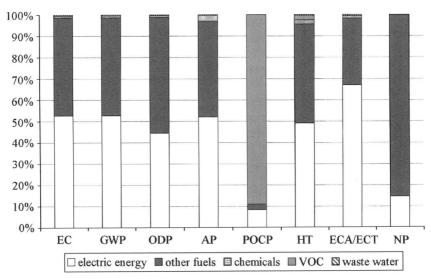

Fig. 17.6   S1 winery impact assessment.

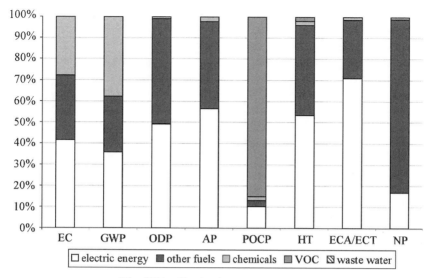

Fig. 17.7 S2 wine impact assessment.

ODP, AP, HT. In the systems S3 and S4 (two white wines) the electric energy contribution is higher than S1 due to the higher electricity needs in the two processes. In the S2 system there is also a contribution of the chemicals to the categories of EC and GWP respectively due to the industrial CO_2 production and CO_2 emissions.

In the four systems, the contribution of the electric energy consumption to the NP impact category is much less, about the 15% compared to a 85% about of other fuels, while in the ECA/ECT category the contribution of the electric energy consumption is about the 70%. In the four systems the category of POCP is dominated by the emission of VOC occurring during the alcoholic fermentation.

17.4.4 Interpretation

The analysis has shown that the profile of the four different wines has principally the same environmental hot spots: vineyard growing operations, due to the use of pesticides and fertilisers, glass bottle production, emissions of VOC in the alcoholic fermentation, electric energy consumption. Taking outside the glass bottle production which is a typical 'background' activity of this system, in the next paragraph some options for the improvement of the wine profile will be identified both in the agricultural step, and in the industrial one.

17.5 Using LCA to improve production

From Fig. 17.8 one can find out the environmental critical points taken from a winery plant environmental analysis (Quaglino and Bettrramo, 2002). By

WINEMAKING ENVIRONMENTAL BURDEN

Activity (in order)				Outputs							Inputs			
Mutual	Red winemaking only	White winemaking only	Sparkling winemaking only	Air emissions	Water emissions	Waste/ by-products	Soil contamination	Toxics	Noise	Smell	Water	Electric energy	Fuel	Raw material and auxiliary
Harvest reception				●	●		■		●		●	●	●	●
Crushing	Destemming	Draining/pressing	Pressing/draining		● ● ●	▶ ▶ ▶	■		● ● ●	● ● ●	● ● ●	● ● ●		● ●
Sulfiting		Defecation	Defecation		●	▶	■		●	●	●	●		● ● ● ●
Alcoholic fermentation	Maceration/ Pumping-over			● ● ●	●	▶	■	■	●	◎	●	●		
Racking	Pressing / Malolactic fermentation / Decanting/ cask filling / Clarifying / Refining/ageing	Stabilization	Stabilization	◎	● ●	▶ ▶ ▶	● ■	■	◎ ● ●	●	● ●	● ● ●		□
Bottling			Second fermentation	●	●	●	●		●		●	●		● ● ●
Packaging			Disgorging	●	● ●	● ●	■		●		●	●	●	● ● ● ●

■ Issues which are related to emergency conditions
▶ There are production waste which could be used as raw materials in distelleries.
◎ Red winemaking only.
□ White winemaking only.

Fig. 17.8 Hot spots in a winery from a site analysis.

analysing the wine environmental burden with a life cycle approach, it has been possible to find out the hot spots of the whole system which should be taken into account in order to minimise the environmental burden of wine. The main problems have been resulted to be the use of pesticides and fertilisers in the viticulture and the emission of VOC in the winery.

17.5.1   Organic agriculture

Organic agriculture and integrated pest management could be an option for the improvement of the wine environmental performance. But, as other studies have shown (Mattson, 1999), the organic agriculture is not a better a priori solution compared to the conventional agriculture. In the case of wine, the main problems are due to the great difference of yields which is on average 40% lower in the organic compared to that of the conventional system, with a consequent higher energy, material and land use for product unit (Nicoletti *et al.*, 2001). Other problems are connected with the type of organic pesticides and fertilisers used: due to its nature, manure is assimilated very slowly by the plant causing relevant nitrogen compounds emissions during its use; moreover, sulphur and copper sulphate have a relevant impact not much during their use but during their production. A relevant reduction in the use of these pesticides and a consequent better environmental profile of the organic system could be reached.

17.5.2   Abatement of the VOC emissions

Carbon dioxide represents the main airborne emission of the winery; it has not been taken into account in the analysis because it is linked to the natural carbon cycle. But, it would be desirable to research for plant solutions which could enable its recovery, in order to use it in the carbonic maceration. With the exception of CO_2, ethanol is the main compound emitted during the alcoholic fermentation. Acetaldehyde, methyl alcohol, n-propyl alcohol, n-butyl alcohol, sec-butyl alcohol, isobutyl alcohol, isoamyl alcohol, and hydrogen sulfide also are emitted but in much smaller quantities. In addition, a large number of other compounds are formed during the fermentation and ageing process as acetates, monoterpenes, higher alcohols, higher acids, aldehydes and ketones, and organosulfides (EPA, 1995).

Fugitive ethanol emissions also occur during the screening of the red wine, pressing of the pomace cap, ageing in barrique and the bottling process. In addition, small amounts of liquified SO_2 are always added to the must prior fermentation, or to the wine after the fermentation is completed; SO_2 emissions can occur during these steps, but they are almost impossible to quantify. There are five potential emissions control systems for VOC – carbon adsorption, water scrubbers, catalytic incineration, condensation, and temperature control – but all the systems have disadvantages in either low control efficiency, cost effectiveness or overall applicability to the wide variety of wineries. The only one which has an emission abatement of about the 98% is wet scrubber, but as the other emission control systems is not currently used during the winemaking.

17.5.3   Recovery of co products

The recovery and reuse of the wine solid co-products – rasps, lees, marc – plays an important role in the wine eco-profile. In life cycle thinking it is possible to skip from the analysis the burden of their disposal and, in industrial ecology terms, they become a raw material for new processes. The LCA enables to see the less environmental burden of the wine in the case the co-products are recovered compared to the case they are disposed of.

17.5.4   Waste water treatment

The winery activities represent a source of significant waste water production, due essentially to the equipment cleaning operations and to the loss occurring during the different operations of raw materials and products movement. The waste water polluting burden has an organic and biodegradable nature, for whose depuration it is possible to use an alternative process to the conventional one called 'activated sludge'. The phyto-depuration makes use of the natural capacity of some acquatic plants to absorb, through the radical apparatus, some substances contained in the waste water or generated by the degradation action of the microorganisms. The plants that are generated by this process could find an easy use as biomass to compost or for energy production. On the contrary, the activated sludge technology requires relevant energy quantities and produces sludges which have to be appropriately treated before their final disposal.

17.6   Future trends

As already pointed out in the previous paragraphs, technology and innovation are very important in the wine sector, first of all for the quality of the final product and also for safety and environmental issues. The most important trends in these directions are relative to plant and to process. The growing use of steel in the equipment and the design of energy saving plants belong at the first category; at the second, the growing use of bio-technology in the vinification and the new refining procedures.

The greater use of steel in plant rather than concrete is absolutely important in order to reduce the use of SO_2, which is nowadays fixed by the EU at 160 mg/L for red wine and 210 mg/L for white and rosè wine. An headache on the morning, after having drunk a glass of wine more, is due to the quantity of SO_2 rather than to the quantity of alcohol metabolised by the body. In winemaking SO_2 has antiseptical and antioxidant functions, whose needs can be minimised by the use of steel equipment.

The design of energy saving equipment is starting to be an important issue in winery, since electricity represents the main input of the process and, of course, its consumption reduction leads to better environmental and cost performances. Very linked to the reduction of the energy consumption in the process is the use of bio-technologies, in terms of selected yeasts or enzymes used in grapes treatments or in wine refining in order to minimise the need of other treatments. Traditional filtration with fossil flours implies the problem of their disposal; consequently, new filtration technologies have been testing, among which the use of the tangential filtration seems to be very promising.

On the contrary, an issue in which the innovation does not seem to go towards the most quality way is the substitution of cork tap with silicone tap, which in the last years has grown quite rapidly. The only reasons which could justify this shifting is the overcoming of the wine 'corked smell' (which sometimes happens due to the presence of some fungi – Mucor, Penicillium, Neurospora – in the cork) and the economical convenience for the winery in using silicone. Going through a quality analysis, it appears clear that cork plays its role in the wine ageing much better than silicon, because of its elastic characteristics which enable it to perfectly stick to the bottle letting the passage of the right micro-quantity of air which permits the ageing of wine and, at the same time, avoids the occurring of oxygenation process. On the contrary, since silicone does not permit any passage of air, it is not appropriate for the ageing of wine; therefore, it could be used just in wines which have to be drunk in a period less than two years. One solution to the cork degradation could be the washing of the cork tap with supercritical carbon dioxide in order to sweep away all the substances and micro-organisms which are extraneous to the cork.

17.7   Sources of further information and advice

The applications of LCA to wine are not very common both in academia or research centres and in wineries or consulting. There are different sources dealing with the environmental impact of the wineries, the guidelines for the application of EMS to the wineries, the inventory of input used in the viticulture. Only for the cork tap there is an available LCA made for the Italian consortium Rilegno (Ecobilancio Italia, 2001). The only LCA approach, but more relative to the greenhouse gases, is the one by the Australian Wine Company Yalumba which is member of the Greenhouse Challenge, Australia (Yalumba Wine Company, 2002).

17.8   References

BRENTRUP F, KüSTERS J, LAMMEL J, KUHLMANN H (2000), 'Methods to estimate on-field nitrogen emission from crop production as an input to LCA studies in the agricultural sector', *Int. J. LCA* 5 (6), 349–357.

DE VITA P (1997), *Course in Enological Mechanics*, Milan, Ulrico Hoepli Editor, (in Italian).

EC (1999), Council Regulation No 1493/1999 of 17 may 1999 on the common organisation of the market in wine, Official Journal of the European Communities, L 179/1, 14.7.1999.

ECETOC (1994), *Ammonia emissions to air in Western Europe*, Technical Report no. 62. European Chemical Industry, Ecology & Toxicology Centre, Brussels.

ECOBILANCIO ITALIA (2001), *Life Cycle Assessment of a cork tap for enological use*, Rome.

EPA (1995), *Emission Factor Documentation for AP-42*, Section 9.12.2. Wines and Brandy.

GUINèE J B, GORèE M, HEIJUNGS R, HUPPES G, KLEIJN R, KONING A DE, OERS L VAN, WEGENER SLEESWIJK A, SUH S, UDO DE HAES H A, BRUIJN H DE, DUIN R VAN, HUIJBREGTS M A J (2002), *Handbook on Life Cycle Assessment. Operational Guide to ISO Standards*, Dordrecht, Kluwer Academic Publishers.

HAUSCHILD M (2000), 'Estimating pesticide emission for LCA of agricultural products'. In Weidema B P, Meeusen M J C (ed.) *Agricultural data for Life Cycle Assessment*, II vol., 64–79, Agricultural Economics Research Institute, The Hague.

HEIJUNGS R, GUINéE J B, HUPPES G, LANKREIJER R M, UDO DE HAES H A, WEGENER SLEESWiJK A, ANSEMS A M, EGGELS P G, VAN DUIN R, GOEDE H P (1992), *Environmental life cycle assessment of products. Guide and Backgrounds*, CML, TNO, B&G, Leiden.

HOUGHTON J T ET AL. (1997), *Greenhouse Gas Inventory Reporting Instructions, Revised 1996 IPCC Guidelines for National Greenhouse Gas Inventories*, Volume 1–3. The Intergovernmental Panel on Climate Change, (IPCC), London.

INTERNATIONAL ORGANIZATION FOR STANDARDIZATION ISO/DIS 14040 (1996), Environmental management - Life cycle assessment - Principles and framework, ISO/TC207/SC5.

MACRAE R, ROBINSON R K, SADLER M J (1993), *Encyclopaedia of Food Science, Food Technology and Nutrition,* London, Academic Press.

MATTSSON B (1999), *Environmental Life Cycle Assessment (LCA) of Agricultural Food Product,* Swedish Institute for Food and Biotechnology, SIK, Gothenburg.

MUCCINELLI M (2002), *Promptuary of pesticides*, (IX edition), Bologna Edagricole, (in Italian).

NAVARRE C (1995), *Enology*, Milan, Ulrico Hoepli Editor, (in Italian).

NICOLETTI G M, NOTARNICOLA B, TASSIELLI G (2001), 'Comparison of conventional and organic wine', Proceedings of the International Conference LCA in Foods, Goteborg 26–27 April 2001.

QUAGLINO A, BELTRAMO R (2002), *Guidelines for the application of environmental management systems in the wineries*, Turin, (in Italian).

RIBAUDO F (1997), *Promptuary of agriculture*, Bologna, Edagricole, (in Italian).

RIBERERAU-GAYON P, DUBOURDIEU D, DONèCHE B, LONVAUD A (2000), *Handbook of Enology*, Chichester, John Wiley & Sons.

YALUMBA WINE COMPANY (2002), Progress Report. Commercial-in-Confidence Report, Angaston, Australia.

17.9 Acknowledgments

The authors thank the oenologist Mr. Nicola Suglia and Mr. Giuseppe Cicco, ITEST, for their precious collaboration.

Index